Lecture Notes in Applied Mathematics and Mechanics

Volume 3

About this Series

The Lecture Notes in Applied Mathematics and Mechanics LAMM are intended for an interdisciplinary readership in the fields of applied mathematics and mechanics. This series is published under the auspices of the International Association of Applied Mathematics and Mechanics (IAAMM; German GAMM).

Topics of interest include for example focus areas of the IAAMM such as: foundations of mechanics, thermodynamics, material theory and modeling, multibody dynamics, structural mechanics, solid mechanics, biomechanics, damage, fracture, multiscale modeling and homogenization, fluid mechanics, gas dynamics, laminar flows and transition, turbulence and reactive flows, interface flows, acoustics, waves, applied analysis, mathematical modeling, calculus of variations, variational principles applied operator theory, evolutionary equations, applied stochastics, systems with uncertainty, dynamical systems, control theory, optimization, applied and numerical linear algebra, analysis and numerics of ordinary and partial differential equations.

Each contribution to the series is intended to be accessible to researchers in mathematics and mechanics and is written in English. The aim of the series is to provide introductory texts for modern developments in applied mathematics and mechanics contributing to crossfertilization. The Lecture Notes are aimed at researchers as well as advanced masters and PhD students in both mechanics and mathematics. Contributions to the series are self-contained and focused on a few central themes. The goal of each contribution is the communication of modern ideas and principles rather than on completeness or detailed proofs. Like lecture notes from a course, a well-chosen example is preferable to an abstract framework that cannot be comprehended without deeper involvement. The typical length of each contribution is between 100 and 300 pages. If the lecture notes represent the proceedings of a summer school with several contributors, a unified, consistent presentation and style are required (e.g., common notation). In exceptional cases, doctoral theses may be accepted, if they fulfill the above-mentioned criteria.

Potential contributors should contact the appropriate editor with a title, table of contents, and a sample chapter. Full manuscripts accepted by the editors will then be peer-reviewed.

More information about this series at http://www.springer.com/series/11915

Adrian Muntean · Jens D.M. Rademacher
Antonios Zagaris
Editors

Macroscopic and Large Scale Phenomena: Coarse Graining, Mean Field Limits and Ergodicity

Springer

Editors
Adrian Muntean
Department of Mathematics and Computer Science, Center for Analysis, Scientific computing and Applications (CASA)
Eindhoven University of Technology
Eindhoven
The Netherlands

Antonios Zagaris
Faculteit Elektrotechniek, Wiskunde en Informatica
Universiteit Twente
Enschede, Overijssel
The Netherlands

Jens D.M. Rademacher
Fachbereich 3 - Mathematik und Informatik
Universität Bremen
Bremen
Germany

ISSN 2197-6724 ISSN 2197-6732 (electronic)
Lecture Notes in Applied Mathematics and Mechanics
ISBN 978-3-319-26882-8 ISBN 978-3-319-26883-5 (eBook)
DOI 10.1007/978-3-319-26883-5

Library of Congress Control Number: 2015956114

Printed on acid-free paper

This Springer imprint is published by SpringerNature
The registered company is Springer International Publishing AG Switzerland

Preface

This book is the offspring of the summer school "Macroscopic and Large Scale Phenomena: Coarse Graining, Mean Field Limits and Ergodicity." The school was held in 2012 at the University of Twente, the Netherlands, under the aegis of the national cluster *Nonlinear Dynamics in Natural Systems* (NDNS+). We assumed the initiative to organize that school to bring certain aspects of particle systems theory to the attention of junior applied mathematicians at large and to applied analysts in particular. To that effect, the main focus was on the modeling, multiscale analysis, and continuum limits of particle systems.

This emphasis is reflected in the book at hand, which collects in print an extension of the material presented in that summer school. Each of the four chapters here is based on a set of lectures delivered at the school, yet all authors have expanded and refined their contributions. The first three of these discuss mean field limits and (evolutionary) Γ-convergence as methods to derive macroscopic model equations from microscopic descriptions. The chapter concluding the volume offers a more abstract perspective, discussing the statistical properties of large-scale dynamics from the standpoint of ergodic and dynamical systems theory. Each chapter concerns one particular aspect of the overall theme and reflects the author's specific viewpoint and focus. We hope that this book can continue to stimulate interaction between these somewhat separate fields, in the spirit of the school that generated it.

François Golse delivers a chapter on the dynamics of large particle systems in the mean field limit. In doing so, he deals with a central issue in statistical mechanics: the derivation of reduced dynamical descriptions of large systems composed of identical microscopic constituents. In such systems, the effect of each individual particle is weak, but the collective action of the particle ensemble generates a nontrivial mean field acting on each and every particle. An early example of such a reduced description is the kinetic theory of gases, which is the mean field limit of molecular dynamics. According to that theory, the state of a system of a large number of identical gas molecules is reflected in the statistical behavior of one typical particle in the system. Golse surveys the most significant tools and methods employed to date to establish mean field limits with mathematical rigor. He illustrates their application actively by means of a variety of examples rooted in physical reality, such as regularized variants of the Vlasov–Poisson or Vlasov–Maxwell systems from plasma physics.

Lucia Scardia focuses on the derivation of continuum limits of discrete models using variational methods. The specific focus is on Γ-convergence, a more recent method that has proved very powerful in tackling problems in material science. Fracture mechanics, spin systems, magneto-mechanics, and dislocations are counted among the many successes of this approach. A common feature of all these problems is that good models are known at the discrete level of atoms and dislocations (microscale), but they are missing at the continuum level (macroscale) which is relevant for engineering applications. The notion of Γ-convergence furnishes a mathematically rigorous way to bridge those scales effectively. Scardia starts with an introduction to this concept, illustrated by concrete examples of discrete-to-continuum upscalings for simple systems. The latter part of the chapter deals with a more advanced application in dislocation theory.

Alexander Mielke's contribution focuses on the multiscale modeling and rigorous analysis of generalized gradient systems. Specifically, he examines the new concept of evolutionary Γ-convergence for systems driven by a dissipation mechanism for a functional, such as an energy or entropy. The focus in this work is on one-parameter families of such vector fields, where the (small) parameter quantifies the multiscale structure of the system in question. In this setting, the central challenge is to describe the limiting behavior of solutions to the detailed model in terms of solutions to a simplified, appropriately chosen, macroscopic one. In this context, the static concept of Γ-convergence provides a framework in which to understand and perform such upscalings. The presentation connects to physical and engineering reality by numerous evocative examples relating, for instance, to periodic homogenization or to the passage from viscous to dry friction.

Martin Göll and Evgeny Verbitskiy conclude this volume with a review of recent advances in a modern branch of ergodic theory. Statistically motivated approaches lie at the heart of ergodic theory and the study of complex dynamical systems. Their numerous offshoots have embraced a diverse array of mathematical fields including number theory, algebra and geometry. The relevant portion of the summer school in Twente discussed fundamental notions of ergodic theory and demonstrated applications of ergodicity in nonlinear dynamics. The chapter in this volume goes much further and reviews recent developments in the study of homoclinic points for certain classes of discrete dynamical systems. From our point of view, the most striking feature is that these relate certain particle systems to algebraic dynamical systems via ergodic properties of, e.g., lattices configurations. The authors motivate the abstract notions and theory by numerous examples, in particular, spatial dynamical systems. Having defined homoclinic points arising through principal algebraic actions, they review existing results and demonstrate the use of the theory in computing the entropy and in studying probabilistic properties of these discrete systems.

We express our gratitude to the Netherlands Organisation for Scientific Research (NWO), the Dutch Applied Mathematics Institute (3TU.AMI), and the Dutch Research School in Mathematics (WONDER) for supporting the summer school financially. We also thank the authors, who kindly agreed to contribute to this volume, as well as the school participants who proved eager to learn about these different approaches. Last but not least, we thank the series editors and publishers for realizing this book.

Eindhoven
Bremen
Twente
June 2015

Adrian Muntean
Jens D.M. Rademacher
Antonios Zagaris

Contents

Contributors

François Golse Centre de mathématiques Laurent Schwartz, Ecole polytechnique, Palaiseau Cedex, France

Martin Göll Mathematical Institute, University of Leiden, Leiden, The Netherlands

Alexander Mielke Weierstraß-Institut für Angewandte Analysis und Stochastik, Berlin, Germany

Lucia Scardia Department of Mathematical Sciences, University of Bath, Bath, UK

Evgeny Verbitskiy Mathematical Institute, University of Leiden, Leiden, The Netherlands; Johann Bernoulli Institute for Mathematics and Computer Science, University of Groningen, Groningen, The Netherlands

Chapter 1
On the Dynamics of Large Particle Systems in the Mean Field Limit

François Golse

In memory of Seiji Ukai (1940–2012)

Abstract This course explains how the usual mean field evolution partial differential equations (PDEs) in Statistical Physics—such as the Vlasov-Poisson system, the vorticity formulation of the two-dimensional Euler equation for incompressible fluids, or the time-dependent Hartree equation in quantum mechanics—can be rigorously derived from the fundamental microscopic equations that govern the evolution of large, interacting particle systems. The emphasis is put on the mathematical methods used in these derivations, such as Dobrushin's stability estimate in the Monge-Kantorovich distance for the empirical measures built on the solution of the N-particle motion equations in classical mechanics, or the theory of BBGKY hierarchies in the case of classical as well as quantum problems. We explain in detail how these different approaches are related; in particular we insist on the notion of chaotic sequences and on the propagation of chaos in the BBGKY hierarchy as the number of particles tends to infinity.

Keywords Mean-field limit · Vlasov-Poisson system · Vorticity formulation of the Euler equation · Empirical measure · Monge-Kantorovich distance · Propagation of chaos · BBGKY hierarchy · Trace-class operators · Time-dependent Hartree equation

The general purpose of all the mean field equations considered below is to describe the dynamics of a very large number of identical particles, assuming that the interaction between particles is known exactly. For instance, at a temperature of 273 K

F. Golse (✉)
Centre de mathématiques Laurent Schwartz, Ecole polytechnique,
91128 Palaiseau Cedex, France
e-mail: francois.golse@polytechnique.edu

A. Muntean et al. (eds.), *Macroscopic and Large Scale Phenomena: Coarse Graining, Mean Field Limits and Ergodicity*, Lecture Notes in Applied Mathematics and Mechanics 3, DOI 10.1007/978-3-319-26883-5_1

and a pressure of 1.01×10^5 Pa, the number of molecules of any ideal gas to be found in a volume of $2.24 \times 10^{-2}\,\mathrm{m}^3$ is the Avogadro number, i.e. 6.02×10^{23}. This is typically what we have in mind when we think of a number of particles so large that the dynamics of each individual particle cannot be exactly determined. Thus, in order to be of any practical interest, these mean field models should be set on (relatively) low dimensional spaces. In any case, they should involve fewer degrees of freedom (of the order of a few units) than the total number of degrees of freedom of the whole particle system.

To be more precise: in classical mechanics, the number of degrees of freedom of a single point particle moving without constraint in the d-dimensional Euclidean space $\mathbf{R}^d$ is d. The *single-particle phase space* is the set $\mathbf{R}^d \times \mathbf{R}^d$ of pairs of all possible positions and momenta of an unconstrained single point particle in $\mathbf{R}^d$.

For a system of N identical point particles moving without constraint in the d-dimensional Euclidean space $\mathbf{R}^d$, the number of degrees of freedom is therefore $2dN$. The space $(\mathbf{R}^d \times \mathbf{R}^d)^N$ of $2N$-tuples of all possible positions and momenta of the N point particles is the *N-particle phase space.*

Thus, the laws of motion of classical mechanics (Newton's second law) written for each molecule of a monatomic gas enclosed in a container form a system of differential equations set on the N-particle phase space, where N is the total number of gas molecules to be found in the container. With N of the order of the Avogadro number, this is of little practical interest. In the kinetic theory of gases, the evolution of the same gas is described by the Boltzmann equation, an integro-differential equation set on the single-particle phase space. Although the kinetic theory of gases will not be discussed in these notes, all the mean field limits considered below will involve the same reduction of the N-particle phase space to the single-particle phase space.

Here is a (by no means exhaustive) list of typical examples of mean field equations:

(a) the particles are the ions and the electrons in a plasma; the interaction is the Coulomb electrostatic force; in that case, the mean field equation is the Vlasov-Poisson system of the kinetic theory of plasmas;

(b) the particles are the nuclei and the electrons in a molecule; the interaction is again the Coulomb electrostatic force; the corresponding mean field model in this context is the Hartree equation or the system of Hartree-Fock equations in atomic physics;

(c) the particles are vortices in an incompressible fluid in space dimension 2; the interaction is given by the Helmholtz potential; the corresponding mean field model is the vorticity formulation of the Euler equations of incompressible fluid mechanics.

All these models are obtained as approximations of the system of equations of motion for each particle in the limit when the number of particles involved tends to infinity. Rigorous justifications of these approximations are based on various mathematical formalisms that are introduced and discussed below.

Excellent references on the issues discussed above are [26, 96, 98].

We conclude this presentation with a quick discussion of the similarities and differences between the material presented in this survey and the other chapters in this volume.

The mean field regime takes into account the pairwise interaction of all the particles in the system. It is of course particularly well suited to situations involving long-range interactions. Other models in physics may involve more local or short-range interactions, such as nearest neighbor interactions, or next-to-nearest neighbor, for instance in the mechanics of solids. These questions are discussed in [93] in terms of Γ-convergence.

The notion of Γ-convergence, due to DeGiorgi, is of considerable importance in the calculus of variations. It is discussed in detail in [77]. Whereas the original setting considered by DeGiorgi and his school corresponded to steady problems, Γ-convergence can be usefully adapted to evolution problems involving gradient flows. This is precisely the topic discussed in [77]. The limiting regimes considered in [77] usually correspond to a reduced, large scale description of the evolution of some field (for instance temperature) in some composite material whose structure is known at the microscopic scale. In all the mean field limits considered here, the small parameter is the reciprocal particle number $1/N$ instead of the ratio of length scales as in most models considered in [77].

Mean field limits are related to the law of large numbers, and this connection is explained in Theorem 1.4.3. In the setting considered in this theorem, chaoticity arises in the choice of the initial states, and is propagated by the dynamics (this is precisely the idea in Dobrushin's estimate discussed below). However, there are very important examples of situations in statistical mechanics where one would expect that chaoticity is not only propagated, but even enhanced by the dynamics. In some sense this is precisely the program laid out in the pioneering work of Krylov [68]. One of Krylov's ideas was to use the mixing property well known in the theory of dynamical systems to account for the notion of relaxation time in the convergence to equilibrium in statistical mechanics. Understanding how randomness emerges from a purely deterministic setting is of course a core issue in statistical mechanics. Since the work of Poincaré on the 3-body problem, the influence of a special kind of geometric situation known as "homoclinic tangles" in producing and enhancing chaoticity has been understood in detail. The dynamical properties of a special class of systems where homoclinic points have some additional algebraic structure are studied in [42]. Although this class of systems does not include the usual dynamics appearing in statistical mechanics, it satisfies some of the properties which one would like to establish on more realistic systems—such as a notion of central limit theorem (see Theorem 13 in [42]) in a purely deterministic context.

1.1 Examples of Mean Field Models in Classical Mechanics

In this section, we consider the examples mentioned above in which the motion of each particle is described in the formalism of classical mechanics, i.e. examples (a) and (c). All these examples have a common structure, to be discussed later, which involves the Liouville equation recalled below.

1.1.1 The Liouville Equation

The Liouville equation governs the evolution of the distribution function for a system of particles subject to an external force field.

The notion of distribution function is fundamental in the kinetic theories of gases and plasmas, and more generally in statistical mechanics. It was introduced by Maxwell in one of his most famous articles.[1]

The distribution function of a system of identical point particles is $f \equiv f(t, x, v)$, that is the number density of particles that are located at the position x and have instantaneous velocity v at time t. In other words, the number of particles to be found at time t in an infinitesimal volume $dxdv$ of the single-particle phase space centered at (x, v) is $\simeq f(t, x, v)dxdv$.

Assume that a particle located at the position x with instantaneous velocity v at time t is subject to some external (or imposed) force field $F(t, x, v)$. As a consequence of Newton's second law of classical mechanics, the distribution function f of the system of particles under consideration satisfies the Liouville equation

$$\partial_t f + v \cdot \nabla_x f + \tfrac{1}{m} \operatorname{div}_v(F(t, x, v)f) = 0,$$

where $m > 0$ is the particle mass. The Liouville equation is a partial differential equation (PDE) of order 1, whose solution can be expressed by the method of characteristics.

To the PDE of order 1 above, one associates the system of ordinary differential equations (ODE)

$$\begin{cases} \dot{X} = V, \\ \dot{V} = \frac{1}{m} F(t, X, V), \end{cases}$$

with the usual notation

$$\dot{\phi}(t) = \frac{d\phi}{dt}(t).$$

These ODEs are referred to as the "equations of characteristics" for the Liouville equation. Denote by $t \mapsto (X(t, t_0, x, v), V(t, t_0, x, v))$ the solution of this ODE system such that

$$X(t_0, t_0, x, v) = x, \quad V(t_0, t_0, x, v) = v\,;$$

the map $(t, t_0, x, v) \mapsto (X(t, t_0, x, v), V(t, t_0, x, v))$ will be henceforth referred to as "the characteristic flow" associated to the Liouville equation above.

[1] *Illustrations of the Dynamical Theory of Gases*, Philosophical Magazine (1860); reprinted in "The Scientific Papers of James Clerk Maxwell", edited by W.D. Niven, Cambridge University Press, 1890; pp. 377–409.

One immediately recognizes in the system of ODEs above the equations of motion of classical mechanics for a particle of mass m subject to the force field $F \equiv F(t, x, v)$ (the first equality being the definition of the velocity, while the second is Newton's second law).

Assume that the force field F is such that the characteristic flow (X, V) is globally defined (i.e. defined on $\mathbf{R} \times \mathbf{R} \times \mathbf{R}^d \times \mathbf{R}^d$). This characteristic flow is used as follows to express the solution of the Cauchy problem

$$\begin{cases} \partial_t f + v \cdot \nabla_x f + \frac{1}{m} \operatorname{div}_v(F(t,x,v)f) = 0, & x, v \in \mathbf{R}^d \times \mathbf{R}^d, \ t \in \mathbf{R}, \\ f\big|_{t=0} = f^{in}. \end{cases}$$

For each test function[2] $\phi \in C_c^1(\mathbf{R}^d \times \mathbf{R}^d)$, and for each $t \in \mathbf{R}$, one has

$$\iint_{\mathbf{R}^d \times \mathbf{R}^d} f(t,x,v)\phi(x,v)dxdv = \iint_{\mathbf{R}^d \times \mathbf{R}^d} f^{in}(x,v)\phi((X,V)(t,0,x,v))dxdv,$$

and this completely determines the distribution function $f(t, x, v)$.

Exercise Deduce from the equality above an explicit formula giving $f(t, x, v)$ in terms of f^{in} and of the characteristic flow (X, V).

For a concise discussion of the method of characteristics (and a solution of the exercise above), see Chap. 1, Sect. 1, in [17].

1.1.2 The Vlasov-Poisson System

Our first example of a mean field kinetic model is the Vlasov-Poisson system used in plasma physics.

Consider a system of identical point particles with mass m and charge q. The electrostatic (Coulomb) force exerted on any such particle located at the position $x \in \mathbf{R}^3$ by another particle located at the position $y \neq x$ is

$$\frac{q^2}{4\pi\varepsilon_0} \frac{x-y}{|x-y|^3},$$

where ε_0 is the dielectric permittivity of vacuum. This force is indeed repulsive, as it is of the form $\lambda(x - y)$ with $\lambda > 0$.

---

[2]For each topological space X and each finite dimensional vector space E on $\mathbf{R}$, we designate by $C(X, E)$ the set of continuous functions defined on X with values in E, and by $C_c(X, E)$ the set of functions belonging to $C(X, E)$ whose support is compact in X. For each $n, k \geq 1$, we denote by $C_c^k(\mathbf{R}^n, E)$ the set of functions of class C^k defined on $\mathbf{R}^n$ with values in E whose support is compact in $\mathbf{R}^n$. We also denote $C(X) := C(X, \mathbf{R})$, $C_c(X) := C_c(X, \mathbf{R})$ and $C_c^k(\mathbf{R}^n) := C_c^k(\mathbf{R}^n, \mathbf{R})$.

More generally, the electrostatic force exerted on a particle located at the position x by a cloud of particles with number density $\rho(t, y)$ (which means that, at time t, approximately $\rho(t, y)dy$ particles are to be found in any infinitesimal volume element dy centered at y) is

$$F(t, x) := \frac{q^2}{4\pi\varepsilon_0} \int_{\mathbf{R}^3} \frac{x - y}{|x - y|^3} \rho(t, y) dy.$$

Recall that

$$G(x) = \frac{1}{4\pi |x|}, \quad x \in \mathbf{R}^3 \setminus \{0\}$$

is the solution of

$$\begin{cases} -\Delta G = \delta_0 & \text{in } \mathscr{D}'(\mathbf{R}^3), \\ G(x) \to 0 & \text{as } |x| \to \infty, \end{cases}$$

where δ_0 designates the Dirac measure at the origin.

Thus the electrostatic force F is given by

$$F(t, x) = qE(t, x),$$

where $E \equiv E(t, x)$ is the electric field, i.e.

$$E(t, x) = -\nabla_x \phi(t, x)$$

with electrostatic potential ϕ given by

$$\phi(t, \cdot) = \frac{q}{\varepsilon_0} G \star \rho(t, \cdot).$$

In particular

$$\begin{aligned} -\Delta\phi(t, \cdot) &= -\frac{q}{\varepsilon_0} \Delta(G \star \rho(t, \cdot)) = -\frac{q}{\varepsilon_0} (\Delta G) \star \rho(t, \cdot) \\ &= \frac{q}{\varepsilon_0} \delta_0 \star \rho(t, \cdot) = \frac{q}{\varepsilon_0} \rho(t, \cdot). \end{aligned}$$

The Vlasov-Poisson system describes the motion of a system of identical charged point particles accelerated by the electrostatic force created by their own distribution of charges—referred to as the "self-consistent electrostatic force". It consists of the Liouville equation governing the evolution of the distribution function, coupled to the Poisson equation satisfied by the self-consistent electrostatic potential as follows:

$$\begin{cases} (\partial_t + v \cdot \nabla_x) f(t,x,v) - \frac{q}{m} \nabla_x \phi(t,x) \cdot \nabla_v f(t,x,v) = 0, \\ \\ -\Delta_x \phi(t,x) = \frac{1}{\varepsilon_0} q \rho_f(t,x), \\ \\ \rho_f(t,x) = \int\limits_{\mathbf{R}^3} f(t,x,v)dv. \end{cases}$$

Exercise Let $f \equiv f(t,x,v)$ and $\phi \equiv \phi(t,x)$ be a solution of the Vlasov-Poisson system. For simplicity, we assume that f and ϕ belong to $C^\infty(\mathbf{R}_+ \times \mathbf{R}^3 \times \mathbf{R}^3)$ and $C^\infty(\mathbf{R}_+ \times \mathbf{R}^3)$ respectively, and that $(x,v) \mapsto f(t,x,v)$ and $x \mapsto \phi(t,x)$ belong to the Schwartz class[3] $\mathscr{S}(\mathbf{R}^3 \times \mathbf{R}^3)$ and $\mathscr{S}(\mathbf{R}^3)$ respectively, for each $t \geq 0$, .
(1) Prove that

$$\partial_t \rho_f(t,x) + \operatorname{div}_x \int\limits_{\mathbf{R}^3} v f(t,x,v)dv = 0,$$

$$\partial_t \int\limits_{\mathbf{R}^3} mvf(t,x,v)dv + \operatorname{div}_x \int\limits_{\mathbf{R}^3} mv^{\otimes 2} f(t,x,v)dv$$
$$+ q\rho_f(t,x)\nabla_x \phi(t,x) = 0,$$

$$\partial_t \int\limits_{\mathbf{R}^3} \tfrac{1}{2} m|v|^2 f(t,x,v)dv + \operatorname{div}_x \int\limits_{\mathbf{R}^3} v \tfrac{1}{2} m|v|^2 f(t,x,v)dv$$
$$+ \nabla_x \phi(t,x) \cdot \int\limits_{\mathbf{R}^3} qvf(t,x,v)dv = 0.$$

These three equalities are respectively the local conservation laws of mass, momentum and energy.
(2) Prove the global conservation of mass (or equivalently of the total number of particles)

$$\frac{d}{dt} \iint\limits_{\mathbf{R}^3 \times \mathbf{R}^3} f(t,x,v)dxdv = 0.$$

(3) Prove that, for each $\psi \in C^2(\mathbf{R}^3)$, one has

$$\Delta\psi\nabla\psi = \operatorname{div}((\nabla\psi)^{\otimes 2}) - \tfrac{1}{2}\nabla|\nabla\psi|^2$$

[3] For each $n \geq 1$, the Schwartz class $\mathscr{S}(\mathbf{R}^n)$ is the set of real-valued C^∞ functions defined on $\mathbf{R}^n$ all of whose partial derivatives are rapidly decreasing at infinity:

$$\mathscr{S}(\mathbf{R}^n) := \{f \in C^\infty(\mathbf{R}^n) \text{ s.t. } |x|^m \partial^\alpha f(x) \to 0 \text{ as } |x| \to \infty \text{ for all } m \geq 1 \text{ and } \alpha \in \mathbf{N}^n\}.$$

and conclude that

$$\frac{d}{dt}\iint\limits_{\mathbf{R}^3\times\mathbf{R}^3} mvf(t,x,v)dxdv = 0.$$

(This is the global conservation of momentum).

(4) Prove the global conservation of energy:

$$\frac{d}{dt}\left(\iint\limits_{\mathbf{R}^3\times\mathbf{R}^3} \tfrac{1}{2}m|v|^2 f(t,x,v)dxdv + \tfrac{1}{2}\varepsilon_0\int\limits_{\mathbf{R}^3} |\nabla_x\phi(t,x)|^2 dx\right) = 0.$$

(Hint: the term

$$\int\limits_{\mathbf{R}^3} \phi(t,x)\,\mathrm{div}_x\left(\int\limits_{\mathbf{R}^3} qvf(t,x,v)dv\right)dx$$

can be simplified by using the result in question (1).)

The Vlasov-Poisson system above is written in the case of a single species of identical charged particles. In reality, a plasma is a globally neutral system, and therefore involves many different species of particles. Denoting by $f_k \equiv f_k(t,x,v)$ for $k = 1,\dots,M$ the distribution function of the kth species in a system of M different kinds of particles, the Vlasov-Poisson describes the motion of this system of particles submitted to the self-consistent electrostatic force resulting from the distribution of charges of all the particles in that system:

$$\begin{cases} (\partial_t + v\cdot\nabla_x)f_k(t,x,v) - \frac{q_k}{m_k}\nabla_x\phi(t,x)\cdot\nabla_v f_k(t,x,v) = 0, \quad k = 1,\dots,M, \\[2ex] -\Delta_x\phi(t,x) = \frac{1}{\varepsilon_0}\sum\limits_{k=1}^{M} q_k\rho_k(t,x), \\[2ex] \rho_k(t,x) = \int\limits_{\mathbf{R}^3} f_k(t,x,v)dv. \end{cases}$$

Here m_k and q_k designate respectively the mass and the charge of particles of the kth species. In practice, considering systems of particles of different species in the Vlasov-Poisson system does not involve additional mathematical difficulties. Therefore, we shall consider only the (unphysical) case of a single species of particles in our mathematical discussion of this system.

There is a huge literature on the Vlasov-Poisson system, which is of considerable importance in plasma physics. The global existence and uniqueness of classical solutions of the Cauchy problem for the Vlasov-Poisson system was obtained by

Ukai-Okabe [102] in the 2-dimensional case, and independently by Lions-Perthame [73] and Pfaffelmoser [86] in the 3-dimensional case. For a presentation of the mathematical analysis of this system, the interested reader is referred to [17, 41, 91].

1.1.3 The Euler Equation for Two-Dimensional Incompressible Fluids

The Euler equation for an incompressible fluid with constant density (set to 1 in the sequel without loss of generality) governs the evolution of the unknown velocity field $u \equiv u(t,x) \in \mathbf{R}^2$ and of the unknown pressure field $p \equiv p(t,x) \in \mathbf{R}$ in the fluid, assumed to fill the Euclidean plane $\mathbf{R}^2$. It takes the form

$$\partial_t u(t,x) + (u \cdot \nabla_x)u(t,x) + \nabla_x p(t,x) = 0, \quad \operatorname{div}_x u(t,x) = 0.$$

The pressure field p can be viewed as the Lagrange multiplier associated to the constraint $\operatorname{div}_x u = 0$.

There is another formulation of the Euler equation in which the pressure field is eliminated. To the velocity field $u \equiv u(t,x) \in \mathbf{R}^2$, one associates its scalar vorticity field $\omega \equiv \omega(t,x)$ defined as follows

$$\omega(t,x) := \partial_{x_1} u_2(t,x) - \partial_{x_2} u_1(t,x).$$

The vorticity field satisfies

$$\partial_t \omega + \operatorname{div}_x(u\omega) = 0.$$

One can check that the Euler equation is (formally) equivalent to the system

$$\begin{cases} \partial_t \omega + \operatorname{div}_x(u\omega) = 0, \\ \operatorname{div}_x u = 0, \\ \omega = \operatorname{div}_x(Ju), \end{cases}$$

where J designates the rotation of an angle $-\frac{\pi}{2}$:

$$J = \begin{pmatrix} 0 & 1 \\ -1 & 0 \end{pmatrix}$$

There is yet another formulation of this system, where the velocity field u is represented in terms of a stream function. Indeed, since $\operatorname{div}_x u = 0$, there exists a stream function $\phi \equiv \phi(t,x)$ such that

$$u(t,x) = J\nabla_x \phi(t,x).$$

The vorticity field ω is related to the stream function ϕ as follows:

$$\omega = \operatorname{div}_x(Ju) = \operatorname{div}_x(J^2\nabla_x\phi) = -\Delta_x\phi,$$

so that the vorticity formulation of the Euler equation becomes

$$\begin{cases} \partial_t\omega + \operatorname{div}_x(\omega J\nabla_x\phi) = 0, \\ -\Delta_x\phi = \omega. \end{cases}$$

In this last system, the unknown is the pair (ω, ϕ); once the stream function ϕ is known, the velocity field u is obtained as the spatial gradient of ϕ rotated of an angle $-\frac{\pi}{2}$.

There is an obvious analogy between this formulation of the Euler equation and the Vlasov-Poisson system: the vorticity field ω is the analogue of the distribution function f in the Vlasov-Poisson system, the stream function is the analogue of the electrostatic potential in the Vlasov-Poisson system, and the velocity field u is the analogue of the electric field in the Vlasov-Poisson system. There is however a subtle difference between both systems: the equations of characteristics associated to the vorticity formulation of the Euler equations are

$$J\dot{X} = -\nabla_x\phi(t, X),$$

while those corresponding with the Vlasov-Poisson system are

$$\ddot{X} = -\frac{q}{m}\nabla_x\phi(t, X),$$

(after eliminating $V = \dot{X}$). The first ODE system is of order 1, while the second is of order 2, because the gradient of the solution of the Poisson equation is a velocity field in the vorticity formulation of the Euler equation, while it is an acceleration field in the case of the Vlasov-Poisson system.

An excellent reference on the Euler equations of incompressible fluids and on issues related to the mean field limit is [76].

1.1.4 The Vlasov-Maxwell System

Observe that magnetic effects are not taken into account in the Vlasov-Poisson system. Yet charged particles in motion generate both an electric and a magnetic field; on the other hand, the electric and magnetic fields both act on charged particles via the Lorentz force. Studying magnetized plasmas (such as tokamak plasmas, or the solar wind, for instance) requires using the Vlasov-Maxwell, instead of the Vlasov-Poisson system.

The unknown in the relativistic Vlasov-Maxwell system is the triple (f, E, B), where $E \equiv E(t,x) \in \mathbf{R}^3$ and $B \equiv B(t,x) \in \mathbf{R}^3$ are respectively the electric and the magnetic fields, while $f \equiv f(t,x,\xi)$ is the particle distribution function. Notice that the velocity variable v is replaced with the momentum ξ in the distribution function. In other words, $f(t,x,\xi)$ is the phase space density of particles located at the position x at time t, with momentum ξ. The momentum and the velocity are related as follows. The relativistic energy of a particle of mass m with momentum ξ is

$$e(\xi) := \sqrt{m^2c^4 + c^2|\xi|^2},$$

where c is the speed of light (notice that $e(0) = mc^2$, the energy at rest of a particle with mass m). Then

$$v(\xi) = \nabla e(\xi) = \frac{c^2\xi}{\sqrt{m^2c^4 + c^2|\xi|^2}}.$$

The (relativistic) Vlasov-Maxwell system takes the form

$$\begin{cases} \partial_t f + v(\xi) \cdot \nabla_x f + q(E + v(\xi) \times B) \cdot \nabla_\xi f = 0, \\ \operatorname{div}_x B = 0, \qquad \partial_t B + \operatorname{rot}_x E = 0, \\ \operatorname{div}_x E = \frac{1}{\varepsilon_0} q\rho_f, \qquad \frac{1}{c^2}\partial_t E - \operatorname{rot}_x B = -\mu_0 q j_f, \\ \rho_f = \int_{\mathbf{R}^3} f d\xi, \qquad j_f = \int_{\mathbf{R}^3} v(\xi) f d\xi. \end{cases}$$

The first equation in this system is the relativistic Liouville equation governing f; the term

$$q(E(t,x) + v(\xi) \times B(t,x))$$

is the Lorentz force field at time t and position x exerted by the electromagnetic field (E, B) on a particle with charge q and momentum ξ. The second and third equations are respectively the equation expressing the absence of magnetic monopoles and the Faraday equation, while the fourth and fifth equations are respectively the Gauss equation (as in electrostatics) and the Maxwell-Ampère equation. Since the source terms in the Maxwell system are the charge density $q\rho_f$ and the current density qj_f, the Lorentz force in the Liouville equation is the self-consistent Lorentz force, i.e. the electromagnetic force generated by the motion of the charged particles accelerated by this force itself.

For the same reason as in the case of the Vlasov-Poisson system, the case of a single species of charged particles is somewhat unrealistic. Physically relevant models in plasma physics involve different species of particles so as to maintain global neutrality of the particle system.

Exercise Following the discussion in the previous section, write the relativistic Vlasov-Maxwell system for a system of M species of particles with masses m_k and charges q_k, for $k = 1, \ldots, M$. Write the local conservation laws of mass, momentum and energy for the resulting system, following the analogous discussion above in the case of the Vlasov-Poisson system.

For more information on the Vlasov-Maxwell system, see [17, 41, 91].

1.2 A General Formalism for Mean Field Limits in Classical Mechanics

We first introduce a formalism for mean field limits in classical mechanics that encompasses all the examples discussed above.

Consider a system of N particles, whose state at time t is defined by phase space coordinates $\hat{z}_1(t), \ldots, \hat{z}_N(t) \in \mathbf{R}^d$. For instance, $\hat{z}_j$ is the position x_j of the jth vortex center in the case of the two dimensional Euler equations for incompressible fluids, and the phase space dimension is $d = 2$. In the case of the Vlasov-Poisson system, the phase space is $\mathbf{R}^3 \times \mathbf{R}^3 \simeq \mathbf{R}^6$, so that $d = 6$, and $\hat{z}_j = (x_j, v_j)$, where x_j and v_j are respectively the position and the velocity of the jth particle.

The interaction between the ith and the jth particle is given by $K(\hat{z}_i, \hat{z}_j)$, where

$$K : \mathbf{R}^d \times \mathbf{R}^d \to \mathbf{R}^d$$

is a map whose properties will be discussed below.

The evolution of $\hat{z}_1(t), \ldots, \hat{z}_N(t) \in \mathbf{R}^d$ is governed by the system of ODEs

$$\frac{d\hat{z}_i}{dt}(t) = \sum_{\substack{j=1\\ j\neq i}}^{N} K(\hat{z}_i(t), \hat{z}_j(t)), \quad i, j = 1, \ldots, N.$$

Problem To describe the behavior of $\hat{z}_1(t), \ldots, \hat{z}_N(t) \in \mathbf{R}^d$ in the large N limit and in some appropriate time scale.

First we need to rescale the time variable, and introduce a new time variable $\hat{t}$ so that, in new time scale, the action on any one of the N particles due to the $N - 1$ other particles is of order 1 as $N \to +\infty$. In other words, the new time variable $\hat{t}$ is chosen so that

$$\frac{d\hat{z}_i}{d\hat{t}} = O(1) \text{ for each } i = 1, \ldots, N \text{ as } N \to \infty.$$

The action on the ith particle of the $N-1$ other particles is

$$\sum_{\substack{j=1\\j\neq i}}^{N} K(\hat{z}_i, \hat{z}_j),$$

and it obviously contains $N-1$ terms of order 1 (assuming each term $K(\hat{z}_i, \hat{z}_j)$ to be of order 1, for instance). Set $\hat{t} = t/N$, then

$$\frac{d\hat{z}_i}{d\hat{t}} = \frac{1}{N}\sum_{\substack{j=1\\j\neq i}}^{N} K(\hat{z}_i, \hat{z}_j).$$

From now on, we drop hats on all variables and consider as our starting point the rescaled problem

$$\dot{z}_i(t) = \frac{1}{N}\sum_{\substack{j=1\\j\neq i}}^{N} K(z_i(t), z_j(t)), \qquad i = 1, \ldots, N.$$

At this point, we introduce an important assumption on the interaction kernel: the action of the jth particle on the ith particle must exactly balance the action of the ith particle on the jth particle. When the interaction is a force, this is precisely Newton's third law of mechanics. Thus we assume that the interaction kernel satisfies

$$K(z, z') = -K(z', z), \qquad z, z' \in \mathbf{R}^d.$$

We have assumed here that the interaction kernel K is defined on the whole $\mathbf{R}^d \times \mathbf{R}^d$ space; in particular, the condition above implies that K vanishes identically on the diagonal, i.e.

$$K(z, z) = 0, \qquad z \in \mathbf{R}^d.$$

Hence the restriction $j \neq i$ can be removed in the summation that appears on the right hand side of the ODEs governing the N-particle dynamics: since $K(z_i(t), z_i(t)) = 0$ for all $i = 1, \ldots, N$, one has

$$\dot{z}_i(t) = \frac{1}{N}\sum_{j=1}^{N} K(z_i(t), z_j(t)) \qquad i = 1, \ldots, N.$$

At this point, we can explain the key idea in the mean field limit: if the points $z_j(t)$ for $j = 1, \ldots, N$ are "distributed at time t under the probability measure $f(t, dz)$" in the large N limit, then, one expects that

$$\frac{1}{N}\sum_{j=1}^{N} K(z_i(t), z_j(t)) \to \int_{\mathbf{R}^d} K(z_i(t), z')f(t, dz') \qquad \text{as } N \to +\infty.$$

This suggests replacing the N-particle system of differential equations with the single differential equation

$$\dot{z}(t) = \int_{\mathbf{R}^d} K(z(t), z')f(t, dz').$$

Here $f(t, dz)$ is unknown, as is $z(t)$, so that it seems that this single differential equation is insufficient to determine both these unknowns.

But one recognizes in the equality above the equation of characteristics for the mean field PDE

$$\partial_t f + \operatorname{div}_z(f\mathscr{K}f) = 0,$$

where the notation $\mathscr{K}$ designates the integral operator defined by the formula

$$\mathscr{K}f(t, z) := \int_{\mathbf{R}^d} K(z, z')f(t, dz').$$

Now, this is a single PDE (in fact an integro-differential equation) for the single unknown f.

A priori f is a time dependent Borel probability measure on $\mathbf{R}^d$, so that the mean field PDE is to be understood in the sense of distributions on $\mathbf{R}^d$. In other words,

$$\frac{d}{dt}\int_{\mathbf{R}^d} \phi(z)f(t, dz) = \int_{\mathbf{R}^d} \mathscr{K}f(t, z) \cdot \nabla\phi(z)f(t, dz)$$

for each test function[4] $\phi \in C_b^1(\mathbf{R}^d)$.

A very important mathematical object in the mathematical theory of the mean field limit is the empirical measure, which is defined below.

[4]For each topological space X and each finite dimensional vector space E on $\mathbf{R}$, we denote by $C_b(X, E)$ the set of continuous functions defined on X with values in E that are bounded on X. For each $n, k \geq 1$, we denote by $C_b^k(\mathbf{R}^n, E)$ the set of functions of class C^k defined on $\mathbf{R}^n$ with values in E all of whose partial derivatives are bounded on $\mathbf{R}^n$: for each norm $|\cdot|_E$ on E, one has

$$C_b^k(\mathbf{R}^n, E) := \{f \in C^k(\mathbf{R}^n, E) \text{ s.t. } \sup_{x\in\mathbf{R}^n} |\partial^\alpha f(x)|_E < \infty \text{ for each } \alpha \in \mathbf{N}^n\}.$$

We also denote $C_b(X) := C_b(X, \mathbf{R})$ and $C_b^k(\mathbf{R}^n) := C_b^k(\mathbf{R}^n, \mathbf{R})$.

Definition 1.2.1 To each N-tuple $Z_N = (z_1, \ldots, z_N) \in (\mathbf{R}^d)^N \simeq \mathbf{R}^{dN}$, one associates its empirical measure

$$\mu_{Z_N} := \frac{1}{N}\sum_{j=1}^{N} \delta_{z_j}.$$

The empirical measure of a N-tuple $Z_N \in (\mathbf{R}^d)^N$ is a Borel probability measure on $\mathbf{R}^d$. As we shall see in the next section, the N-tuple

$$t \mapsto Z_N(t) = (z_1(t), \ldots, z_N(t))$$

is a solution of the N-particle ODE system

$$\dot{z}_i(t) = \frac{1}{N}\sum_{j=1}^{N} K(z_i(t), z_j(t)), \quad i = 1, \ldots, N$$

if and only if the empirical measure $\mu_{Z_N(t)}$ is a solution of the mean field PDE

$$\partial_t \mu_{Z_N(t)} + \mathrm{div}_z(\mu_{Z_N(t)} \mathscr{K} \mu_{Z_N(t)}) = 0.$$

We conclude this section with a few exercises where the reader can verify that the formalism introduced here encompasses the two main examples of mean field theories presented above, i.e. the two dimensional Euler equation and the Vlasov-Poisson system.

Exercise

(1) Compute $\Delta \ln |x|$ in the sense of distributions on $\mathbf{R}^2$ (answer: $2\pi\delta_0$).
(2) Define

$$K(x, x') := -\tfrac{1}{2\pi}\frac{J(x-x')}{|x-x'|^2}, \quad x \neq x' \in \mathbf{R}^2,$$

where J designates the rotation of an angle $-\frac{\pi}{2}$:

$$J = \begin{pmatrix} 0 & 1 \\ -1 & 0 \end{pmatrix}$$

For each $\omega \equiv \omega(t, x)$ belonging to $C_b^1(\mathbf{R}_+ \times \mathbf{R}^2)$ such that $\mathrm{supp}(\omega(t, \cdot))$ is compact for each $t \geq 0$, prove that the vector field u defined by

$$u(t, x) := \int_{\mathbf{R}^2} K(x, x')\omega(t, x')dx'$$

is of class C_b^1 on $\mathbf{R}_+ \times \mathbf{R}^2$ and satisfies

$$\operatorname{div}_x u(t,x) = 0, \quad \operatorname{div}_x(Ju)(t,x) = \omega(t,x).$$

(3) Conclude that the two dimensional Euler equation for incompressible fluids can be put in the formalism described in the present section, except for the fact that the interaction kernel K is singular on the diagonal of $\mathbf{R}^2 \times \mathbf{R}^2$.

Exercise Let (f, ϕ) be a solution of the Vlasov-Poisson system satisfying the same assumptions as in the exercise of Sect. 1.1.2. Assume further that

$$\iint_{\mathbf{R}^3\times\mathbf{R}^3} f(0,x,v)dxdv = 1, \quad \text{and} \quad \iint_{\mathbf{R}^3\times\mathbf{R}^3} vf(0,x,v)dxdv = 0.$$

(1) Prove that

$$\iint_{\mathbf{R}^3\times\mathbf{R}^3} f(t,x,v)dxdv = 1 \quad \text{and} \quad \iint_{\mathbf{R}^3\times\mathbf{R}^3} vf(t,x,v)dxdv = 0 \quad \text{for all } t \geq 0.$$

(2) Set $z = (x, v)$ and

$$K(z,z') = K(x,v,x',v') := \left(v - v', \frac{q^2}{4\pi\varepsilon_0 m}\frac{x-x'}{|x-x'|^3}\right).$$

Prove that

$$\iint_{\mathbf{R}^3\times\mathbf{R}^3} K(x,v,x',v')f(t,x',v')dx'dv' = (v, -\tfrac{q}{m}\nabla_x\phi(t,x)),$$

where

$$-\Delta_x\phi(t,x) = \frac{q}{\varepsilon_0}\int_{\mathbf{R}^3} f(t,x,v)dv.$$

(3) Conclude that the Vlasov-Poisson system can be put in the formalism described in the present section, except for the fact that the interaction kernel K is singular on the set $\{(x,v,x',v') \in (\mathbf{R}^3)^4 \text{ s.t. } x = x'\}$.

1.3 The Mean Field Characteristic Flow

Henceforth we assume that the interaction kernel $K : \mathbf{R}^d \times \mathbf{R}^d \to \mathbf{R}^d$ satisfies the following assumptions.

First K is skew-symmetric:

$$(HK1) \qquad K(z,z') = -K(z',z) \quad \text{for all } z, z' \in \mathbf{R}^d.$$

Besides, $K \in C^1(\mathbf{R}^d \times \mathbf{R}^d; \mathbf{R}^d)$, with bounded partial derivatives of order 1. In other words, there exists a constant $L \geq 0$ such that

$$(HK2)\ \sup_{z,z' \in \mathbf{R}^d} |\nabla_z K(z,z')| \leq L, \quad \text{and } \sup_{z,z' \in \mathbf{R}^d} |\nabla_{z'} K(z,z')| \leq L.$$

Applying the mean value theorem shows that assumption (HK2) implies that K is Lipschitz continuous in z uniformly in z' (and conversely):

$$\begin{cases} \sup_{z' \in \mathbf{R}^d} |K(z_1,z') - K(z_2,z')| \leq L|z_1 - z_2|, \\ \sup_{z \in \mathbf{R}^d} |K(z,z_1) - K(z,z_2)| \leq L|z_1 - z_2|. \end{cases}$$

Assumption (HK2) also implies that K grows at most linearly at infinity:

$$|K(z,z')| \leq L(|z| + |z'|), \quad z, z' \in \mathbf{R}^d.$$

Notice also that the integral operator $\mathcal{K}$ can be extended to the set of Borel probability measures[5] on $\mathbf{R}^d$ with finite moment of order 1, i.e.

$$\mathcal{P}_1(\mathbf{R}^d) := \left\{ p \in \mathcal{P}(\mathbf{R}^d) \text{ s.t.} \int_{\mathbf{R}^d} |z| p(dz) < \infty \right\},$$

in the obvious manner, i.e.

$$\mathcal{K} p(z) := \int_{\mathbf{R}^d} K(z,z') p(dz').$$

The extended operator $\mathcal{K}$ so defined maps $\mathcal{P}_1(\mathbf{R}^d)$ into the class $\mathrm{Lip}(\mathbf{R}^d; \mathbf{R}^d)$ of Lipschitz continuous vector fields on $\mathbf{R}^d$.

With the assumptions above, one easily arrives at the existence and uniqueness theory for the N-body ODE system.

Theorem 1.3.1 *Assume that the interaction kernel $K \in C^1(\mathbf{R}^d \times \mathbf{R}^d, \mathbf{R}^d)$ satisfies assumptions (HK1-HK2). Then*

(a) for each $N \geq 1$ and each N-tuple $Z_N^{in} = (z_1^{in}, \ldots, z_N^{in})$, the Cauchy problem for the N-particle ODE system

[5] Henceforth, the set of Borel probability measures on $\mathbf{R}^d$ will be denoted by $\mathcal{P}(\mathbf{R}^d)$.

$$\begin{cases} \dot{z}_i(t) = \dfrac{1}{N}\displaystyle\sum_{j=1}^{N} K(z_i(t), z_j(t)), & i = 1, \ldots, N, \\ z_i(0) = z_i^{in}, \end{cases}$$

has a unique solution of class C^1 on $\mathbf{R}$

$$t \mapsto Z_N(t) = (z_1(t), \ldots, z_N(t)) =: T_t Z_N^{in} ;$$

(b) the empirical measure $f(t, dz) := \mu_{T_t Z_N^{in}}$ is a weak solution of the Cauchy problem for the mean field PDE

$$\begin{cases} \partial_t f + \operatorname{div}_z(f \mathscr{K} f) = 0, \\ f\big|_{t=0} = f^{in}. \end{cases}$$

Statement (a) follows from the Cauchy-Lipschitz theorem. Statement (b) follows from the method of characteristics for the transport equation. For the sake of being complete, we sketch the main steps in the proof of statement (b), and leave the details as an exercise to be treated by the reader.

Exercise Let $b \equiv b(t, y) \in C([0, \tau]; \mathbf{R}^d)$ be such that $D_y b \in C([0, \tau]; \mathbf{R}^d)$ and

$$(H) \qquad |b(t, y)| \le \kappa(1 + |y|)$$

for all $t \in [0, \tau]$ and $y \in \mathbf{R}^d$, where κ is a positive constant.
(1) Prove that, for each $t \in [0, \tau]$, the Cauchy problem for the ODE

$$\begin{cases} \dot{Y}(s) = b(s, Y(s)), \\ Y(t) = y, \end{cases}$$

has a unique solution $s \mapsto Y(s, t, y)$. What is the maximal domain of definition of this solution? What is the regularity of the map Y viewed as a function of the 3 variables s, t, y?
(2) What is the role of assumption (H)?
(3) Prove that, for each $t_1, t_2, t_3 \in [0, \tau]$ and $y \in \mathbf{R}^d$, one has

$$Y(t_3, t_2, Y(t_2, t_1, y)) = Y(t_3, t_1, y).$$

(4) Compute

$$\partial_t Y(s, t, y) + b(t, y) \cdot \nabla_y Y(s, t, y).$$

(5) Let $f^{in} \in C^1(\mathbf{R}^d)$. Prove that the Cauchy problem for the transport equation

$$\begin{cases} \partial_t f(t,y) + b(t,y) \cdot \nabla_y f(t,y) = 0, \\ f\big|_{t=0} = f^{in}, \end{cases}$$

has a unique solution $f \in C^1([0,\tau] \times \mathbf{R}^d)$, and that this solution is given by the formula

$$f(t,y) = f^{in}(Y(0,t,y)).$$

(6) Let μ^{in} be a Borel probability measure on $\mathbf{R}^d$. Prove that the push-forward measure[6]

$$\mu(t) := Y(t,0,\cdot)\#\mu^{in}$$

is a weak solution of

$$\begin{cases} \partial_t \mu + \operatorname{div}_y(\mu b) = 0, \\ \mu\big|_{t=0} = \mu^{in}. \end{cases}$$

Hint: for $\phi \in C_c^1(\mathbf{R}^d)$, compute

$$\frac{d}{dt} \int_{\mathbf{R}^d} \phi(Y(t,0,y))\mu^{in}(dy).$$

(7) Prove that the unique weak solution[7] $\mu \in C([0,\tau], w - \mathscr{P}(\mathbf{R}^d))$ of the Cauchy problem considered in (6) is the push-forward measure defined by the formula

$$\mu(t) := Y(t,0,\cdot)\#\mu^{in}$$

for each $t \in [0,\tau]$. (Hint: for $\phi \in C_c^1(\mathbf{R}^d)$, compute

---

[6]Given two measurable spaces $(X, \mathscr{A})$ and $(Y, \mathscr{B})$, a measurable map $\Phi : (X, \mathscr{A}) \to (Y, \mathscr{B})$ and a measure m on $(X, \mathscr{A})$, the push-forward of m under Φ is the measure on $(Y, \mathscr{B})$ defined by the formula

$$\Phi\#m(B) = m(\Phi^{-1}(B)), \quad \text{for all } B \in \mathscr{B}.$$

[7]We designate by $w - \mathscr{P}(\mathbf{R}^d)$ the set $\mathscr{P}(\mathbf{R}^d)$ equipped with the weak topology of probability measures, i.e. the topology defined by the family of semi-distances

$$d_\phi(\mu,\nu) := \left| \int_{\mathbf{R}^d} \phi(z)\mu(dz) - \int_{\mathbf{R}^d} \phi(z)\nu(dz) \right|$$

as ϕ runs through $C_b(\mathbf{R}^d)$.

$$\frac{d}{dt}\langle Y(0,t,\cdot)\#\mu(t),\phi\rangle$$

in the sense of distributions on $(0,\tau)$.)

For a solution of this exercise, see Chap. 1, Sect. 1 of [17].

Our next step is to formulate and solve a new problem that will contain both the N-particle ODE system in the mean field scaling and the mean field PDE.

Theorem 1.3.2 *Assume that the interaction kernel $K \in C^1(\mathbf{R}^d \times \mathbf{R}^d, \mathbf{R}^d)$ satisfies assumptions (HK1-HK2). For each $\zeta^{in} \in \mathbf{R}^d$ and each Borel probability measure $\mu^{in} \in \mathscr{P}_1(\mathbf{R}^d)$, there exists a unique solution denoted by*

$$\mathbf{R} \ni t \mapsto Z(t,\zeta^{in},\mu^{in}) \in \mathbf{R}^d$$

of class C^1 of the problem

$$\begin{cases} \partial_t Z(t,\zeta^{in},\mu^{in}) = (\mathscr{K}\mu(t))(Z(t,\zeta^{in},\mu^{in})), \\ \mu(t) = Z(t,\cdot,\mu^{in})\#\mu^{in}, \\ Z(0,\zeta^{in},\mu^{in}) = \zeta^{in}. \end{cases}$$

Notice that the ODE governing the evolution of $t \mapsto Z(t,\zeta^{in},\mu^{in})$ is set in the single-particle phase space $\mathbf{R}^d$, and not in the N-particle phase space, as is the case of the ODE system studied in Theorem 1.3.1.

Obviously, the ODE appearing in Theorem 1.3.2 is precisely the equation of characteristics for the mean field PDE. Henceforth, we refer to this ODE as the equations of "mean field characteristics", and to its solution Z as the "mean field characteristic flow".

How the mean field characteristic flow Z and the flow T_t associated to the N-particle ODE system are related is explained in the next proposition.

Proposition 1.3.1 *Assume that the interaction kernel $K \in C^1(\mathbf{R}^d \times \mathbf{R}^d, \mathbf{R}^d)$ satisfies assumptions (HK1-HK2). For each $Z_N^{in} = (z_1^{in},\ldots,z_N^{in})$, the solution*

$$T_t Z_N^{in} = (z_1(t),\ldots,z_N(t))$$

of the N-body problem and the mean field characteristic flow $Z(t,\zeta^{in},\mu^{in})$ satisfy

$$z_i(t) = Z(t,z_i^{in},\mu_{Z_N^{in}}), \quad i=1,\ldots,N,$$

for all $t \in \mathbf{R}$.

Proof (Proof of Proposition 1.3.1) Define

$$\zeta_i(t) := Z(t,z_i^{in},\mu_{Z_N^{in}}), \quad i=1,\ldots,N.$$

Then[8]

$$\mu(t) = Z(t, \cdot, \mu_{Z_N^{in}})\#\mu_{Z_N^{in}} = \frac{1}{N}\sum_{j=1}^{N}\delta_{\zeta_j(t)}$$

for all $t \in \mathbf{R}$. Therefore, ζ_i satisfies

$$\dot{\zeta}_i(t) = (\mathscr{K}\mu(t))(\zeta_i(t)) = \frac{1}{N}\sum_{j=1}^{N} K(\zeta_i(t), \zeta_j(t)), \quad i = 1, \ldots, N,$$

for all $t \in \mathbf{R}$. Moreover

$$\zeta_i(0) = Z(0, z_i^{in}, \mu^{in}) = z_i^{in}, \quad i = 1, \ldots, N.$$

Therefore, by uniqueness of the solution of the N-particle equation (Theorem 1.3.1), one has

$$\zeta_i(t) = z_i(t),$$

for all $i = 1, \ldots, N$ and all $t \in \mathbf{R}$.

The proof of Theorem 1.3.2 is a simple variant of the proof of the Cauchy-Lipschitz theorem.

Proof (Proof of Theorem 1.3.2) Let $\mu^{in} \in \mathscr{P}_1(\mathbf{R}^d)$, and denote

$$C_1 := \int_{\mathbf{R}^d} |z|\mu^{in}(dz).$$

Let

$$X := \left\{ v \in C(\mathbf{R}^d; \mathbf{R}^d) \text{ s.t. } \sup_{z \in \mathbf{R}^d} \frac{|v(z)|}{1 + |z|} < \infty \right\},$$

[8]The reader should be aware of the following subtle point. In classical references on distribution theory, such as [57], the Dirac mass is viewed as a distribution, therefore as an object that generalizes the notion of function. There is a notion of pull-back of a distribution under a C^∞ diffeomorphism such that the pull-back of the Dirac mass at y_0 with a C^∞ diffeomorphism $\chi : \mathbf{R}^N \to \mathbf{R}^N$ satisfying $\chi(x_0) = y_0$ is

$$\delta_{y_0} \circ \chi = |\det(D\chi(x_0))|^{-1}\delta_{x_0}.$$

This notion is not to be confused with the push-forward under χ of the Dirac mass at δ_{x_0} viewed as a probability measure, which, according to the definition in the previous footnote is

$$\chi\#\delta_{x_0} = \delta_{y_0}.$$

In particular

$$\chi\#\delta_{x_0} \neq \delta_{x_0} \circ \chi^{-1}$$

unless χ has Jacobian determinant 1 at x_0.

which is a Banach space for the norm

$$\|v\|_X := \sup_{z\in\mathbf{R}^d} \frac{|v(z)|}{1+|z|}.$$

By assumption (HK2) on the interaction kernel K, for each $v, w \in X$, one has

$$\begin{aligned}
&\left|\int_{\mathbf{R}^d} K(v(z), v(z'))\mu^{in}(dz') - \int_{\mathbf{R}^d} K(w(z), w(z'))\mu^{in}(dz')\right| \\
&\qquad \le L\int_{\mathbf{R}^d} (|v(z)-w(z)| + |v(z')-w(z')|)\mu^{in}(dz') \\
&\le L\|v-w\|_X(1+|z|) + L\|v-w\|_X \int_{\mathbf{R}^d} (1+|z'|)\mu^{in}(dz') \\
&\qquad = L\|v-w\|_X(1+|z|+1+C_1) \\
&\qquad \le L\|v-w\|_X(2+C_1)(1+|z|).
\end{aligned}$$

Define a sequence $(Z_n)_{n\ge 0}$ by induction, as follows:

$$\begin{cases} Z_{n+1}(t,\zeta) = \zeta + \displaystyle\int_0^t \int_{\mathbf{R}^d} K(Z_n(t,\zeta), Z_n(t,\zeta'))\mu^{in}(d\zeta')ds, & n \ge 0, \\ Z_0(t,\zeta) = \zeta. \end{cases}$$

One checks by induction with the inequality above that, for each $n \in \mathbf{N}$,

$$\|Z_{n+1}(t,\cdot) - Z_n(t,\cdot)\|_X \le \frac{((2+C_1)L|t|)^n}{n!}\|Z_1(t,\cdot) - Z_0(t,\cdot)\|_X.$$

Since

$$\begin{aligned}
|Z_1(t,\zeta) - \zeta| &= \left|\int_0^t \int_{\mathbf{R}^d} K(\zeta,\zeta')\mu^{in}(d\zeta')ds\right| \\
&\le \int_0^{|t|}\int_{\mathbf{R}^d} L(|\zeta| + |\zeta'|)\mu^{in}(d\zeta')ds \\
&= \int_0^{|t|} L(|\zeta| + C_1)ds \le L(1+C_1)(1+|\zeta|)|t|,
\end{aligned}$$

one has

$$\|Z_{n+1}(t,\cdot) - Z_n(t,\cdot)\|_X \le \frac{((2+C_1)L|t|)^{n+1}}{n!}.$$

Thus, for each $\tau > 0$,

$$Z_n(t,\cdot) \to Z(t,\cdot) \quad \text{in } X \text{ uniformly on } [-\tau,\tau],$$

where $Z \in C(\mathbf{R}; X)$ satisfies

$$Z(t,\zeta) = \zeta + \int_0^t \int_{\mathbf{R}^d} K(Z(s,\zeta), Z(s,\zeta'))\mu^{in}(d\zeta')ds$$

for all $t \in \mathbf{R}$ and all $\zeta \in \mathbf{R}^d$.

If Z and $\tilde{Z} \in C(\mathbf{R}; X)$ satisfy the integral equation above, then

$$Z(t,\zeta) - \tilde{Z}(t,\zeta) = \int_{\mathbf{R}^d} (K(Z(s,\zeta), Z(s,\zeta')) - K(\tilde{Z}(s,\zeta), \tilde{Z}(s,\zeta')))\mu^{in}(d\zeta'),$$

so that, for all $t \in \mathbf{R}$, one has

$$\|Z(t,\cdot) - \tilde{Z}(t,\cdot)\|_X \le L(2+C_1)\left| \int_0^t \|Z(s,\cdot) - \tilde{Z}(s,\cdot)\|_X ds \right|.$$

This implies that

$$\|Z(t,\cdot) - \tilde{Z}(t,\cdot)\|_X = 0$$

by Gronwall's inequality, so that $Z = \tilde{Z}$. Hence the integral equation has only one solution $Z \in C(\mathbf{R}; X)$.

Since $Z \in C(\mathbf{R}_+; X)$, $K \in C^1(\mathbf{R}^d \times \mathbf{R}^d, \mathbf{R}^d)$ satisfies (HK2) and $\mu^{in} \in \mathscr{P}_1(\mathbf{R}^d)$, the function

$$s \mapsto \int_{\mathbf{R}^d} K(Z(s,\zeta), Z(s,\zeta'))\mu^{in}(d\zeta')$$

is continuous on $\mathbf{R}$.

Using the integral equation shows that the function $t \mapsto Z(t,\zeta)$ is of class C^1 on $\mathbf{R}$ and satisfies

$$\begin{cases} \partial_t Z(t,\zeta) = \displaystyle\int_{\mathbf{R}^d} K(Z(t,\zeta), Z(t,\zeta'))\mu^{in}(d\zeta'), \\ Z(0,\zeta) = \zeta. \end{cases}$$

Substituting $z' = Z(t, \zeta')$ in the integral above, one has

$$\int_{\mathbf{R}^d} K(Z(t,\zeta), Z(t,\zeta'))\mu^{in}(d\zeta') = \int_{\mathbf{R}^d} K(Z(t,\zeta), z')Z(t,\cdot)\#\mu^{in}(dz')$$

so that the element Z of $C(\mathbf{R}; X)$ so constructed is the unique solution of the mean field characteristic equation.

References for this and the previous section are [20, 81].

1.4 Dobrushin's Stability Estimate and the Mean Field Limit

1.4.1 The Monge-Kantorovich Distance

For each $r > 1$, we denote by $\mathscr{P}_r(\mathbf{R}^d)$ the set of Borel probability measures on $\mathbf{R}^d$ with a finite moment of order r, i.e. satisfying

$$\int_{\mathbf{R}^d} |z|^r \mu(dz) < \infty.$$

Given $\mu, \nu \in \mathscr{P}_r(\mathbf{R}^d)$, we define $\Pi(\mu, \nu)$ to be the set of Borel probability measures π on $\mathbf{R}^d \times \mathbf{R}^d$ with first and second marginals μ and ν respectively. Equivalently, for each $\pi \in \mathscr{P}(\mathbf{R}^d \times \mathbf{R}^d)$,

$$\pi \in \Pi(\mu,\nu) \Leftrightarrow \iint_{\mathbf{R}^d\times\mathbf{R}^d} (\phi(x) + \psi(y))\pi(dxdy) = \int_{\mathbf{R}^d} \phi(x)\mu(dx) + \int_{\mathbf{R}^d} \psi(y)\nu(dy)$$

for each $\phi, \psi \in C(\mathbf{R}^d)$ such that $\phi(z) = O(|z|^r)$ and $\psi(z) = O(|z|^r)$ as $|z| \to \infty$.

Probability measures belonging to $\Pi(\mu, \nu)$ are sometimes referred to as "couplings of μ and ν".

Exercise Check that, if μ and $\nu \in \mathscr{P}_r(\mathbf{R}^d)$ for $r > 0$, then $\Pi(\mu, \nu) \subset \mathscr{P}_r(\mathbf{R}^d \times \mathbf{R}^d)$.

With these elements of notation, we now introduce the notion of Monge-Kantorovich distance.

Definition 1.4.1 For each $r \geq 1$ and each $\mu, \nu \in \mathscr{P}_r(\mathbf{R}^d)$, the Monge-Kantorovich distance $\text{dist}_{MK,r}(\mu, \nu)$ between μ and ν is defined by the formula

$$\text{dist}_{MK,r}(\mu,\nu) = \inf_{\pi\in\Pi(\mu,\nu)} \left(\iint_{\mathbf{R}^d\times\mathbf{R}^d} |x-y|^r \pi(dxdy) \right)^{1/r}.$$

These distances also go by the name of "Kantorovich-Rubinstein distances" or "Wasserstein distances"—although the minimization problem in the right hand side of the formula defining $\operatorname{dist}_{MK,r}$ had been considered for the first time by Monge[9] and systematically studied by Kantorovich.

We shall use the Monge-Kantorovich distances only as a convenient tool for studying the stability of the mean field characteristic flow. Therefore, we shall not attempt to present the mathematical theory of these distances and refer instead to the C. Villani's books [103, 104] for a very detailed discussion of this topic.

However, it is useful to know the following property that is special to the case $r = 1$.

Proposition 1.4.1 *The Monge-Kantorovich distance with exponent* 1 *is also given by the formula*

$$\operatorname{dist}_{MK,1}(\mu,\nu) = \sup_{\substack{\phi\in\operatorname{Lip}(\mathbf{R}^d)\\ \operatorname{Lip}(\phi)\le 1}} \left| \int_{\mathbf{R}^d} \phi(z)\mu(dz) - \int_{\mathbf{R}^d} \phi(z)\nu(dz) \right|,$$

with the notation

$$\operatorname{Lip}(\phi) := \sup_{x\neq y\in\mathbf{R}^d} \frac{|\phi(x)-\phi(y)|}{|x-y|}.$$

for the Lipschitz constant of ϕ.

The proof of this proposition is based on a duality argument in optimization: see for instance Theorems 1.14 and 7.3(i) in [103].

1.4.2 Dobrushin's Estimate

As explained in Proposition 1.3.1, the mean field characteristic flow contains all the relevant information about both the mean field PDE and the N-particle ODE system.

Dobrushin's approach to the mean field limit is based on the idea of proving the stability of the mean field characteristic flow $Z(t,\zeta^{in},\mu^{in})$ in *both* the initial position in phase space ζ^{in} and the initial distribution μ^{in}. As we shall see, the Monge-Kantorovich distance is the best adapted mathematical tool to measure this stability.

Dobrushin's idea ultimately rests on the following key computation. Let $\zeta_1^{in},\zeta_2^{in}\in\mathbf{R}^d$, and let $\mu_1^{in},\mu_2^{in}\in\mathscr{P}_1(\mathbf{R}^d)$. Then

---

[9]Monge's original problem was to minimize over the class of all Borel measurable transportation maps $T:\mathbf{R}^d\to\mathbf{R}^d$ such that $T\#\mu=\nu$ the transportation cost $\int_{\mathbf{R}^d}|x-T(x)|\mu(dx)$.

$$Z(t,\zeta_1,\mu_1^{in})-Z(t,\zeta_2,\mu_2^{in})=\zeta_1-\zeta_2$$
$$+\int_0^t\int_{\mathbf{R}^d}K(Z(s,\zeta_1,\mu_1^{in}),z')\mu_1(s,dz')ds$$
$$-\int_0^t\int_{\mathbf{R}^d}K(Z(s,\zeta_2,\mu_2^{in}),z')\mu_2(s,dz')ds.$$

Since $\mu_j(t)=Z(t,\cdot,\mu_j^{in})\#\mu_j^{in}$ for $j=1,2$, each inner integral on the right hand side of the equality above can be expressed as follows:

$$\int_{\mathbf{R}^d}K(Z(s,\zeta_j,\mu_j^{in}),z')\mu_j(s,dz')$$
$$=\int_{\mathbf{R}^d}K(Z(s,\zeta_j,\mu_j^{in}),Z(s,\zeta_j',\mu_j^{in}))\mu_j^{in}(d\zeta_j')$$

for $j=1,2$. Therefore, for each coupling $\pi^{in}\in\mathscr{P}_1(\mu_1^{in},\mu_2^{in})$, one has

$$\int_{\mathbf{R}^d}K(Z(s,\zeta_1,\mu_1^{in}),Z(s,\zeta_1',\mu_1^{in}))\mu_1^{in}(d\zeta_1')$$
$$-\int_{\mathbf{R}^d}K(Z(s,\zeta_2,\mu_2^{in}),Z(s,\zeta_2',\mu_2^{in}))\mu_2^{in}(d\zeta_2')$$
$$=\iint_{\mathbf{R}^d\times\mathbf{R}^d}(K(Z(s,\zeta_1,\mu_1^{in}),Z(s,\zeta_1',\mu_1^{in}))$$
$$-K(Z(s,\zeta_2,\mu_2^{in}),Z(s,\zeta_2',\mu_2^{in})))\pi^{in}(d\zeta_1',d\zeta_2'),$$

so that

$$Z(t,\zeta_1,\mu_1^{in})-Z(t,\zeta_2,\mu_2^{in})=\zeta_1-\zeta_2$$
$$+\int_0^t\iint_{\mathbf{R}^d\times\mathbf{R}^d}(K(Z(s,\zeta_1,\mu_1^{in}),Z(s,\zeta_1',\mu_1^{in}))$$
$$-K(Z(s,\zeta_2,\mu_2^{in}),Z(s,\zeta_2',\mu_2^{in})))\pi^{in}(d\zeta_1',d\zeta_2')ds.$$

This last equality is the key observation in Dobrushin's argument, which explains the role of couplings of μ_1^{in} and μ_2^{in} in this problem, and therefore why it is natural to use the Monge-Kantorovich distance.

After this, the end of the argument is plain sailing. By assumption (HK2) on the interaction kernel K, for all $a,a',b,b'\in\mathbf{R}^d$, one has

$$\begin{aligned}|K(a,a') - K(b,b')| &\le |K(a,a') - K(b,a')| + |K(b,a') - K(b,b')| \\ &\le L|a-b| + L|a'-b'|.\end{aligned}$$

Therefore

$$\begin{aligned}&|Z(t,\zeta_1,\mu_1^{in}) - Z(t,\zeta_2,\mu_2^{in})| \\ &\quad\le |\zeta_1 - \zeta_2| + L\int_0^t |Z(s,\zeta_1,\mu_1^{in}) - Z(s,\zeta_2,\mu_2^{in})|ds \\ &\quad + L\int_0^t \iint_{\mathbf{R}^d\times\mathbf{R}^d} |Z(s,\zeta_1',\mu_1^{in}) - Z(s,\zeta_2',\mu_2^{in})|\pi^{in}(d\zeta_1' d\zeta_2')ds.\end{aligned}$$

It is convenient at this point to introduce the notation

$$D[\pi](s) := \iint_{\mathbf{R}^d\times\mathbf{R}^d} |Z(s,\zeta_1',\mu_1^{in}) - Z(s,\zeta_2',\mu_2^{in})|\pi(d\zeta_1' d\zeta_2')$$

for each $\pi \in \mathscr{P}_1(\mathbf{R}^d \times \mathbf{R}^d)$. Thus, the previous inequality becomes

$$\begin{aligned}|Z(t,\zeta_1,\mu_1^{in}) - Z(t,\zeta_2,\mu_2^{in})| &\le |\zeta_1 - \zeta_2| \\ &+ L\int_0^t |Z(s,\zeta_1,\mu_1^{in}) - Z(s,\zeta_2,\mu_2^{in})|ds + L\int_0^t D[\pi^{in}](s)ds.\end{aligned}$$

Integrating both sides of the inequality above with respect to $\pi^{in}(d\zeta_1 d\zeta_2)$ leads to

$$\begin{aligned}D[\pi^{in}](t) &\le \iint_{\mathbf{R}^d\times\mathbf{R}^d} |\zeta_1 - \zeta_2|\pi^{in}(d\zeta_1 d\zeta_2) + 2L\int_0^t D[\pi^{in}](s)ds \\ &= D[\pi^{in}](0) + 2L\int_0^t D[\pi^{in}](s)ds.\end{aligned}$$

By Gronwall's inequality, we conclude that, for all $t \in \mathbf{R}$, one has

$$D[\pi^{in}](t) \le D[\pi^{in}](0)e^{2L|t|}.$$

Now we can state Dobrushin's stability theorem.

Theorem 1.4.1 (Dobrushin) *Assume that* $K \in C^1(\mathbf{R}^d \times \mathbf{R}^d, \mathbf{R}^d)$ *satisfies (HK1-HK2). Let* $\mu_1^{in}, \mu_2^{in} \in \mathscr{P}_1(\mathbf{R}^d)$. *For all* $t \in \mathbf{R}$, *let*

$$\begin{cases} \mu_1(t) = Z(t, \cdot, \mu_1^{in})\#\mu_1^{in}, \\ \mu_2(t) = Z(t, \cdot, \mu_2^{in})\#\mu_2^{in}, \end{cases}$$

where Z is the mean field characteristic flow defined in Theorem 1.3.2.

Then, for all $t \in \mathbf{R}$, *one has*

$$\mathrm{dist}_{MK,1}(\mu_1(t), \mu_2(t)) \le e^{2L|t|}\,\mathrm{dist}_{MK,1}(\mu_1^{in}, \mu_2^{in}).$$

Proof We have seen that, for all $\mu_1^{in}, \mu_2^{in} \in \mathscr{P}_1(\mathbf{R}^d)$ and all $\pi^{in} \in \Pi(\mu_1^{in}, \mu_2^{in})$, one has

$$D[\pi^{in}](t) \le D[\pi^{in}](0)e^{2L|t|}$$

for all $t \in \mathbf{R}$.

Since $Z(t, \cdot, \mu_j^{in})\#\mu_j^{in} = \mu_j(t)$ for $j = 1, 2$, the map

$$\Phi_t : (\zeta_1, \zeta_2) \mapsto (Z(t, \zeta_1, \mu_1^{in}), Z(t, \zeta_2, \mu_2^{in}))$$

satisfies

$$\Phi_t\#\pi^{in} = \pi(t) \in \Pi(\mu_1(t), \mu_2(t))$$

for all $t \in \mathbf{R}$, since $\pi^{in} \in \Pi(\mu_1^{in}, \mu_2^{in})$.

Thus

$$\begin{aligned} \mathrm{dist}_{MK,1}(\mu_1(t), \mu_2(t)) &= \inf_{\pi \in \Pi(\mu_1(t),\mu_2(t))} \iint_{\mathbf{R}^d\times\mathbf{R}^d} |\zeta_1 - \zeta_2|\pi(d\zeta_1 d\zeta_2) \\ &\le \inf_{\pi^{in} \in \Pi(\mu_1^{in},\mu_2^{in})} \iint_{\mathbf{R}^d\times\mathbf{R}^d} |Z(t, \zeta_1, \mu_1^{in}) - Z(t, \zeta_2, \mu_2^{in})|\pi^{in}(d\zeta_1 d\zeta_2) \\ &= \inf_{\pi^{in} \in \Pi(\mu_1^{in},\mu_2^{in})} D[\pi^{in}](t) \le e^{2L|t|} \inf_{\pi^{in} \in \Pi(\mu_1^{in},\mu_2^{in})} D[\pi^{in}](0) \\ &= e^{2L|t|}\,\mathrm{dist}_{MK,1}(\mu_1^{in}, \mu_2^{in}) \end{aligned}$$

which concludes the proof.

The discussion in this section is inspired from [29]; see also [76]. The interested reader is also referred to the very interesting paper [74] where Monge-Kantorovich distances with exponents different from 1 are used in the same context—see also [52].

1.4.3 The Mean Field Limit

The mean field limit of the N-particle system is a consequence of Dobrushin's stability theorem, as explained below.

Theorem 1.4.2 *Assume that the interaction kernel $K \in C^1(\mathbf{R}^d \times \mathbf{R}^d)$ and satisfies assumptions (HK1-HK2). Let f^{in} be a probability density on $\mathbf{R}^d$ such that*

$$\int_{\mathbf{R}^d} |z| f^{in}(z)dz < \infty.$$

Then the Cauchy problem for the mean field PDE

$$\begin{cases} \partial_t f(t,z) + \operatorname{div}_z(f(t,z)\mathscr{K}f(t,z)) = 0, \quad z \in \mathbf{R}^d,\ t \in \mathbf{R}, \\ f\big|_{t=0} = f^{in} \end{cases}$$

has a unique weak solution $f \in C(\mathbf{R}; L^1(\mathbf{R}^d))$.

For each $N \geq 1$, let $\mathfrak{z}(N) = (z^{in}_{1,N}, \ldots, z^{in}_{N,N}) \in (\mathbf{R}^d)^N$ be such that

$$\mu_{\mathfrak{z}(N)} = \frac{1}{N}\sum_{=1}^{N} \delta_{z^{in}_{j,N}}$$

satisfies

$$\operatorname{dist}_{MK,1}(\mu_{\mathfrak{z}(N)}, f^{in}) \to 0 \quad \text{as } N \to \infty.$$

Let $t \mapsto T_t\mathfrak{z}(N) = (z_{1,N}(t), \ldots, z_{N,N}(t)) \in (\mathbf{R}^d)^N$ be the solution of the N-particle ODE system with initial data $\mathfrak{z}(N)$, i.e.

$$\begin{cases} \dot{z}_i(t) = \dfrac{1}{N}\displaystyle\sum_{j=1}^{N} K(z_i(t), z_j(t)), \quad i = 1, \ldots, N, \\ z_i(0) = z_i^{in}. \end{cases}$$

Then[10]

$$\mu_{T_t\mathfrak{z}(N)} \rightharpoonup f(t,\cdot)\mathscr{L}^d \text{ as } N \to \infty$$

in the weak topology of probability measures, with convergence rate

$$\operatorname{dist}_{MK,1}(\mu_{T_t\mathfrak{z}(N)}, f(t,\cdot)\mathscr{L}^d) \leq e^{2L|t|} \operatorname{dist}_{MK,1}(\mu_{\mathfrak{z}(N)}, f^{in}) \to 0$$

as $N \to \infty$ for each $t \in \mathbf{R}$.

---

[10]The notation $\mathscr{L}^d$ designates the Lebesgue measure on $\mathbf{R}^d$.

Proof By Theorem 1.3.2 and questions 6 and 7 in the exercise on the method of characteristics before Theorem 1.3.2, one has

$$f(t,\cdot)\mathscr{L}^d = Z(t,\cdot,f^{in}\mathscr{L}^d)\#f^{in}\mathscr{L}^d$$

for all $t \in \mathbf{R}$. This implies in particular the uniqueness of the solution of the Cauchy problem in $C(\mathbf{R}; L^1(\mathbf{R}^d))$ for the mean field PDE.

By Proposition 1.3.1,

$$\mu_{T_t\mathfrak{Z}(N)} = Z(t,\cdot,\mu_{\mathfrak{Z}(N)})\#\mu_{\mathfrak{Z}(N)}$$

for all $t \in \mathbf{R}$.

By Dobrushin's stability estimate,

$$\mathrm{dist}_{MK,1}(\mu_{T_t\mathfrak{Z}(N)}, f(t,\cdot)\mathscr{L}^d) \le e^{2L|t|}\,\mathrm{dist}_{MK,1}(\mu_{\mathfrak{Z}(N)}, f^{in})$$

for all $t \in \mathbf{R}$, and since we have chosen $\mathfrak{Z}(N)$ so that

$$\mathrm{dist}_{MK,1}(\mu_{\mathfrak{Z}(N)}, f^{in}) \to 0$$

as $N \to \infty$, we conclude that

$$\mathrm{dist}_{MK,1}(\mu_{T_t\mathfrak{Z}(N)}, f(t,\cdot)\mathscr{L}^d) \to 0$$

as $N \to \infty$ for each $t \in \mathbf{R}$.

As for weak convergence, pick $\phi \in \mathrm{Lip}(\mathbf{R}^d)$; then

$$\begin{aligned}\left|\int_{\mathbf{R}^d} \phi(z)\mu_{T_t\mathfrak{Z}(N)}(dz) - \int_{\mathbf{R}^d} \phi(z)f(t,z)dz\right| &= \left|\iint_{\mathbf{R}^d\times\mathbf{R}^d} (\phi(x)-\phi(y))\pi(dxdy)\right| \\ &\le \iint_{\mathbf{R}^d\times\mathbf{R}^d} |\phi(x)-\phi(y)|\pi(dxdy) \\ &\le \mathrm{Lip}(\phi)\iint_{\mathbf{R}^d\times\mathbf{R}^d} |x-y|\pi(dxdy)\end{aligned}$$

for each $\pi \in \Pi(\mu_{T_t\mathfrak{Z}(N)}, f(t,\cdot)\mathscr{L}^d)$. Thus

$$
\begin{aligned}
&\left|\int_{\mathbf{R}^d} \phi(z)\mu_{T_t\mathfrak{Z}(N)}(dz) - \int_{\mathbf{R}^d} \phi(z)f(t,z)dz\right| \\
&\quad \leq \operatorname{Lip}(\phi) \inf_{\pi\in\Pi(\mu_{T_t\mathfrak{Z}(N)},f(t,\cdot)\mathscr{L}^d)} \iint_{\mathbf{R}^d\times\mathbf{R}^d} |x-y|\pi(dxdy) \\
&\quad = \operatorname{Lip}(\phi)\operatorname{dist}_{MK,1}(\mu_{T_t\mathfrak{Z}(N)}, f(t,\cdot)\mathscr{L}^d) \to 0
\end{aligned}
$$

for each $t \in \mathbf{R}$ as $N \to \infty$. (Notice that the inequality above is an obvious consequence of the definition of $\operatorname{dist}_{MK,1}$, so that the equality in Proposition 1.4.1 is not needed here.)

This is true in particular for each $\phi \in C_c^1(\mathbf{R}^d)$, and since $C_c^1(\mathbf{R}^d)$ is dense in $C_c(\mathbf{R}^d)$, we conclude that

$$
\int_{\mathbf{R}^d} \phi(z)\mu_{T_t\mathfrak{Z}(N)}(dz) \to \int_{\mathbf{R}^d} \phi(z)f(t,z)dz
$$

as $N \to \infty$ for each $\phi \in C_c(\mathbf{R}^d)$. Since

$$
\int_{\mathbf{R}^d} \mu_{T_t\mathfrak{Z}(N)}(dz) = \int_{\mathbf{R}^d} f(t,z)dz = 1
$$

for all $t \in \mathbf{R}$, we conclude that the convergence above holds for each $\phi \in C_b(\mathbf{R}^d)$, which means that

$$
\mu_{T_t\mathfrak{Z}(N)} \to f(t,\cdot)\mathscr{L}^d
$$

as $N \to \infty$ in the weak topology of probability measures, by applying Theorem 6.8 in Chap. II of [75], sometimes referred to as the "portmanteau theorem".

The theorem above is the main result on the mean field limit in [20, 29, 81].

1.4.4 On the Choice of the Initial Data

In practice, using Theorem 1.4.2 as a rigorous justification of the mean field limit requires being able to generate N-tuples of the form $\mathfrak{Z}(N) = (z_{1,N}^{in}, \ldots, z_{N,N}^{in}) \in (\mathbf{R}^d)^N$ such that

$$
\mu_{\mathfrak{Z}(N)} = \frac{1}{N}\sum_{=1}^{N} \delta_{z_{j,N}^{in}}
$$

satisfies

$$
\operatorname{dist}_{MK,1}(\mu_{\mathfrak{Z}(N)}, f^{in}) \to 0 \quad \text{as } N \to \infty.
$$

Assume that f^{in} is a probability density on $\mathbf{R}^d$ such that

$$\int_{\mathbf{R}^d} |z|^2 f(z)dz < \infty.$$

Let $\Omega := (\mathbf{R}^d)^{\mathbf{N}^*}$, the set of sequences of points in $\mathbf{R}^d$ indexed by $\mathbf{N}^*$. Let $\mathscr{F}$ be the σ-algebra on Ω generated by cylinders, i.e. by sets of the form

$$\prod_{n\geq 1} B_n \quad \text{with } B_n \text{ Borel set in } \mathbf{R}^d$$
$$\text{and } B_n = \mathbf{R}^d \text{ for all but finitely many } n.$$

Finally, we endow the measurable space $(\Omega, \mathscr{F})$ with the factorized probability measure $\mathbf{P} := (f^{in})^{\otimes\infty}$, defined on the set of cylinders of Ω by the formula

$$\mathbf{P}\left(\prod_{n\geq 1} B_n\right) = \prod_{n\geq 1} f^{in}(B_n).$$

(Notice that $f^{in}(B_n) = 1$ for all but finitely many n, since $B_n = \mathbf{R}^d$ except for finitely many n.)

Theorem 1.4.3 *For each* $\mathbf{z}^{in} = (z_k^{in})_{k\geq 1} \in \Omega$, *let* $Z_N^{in} = (z_1^{in}, \dots, z_N^{in})$. *Then*

$$\mathrm{dist}_{MK,1}(\mu_{Z_N^{in}}, f^{in}\mathscr{L}^d) \to 0$$

as $N \to \infty$ *for* $\mathbf{P}$-*a.e.* $\mathbf{z}^{in} \in \Omega$.

Proof For $\phi \in C_c(\mathbf{R}^d)$ or $\phi(z) = |z|$, consider the sequence of random variables on $(\Omega, \mathscr{F})$ defined by

$$Y_n(\mathbf{z}) = \phi(z_n),$$

where

$$\mathbf{z} := (z_1, \dots, z_n, \dots) \in \Omega.$$

The random variables Y_n are identically distributed, since

$$\mathbf{P}(Y_n \geq a) = \int_{\mathbf{R}^d} \mathbf{1}_{\phi(z)\geq a} f^{in}(z)dz$$

is independent of n.

The random variables Y_n are also independent, since for all $N \geq 1$ and all $g_1, \dots, g_N \in C_b(\mathbf{R})$, one has

$$\mathbf{E}^{\mathbf{P}}(g_1(Y_1)\dots g_N(Y_N)) = \prod_{k=1}^{N} \int_{\mathbf{R}^d} g_k(\phi(z)) f^{in}(z)dz = \prod_{k=1}^{N} \mathbf{E}^{\mathbf{P}}(g_k(Y_k)).$$

Finally, the random variables Y_n have finite variance since

$$\mathbf{E}^{\mathbf{P}}(|Y_n|^2) = \int_{\mathbf{R}^d} |z|^2 f^{in}(z)dz < \infty.$$

By the strong law of large numbers (see Theorem 3.27 in [21]), one has

$$\left\langle \frac{1}{N}\sum_{k=1}^{N} \delta_{z_k}, \phi \right\rangle = \frac{1}{N}\sum_{k=1}^{N} Y_k \to \mathbf{E}^{\mathbf{P}}(Y_1) = \int_{\mathbf{R}^d} \phi(z) f^{in}(z)dz$$

for $\mathbf{P}$-a.e. $\mathbf{z}$ as $N \to \infty$.

Since $C_c(\mathbf{R}^d)$ is separable, one can assume that the $\mathbf{P}$-negligible set is the same for all $\phi \in C_c(\mathbf{R}^d)$, and take its union with the one corresponding to $\phi(z) = |z|$. This means precisely that

$$\frac{1}{N}\sum_{k=1}^{N} \delta_{z_k} \to f^{in}\mathscr{L}^d$$

weakly in $\mathscr{P}_1(\mathbf{R}^d)$ for $\mathbf{P}$-a.e. $\mathbf{z} \in \Omega$. One concludes the proof with the lemma below.

Lemma 1.4.1 *The Monge-Kantorovich distance* $\text{dist}_{MK,1}$ *metricizes the topology of weak convergence on* $\mathscr{P}_1(\mathbf{R}^d)$*. In other words, given a sequence* $(\mu_n)_{n\geq 1}$ *of elements of* $\mathscr{P}_1(\mathbf{R}^d)$ *and* $\mu \in \mathscr{P}_1(\mathbf{R}^d)$*, the two following statements are equivalent:*

(1) $\text{dist}_{MK,1}(\mu_n, \mu) \to 0$ *as* $n \to \infty$*;*

(2) $\mu_n \to \mu$ *weakly in* $\mathscr{P}(\mathbf{R}^d)$ *as* $n \to \infty$ *and*

$$\sup_n \int_{\mathbf{R}^d} |z|\mathbf{1}_{|z|\geq R}\mu_n(dz) \to 0 \quad \text{as } R \to \infty.$$

For a proof of Lemma 1.4.1, see [103].

Exercise The reader is invited to verify the fact that one can choose the $\mathbf{P}$-negligible set that appears in the proof of Theorem 1.4.3 to be the same for all $\phi \in C_c(\mathbf{R}^d)$ and for $\phi(z) = |z|$. Here is an outline of the argument.

(a) Let $R > 0$; let E_R be the set of real-valued continuous functions defined on $[-R, R]^d$ that vanish identically on $\partial[-R, R]^d$, equipped with the sup-norm

$$\|\phi\| := \sup_{x\in[-R,R]^d} |\phi(x)|.$$

Prove that E_R is a separable Banach space.

Denote by $\mathcal{N}_\phi$ be the set of $\mathbf{z} \in \Omega$ such that

$$\left\langle \frac{1}{N}\sum_{k=1}^{N} \delta_{z_k}, \phi \right\rangle$$

does not converge to

$$\int_{\mathbf{R}^d} \phi(z) f^{in}(z)dz$$

as $N \to \infty$. Let $R > 0$ and let $(\phi_n)_{n\geq 1}$ be a dense sequence of elements of E_R, extended by 0 to $\mathbf{R}^d$. Define

$$\mathcal{N}_R := \bigcup_{n\geq 1} \mathcal{N}_{\phi_n}.$$

(b) Prove that

$$\left\langle \frac{1}{N}\sum_{k=1}^{N} \delta_{z_k}, \phi \right\rangle \to \int_{\mathbf{R}^d} \phi(z) f^{in}(z)dz$$

as $N \to \infty$ for all $\phi \in E_R$ and all $\mathbf{z} \notin \mathcal{N}_R$. (Hint: pick $\phi \in E_R$ and $\varepsilon > 0$, and choose $m := m(\phi, \varepsilon)$ such that $\|\phi - \phi_m\| < \varepsilon$. With the decomposition

$$\begin{aligned}\left\langle \frac{1}{N}\sum_{k=1}^{N} \delta_{z_k}, \phi \right\rangle - \int_{\mathbf{R}^d} \phi(z) f^{in}(z)dz &= \left\langle \frac{1}{N}\sum_{k=1}^{N} \delta_{z_k}, \phi - \phi_m \right\rangle \\ &+ \left\langle \frac{1}{N}\sum_{k=1}^{N} \delta_{z_k}, \phi_m \right\rangle - \int_{\mathbf{R}^d} \phi_m(z) f^{in}(z)dz \\ &+ \int_{\mathbf{R}^d} (\phi_m(z) - \phi(z)) f^{in}(z)dz,\end{aligned}$$

prove that

$$\left| \left\langle \frac{1}{N}\sum_{k=1}^{N} \delta_{z_k}, \phi \right\rangle - \int_{\mathbf{R}^d} \phi(z) f^{in}(z)dz \right| < 3\varepsilon$$

for all $\mathbf{z} \notin \mathcal{N}_R$ provided that $N \geq N_0 = N_0(\varepsilon, \phi)$.)

(c) Complete the proof of Theorem 1.4.3.

Thus, using Theorem 1.4.2 to prove the mean field limit requires choosing

$$\mathfrak{Z}(N) = (z_{1,N}^{in}, \ldots, z_{N,N}^{in}) \in (\mathbf{R}^d)^N$$

for each $N \geq 1$ so that

$$\mathrm{dist}_{MK,1}(\mu_{\mathfrak{Z}(N)}, f^{in}\mathscr{L}^d) \to 0 \quad \text{as } N \to \infty.$$

Theorem 1.4.3 provides us with a strategy for making this choice, which is to draw an infinite sequence z_j^{in} at random and independently with distribution $f^{in}\mathscr{L}^d$, and to set $z_{j,N}^{in} := z_j^{in}$. This strategy avoids the unpleasant task of having to change the first terms in $\mathfrak{Z}(N)$ as $N \to \infty$.

Since Dobrushin's estimate bounds $\mathrm{dist}_{MK,1}(f(t,\cdot)\mathscr{L}^d, \mu_{T_t\mathfrak{Z}(N)})$ in terms of

$$\mathrm{dist}_{MK,1}(f^{in}\mathscr{L}^d, \mu_{\mathfrak{Z}(N)}),$$

having an explicit bound on $\mathrm{dist}_{MK,1}(f^{in}\mathscr{L}^d, \mu_{\mathfrak{Z}(N)})$ would provide us with a quantitative error estimate for the mean field limit. Such a bound will be given below—see Theorem 1.6.3.

More details on the topics discussed in the present section are to be found in [20], as well as a precise statement concerning the behavior of fluctuations around the mean field limit—in some sense, the asymptotic behavior at next order after the mean field limit (see Theorem 3.5 in [20]).

1.5 The BBGKY Hierarchy and the Mean Field Limit

In the previous derivation of the mean field limit of the N-particle system with interaction kernel K satisfying assumptions (HK1)-(HK2), we benefited from a happy circumstance, i.e. the fact that the empirical measure built on any solution of the N-particle ODE system is an *exact* solution of the mean field PDE. This is why the mean field limit was reduced to the stability of the solution of the mean field PDE in terms of its initial data, which follows from Dobrushin's stability estimate.

However, there are other situations in statistical mechanics where the empirical measure built on solutions of the N-particle ODE system may not be an exact solution of the target equation—the best known example of this being the Boltzmann equation of the kinetic theory of gases. There are various examples of such situations where the mean field limit can nevertheless be justified rigorously—see for instance [79, 100] in the case of random dynamics.

There are also situations where there is no clear notion of empirical measure—think for instance to the N-body problem in quantum mechanics: in that case, it is impossible to exactly localize any one of the N particles in phase space, according to the Heisenberg uncertainty principle.

In the present section, we present another approach to the mean field limit of N-particle systems, that is, in some sense, more systematic than the method based on empirical measure and that can be applied to a greater variety of situations (including quantum models, as we shall see later).

1.5.1 N-Particle Distributions

The state at time t of a system of N identical particles located at the positions $z_1(t), \ldots, z_N(t)$ in the single-particle phase space $\mathbf{R}^d$ was described in the previous section by means of the empirical measure

$$\mu_{Z_N(t)} := \frac{1}{N} \sum_{i=1}^{N} \delta_{z_i(t)},$$

where $Z_N(t) := (z_1(t), \ldots, z_N(t))$. The empirical measure is a probability measure in the single-particle phase space $\mathbf{R}^d$ as mentioned above. This measure is parametrized by the element $Z_N(t)$ of the N-particle phase space $(\mathbf{R}^d)^N$.

Another way of describing the state of the same system of N particles at time t is to use its N-particle distribution function, that is

$$F_N(t, z_1, \ldots, z_N).$$

More generally, one could think of the N-particle distribution as being a probability measure on the N-body phase space $(\mathbf{R}^d)^N$

$$F_N(t, dz_1 \ldots dz_N).$$

The meaning of this N-particle distribution is as follows. Let $A_j \subset \mathbf{R}^d$ be Borel measurable sets for $j = 1, \ldots, N$; then, the joint probability at time t to have particle 1 in A_1, particle 2 in A_2...and particle N in A_N is

$$\int_{A_1} \int_{A_2} \ldots \int_{A_N} F(t, dz_1 dz_2 \ldots dz_N).$$

Now, we are interested in situations where all the particles in the N-particle system considered are identical. (For example, all electrons in the universe are identical; ions of any given species in a plasma are identical too.) Therefore, for any permutation $\sigma \in \mathfrak{S}_N$, the joint probability of having particle 1 in A_1, particle 2 in A_2...and particle N in A_N is equal to the joint probability of having particle 1 in $A_{\sigma^{-1}(1)}$, particle 2 in $A_{\sigma^{-1}(2)}$...and particle N in $A_{\sigma^{-1}(N)}$. This is indeed obvious since it is impossible to distinguish particle 1 from particle $\sigma^{-1}(1)$, particle 2 from particle $\sigma^{-1}(2)$...and particle N from particle $\sigma^{-1}(N)$. Thus

$$\int_{A_1}\int_{A_2}\dots\int_{A_N} F(t, dz_1 dz_2 \dots dz_N)$$
$$= \int_{A_{\sigma^{-1}(1)}}\int_{A_{\sigma^{-1}(2)}}\dots\int_{A_{\sigma^{-1}(N)}} F(t, dz_1 dz_2 \dots dz_N).$$

Equivalently

$$\int_{(\mathbf{R}^d)^N} \mathbf{1}_{A_1\times A_2\times\dots\times A_N}(z_1, z_2, \dots, z_N) F(t, dz_1 dz_2 \dots dz_N)$$
$$= \int_{(\mathbf{R}^d)^N} \mathbf{1}_{A_1\times A_2\times\dots\times A_N}(z_{\sigma(1)}, z_{\sigma(2)}, \dots, z_{\sigma(N)}) F(t, dz_1 dz_2 \dots dz_N).$$

For $\sigma \in \mathfrak{S}_N$, define

$$S_\sigma : (z_1, \dots, z_N) \mapsto (z_{\sigma(1)}, \dots, z_{\sigma(N)}).$$

The equality above is recast as

$$\int_{(\mathbf{R}^d)^N} \mathbf{1}_{A_1\times A_2\times\dots\times A_N}(z_1, z_2, \dots, z_N) F(t, dz_1 dz_2 \dots dz_N)$$
$$= \int_{(\mathbf{R}^d)^N} \mathbf{1}_{A_1\times A_2\times\dots\times A_N} \circ S_\sigma(z_1, \dots, z_N) F(t, dz_1 dz_2 \dots dz_N)$$

for all $A_1, A_2, \dots, A_N$ Borel subsets of $\mathbf{R}^d$. This is equivalent to the equality

$$S_\sigma \# F(t, \cdot) = F(t, \cdot)$$

for all $\sigma \in \mathfrak{S}_N$.

Obviously, when F is a probability density instead of a probability measure, the condition

$$S_\sigma \# F(t, \cdot)\mathscr{L}^{dN} = F(t, \cdot)\mathscr{L}^{dN}$$

for all $\sigma \in \mathfrak{S}_N$ is equivalent to the condition

$$F(t, z_{\sigma(1)}, \dots, z_{\sigma(N)}) = F(t, S_\sigma(z_1, \dots, z_N)) = F(t, z_1, \dots, z_N)$$

for all $\sigma \in \mathfrak{S}_N$ and for all $z_1, \ldots, z_N \in \mathbf{R}^d$. In other words, the function

$$(z_1, \ldots, z_N) \mapsto F(t, z_1, \ldots, z_N)$$

is symmetric.

1.5.2 Marginal Distributions of Symmetric N-Particle Distributions

There is however one serious difficulty in considering N-particle distributions in the context of the mean field limit. Indeed, this limit assumes that $N \to \infty$, so that one would have to deal with "functions of infinitely many variables" in this limit, which does not make much sense at first sight.[11]

A traditional way of circumventing this difficulty is by considering the string of marginal distributions of the N-particle distribution. Before giving precise definitions, let us explain the idea in simple geometrical terms.

Consider a sphere centered at the origin in the 3 dimensional Euclidean space. The only missing information in order to completely define this sphere is its diameter. In other words, the sphere is completely determined as soon as one knows its orthogonal projection on any axis passing through the origin. However, if one does not know a priori that the object is a sphere, its orthogonal projections on each axis passing through the origin is not enough in order to reconstruct completely the object, since they will not distinguish between a ball centered at the origin and its boundary that is a sphere of equal radius.

The situation that we consider here is slightly more complicated, since the group of symmetries is not the orthogonal group, but the group generated by the reflections exchanging two coordinate axis in the N dimensional Euclidean space. Knowing the orthogonal projection of a set that is invariant under the action of this group on the first coordinate axis is again not sufficient as it will not distinguish between a sphere of radius r centered at the origin and the (hyper)cube of side $2r$ centered at the origin with edges parallel to the coordinate axis. Knowing the orthogonal projection on any one of the planes defined by two coordinate axis removes this ambiguity.

Considering marginals of an N-particle distribution is the analogous operation on probability measures. Denote by $\mathscr{P}_{sym}((\mathbf{R}^d)^N)$ the set of symmetric probability measures on the N-particle phase space, i.e.

$$\mathscr{P}_{sym}((\mathbf{R}^d)^N) := \{P \in \mathscr{P}((\mathbf{R}^d)^N) \mid S_\sigma \# P = P \text{ for all } \sigma \in \mathfrak{S}_N\},$$

[11] This last statement is not completely correct, as P.-L. Lions recently proposed a well defined mathematical object that would play the role of a "symmetric function of infinitely many variables that is slowly varying in each variable": see [72] and Sect. 1.7.3.

where we recall that S_σ is the transformation on $(\mathbf{R}^d)^N$ defined by

$$S_\sigma(z_1, \ldots, z_N) = (z_{\sigma(1)}, \ldots, z_{\sigma(N)})$$

for all $\sigma \in \mathfrak{S}_N$ and all $z_1, \ldots, z_N \in \mathbf{R}^d$.

Definition 1.5.1 For each $N \in \mathbf{N}^*$, each $P_N \in \mathscr{P}_{sym}((\mathbf{R}^d)^N)$ and each $k \in \{1, \ldots, N\}$, the k-particle marginal of P_N is the element of $\mathscr{P}_{sym}((\mathbf{R}^d)^k)$ defined by the formula

$$\int_{(\mathbf{R}^d)^k} \phi(z_1, \ldots, z_k) P_{N:k}(dz_1 \ldots dz_k) = \int_{(\mathbf{R}^d)^N} \phi(z_1, \ldots, z_k) P_N(dz_1 \ldots dz_N)$$

for each test function $\phi \in C_b((\mathbf{R}^d)^k)$. We shall systematically use the convention

$$P_{N:k} = 0 \quad \text{whenever } j > N.$$

If

$$P_N(dz_1 \ldots dz_N) = F_N(z_1, \ldots, z_N) dz_1 \ldots dz_N$$

where F_N is a symmetric probability density on $(\mathbf{R}^d)^N$, then, for each $k = 1, \ldots, N$, one has

$$P_{N:k}(dz_1 \ldots dz_k) = F_{N:k}(z_1, \ldots, z_k) dz_1 \ldots dz_k$$

with

$$F_{N:k}(z_1, \ldots, z_k) = \int_{(\mathbf{R}^d)^{N-k}} F_N(z_1, \ldots, z_N) dz_{k+1} \ldots dz_N.$$

Obviously, $F_{N:k}$ is also a symmetric probability density on $(\mathbf{R}^d)^k$.

The following elementary exercise confirms the analogy between the orthogonal projections of a subset of the Euclidean space on the subspaces generated by the coordinate axis and the marginal distributions associated to a symmetric probability on the N-particle phase space.

Exercise Consider for each $N \in \mathbf{N}^*$ and each $k = 1, \ldots, N$ the orthogonal projection

$$\mathbf{P}_N^k : (\mathbf{R}^d)^N \ni (z_1, \ldots, z_N) \mapsto (z_1, \ldots, z_k) \in (\mathbf{R}^d)^k.$$

Check that, for each $P_N \in \mathscr{P}_{sym}((\mathbf{R}^d)^N)$, one has

$$P_{N:k} = \mathbf{P}_N^k \# P_N,$$

and that

$$(P_{N:k})_{:j} = P_{N:j} \quad \text{for all } j, k \text{ such that } 1 \leq j \leq k \leq N.$$

An important example of symmetric N-particle distributions is the case of *factorized distributions*. Given a probability density f on $\mathbf{R}^d$, we consider for each $N \in \mathbf{N}^*$ the N-particle probability density F_N defined by the formula

$$F_N(z_1, \ldots, z_N) := \prod_{j=1}^{N} f(z_j).$$

This N-particle probability density is denoted as follows:

$$F_N = f^{\otimes N}.$$

Obviously $F_N = f^{\otimes N}$ is a symmetric N-particle probability distribution, and its marginals are also factorized distributions, since

$$F_N = f^{\otimes N} \Rightarrow F_{N:k} = f^{\otimes k}$$

for all $k = 1, \ldots, N$.

There is a very nice characterization of factorized distributions in terms of entropy. We shall not use it in the sequel. Nevertheless, it is important to know it, and we leave it as an exercise.

Exercise For each probability density f on $\mathbf{R}^d$ and each $N \in \mathbf{N}^*$, define

$$E_N(f) = \{F_N \text{ symmetric probability density on } (\mathbf{R}^d)^N \text{ s.t. } F_{N:1} = f\}.$$

We want to prove that $F_N = f^{\otimes N}$ realizes

$$\inf_{F_N \in E_N(f)} \int_{(\mathbf{R}^d)^N} F_N \ln F_N(z_1, \ldots, z_N) dz_1 \ldots dz_N.$$

(1) Let $a = (a_1, \ldots, a_d) \in (\mathbf{R}_+^*)^d$ satisfy $a_1 + \ldots + a_d = 1$, and consider

$$M(a) := \{A = A^T \in M_d(\mathbf{R}_+^*) \text{ s.t. } (1, \ldots, 1) \cdot A = a\}.$$

Find the critical points of the function

$$H : M(a) \ni A \mapsto \sum_{i,j=1}^{d} A_{ij} \ln A_{ij} \in \mathbf{R}.$$

(2) Prove that, for each $x, y > 0$

$$\phi(x, y) := x \ln\left(\frac{x}{y}\right) - x + y \geq 0,$$

with equality if and only if $x = y$.
(3) Express in terms of $\phi(A_{ij}, a_i a_j)$ the quantity

$$\sum_{i,j=1}^{d} (A_{ij} \ln A_{ij} - a_i a_j \ln(a_i a_j)).$$

(4) Find

$$\inf_{A \in M(a)} H(A).$$

(5) Using the intuition provided by questions 1–4, solve the minimization problem

$$\inf_{F_N \in E_N(f)} \int_{(\mathbf{R}^d)^N} F_N \ln F_N(z_1, \dots, z_N) dz_1 \dots dz_N.$$

References for this and the previous section are Chap. 3 in [17] and Chap. 3 in [26].

1.5.3 The N-Particle Liouville Equation

We have explained above how the state of a system of N identical particles is described by a symmetric probability measure on the N-particle phase space. Our next task is to define the evolution of such a probability measure, knowing that the positions of the particles in phase space are governed by the system of N-particle ODEs

$$\begin{cases} \dot{z}_i(t) = \dfrac{1}{N} \displaystyle\sum_{j=1}^{N} K(z_i(t), z_j(t)), \quad i = 1, \dots, N, \\ z_i(0) = z_i^{in}. \end{cases}$$

As explained in Theorem 1.3.1, whenever $K \in C^1(\mathbf{R}^d \times \mathbf{R}^d, \mathbf{R}^d)$ satisfies assumptions (HK1-HK2), the system of ODEs above generates a flow on the N-particle phase space $(\mathbf{R}^d)^N$ denoted by T_t and defined by the formula

$$T_t(z_1^{in}, \dots, z_N^{in}) := (z_1(t), \dots, z_N(t))$$

for all $t \in \mathbf{R}$.

Given a N-particle symmetric probability measure $F_N^{in} \in \mathscr{P}_{sym}((\mathbf{R}^d)^N)$, we set

$$F_N(t) := T_t \# F_N^{in}, \qquad t \in \mathbf{R}.$$

This formula defines $F_N(t)$ as the unique weak solution in $C(\mathbf{R}; w-\mathscr{P}((\mathbf{R}^d)^N))$ of the Cauchy problem for the N-particle Liouville equation

$$\begin{cases} \partial_t F_N + \dfrac{1}{N}\displaystyle\sum_{i,j=1}^{N} \operatorname{div}_{z_i}(F_N K(z_i, z_j)) = 0, \quad z_1, \ldots, z_N \in \mathbf{R}^d, \ t \in \mathbf{R}, \\ F_N\big|_{t=0} = F_N^{in}. \end{cases}$$

Exercise Check this, by using the method of characteristics as in the exercise following the statement of Theorem 1.3.1. (For the solution, see Chap. 1 in [17]).

As explained above, it is important that the probability measure in the N-particle phase space describing the state of a system of identical particles should be symmetric. Whether this symmetry property is propagated by the flow of the N-particle Liouville equation is therefore a very natural question. The answer to that question is given by the following proposition.

Proposition 1.5.1 *Assume that the interaction kernel $K \in C^1(\mathbf{R}^d \times \mathbf{R}^d, \mathbf{R}^d)$ satisfies the assumptions (HK1)-(HK2) so that the N-particle ODE system defines a unique flow T_t on $(\mathbf{R}^d)^N$ as proved in Theorem 1.3.1. For each $\sigma \in \mathfrak{S}_N$ we denote as above by S_σ the transformation on $(\mathbf{R}^d)^N$ defined by*

$$S_\sigma(z_1, \ldots, z_N) = (z_{\sigma(1)}, \ldots, z_{\sigma(N)})$$

for all $z_1, \ldots, z_N \in \mathbf{R}^d$.

(1) For all $\sigma \in \mathfrak{S}_N$ and all $t \in \mathbf{R}$, one has

$$T_t S_\sigma = S_\sigma T_t,$$

viz.

$$t \mapsto (z_{\sigma(1)}(t), \ldots, z_{\sigma(N)}(t)) \textit{ is the solution of the N-particle ODE system}$$
$$\textit{with initial condition } (z^{in}_{\sigma(1)}, \ldots, z^{in}_{\sigma(N)})\,;$$

(2) For each $F_N^{in} \in \mathscr{P}((\mathbf{R}^d)^N)$ and all $t \in \mathbf{R}$, the probability measure

$$F_N(t) := T_t \# F_N^{in}$$

is symmetric if F_N^{in} is symmetric, i.e.

$$S_\sigma \# F_N(t) = F_N(t) \quad \textit{for all } t \in \mathbf{R},$$

if

$$S_\sigma\#F_N^{in} = F_N^{in}.$$

Proof An elementary computation shows that

$$t \mapsto (z_{\sigma(1)}(t), \ldots, z_{\sigma(N)}(t))$$

satisfies the same N-particle ODE system as

$$t \mapsto (z_1(t), \ldots, z_N(t)).$$

By uniqueness of the solution of the Cauchy problem for this ODE system (Theorem 1.3.1), this is therefore the unique solution of that problem with initial data $(z_{\sigma(1)}^{in}, \ldots, z_{\sigma(N)}^{in})$. In other words,

$$T_t S_\sigma(z_1^{in}, \ldots, z_N^{in}) = (z_{\sigma(1)}(t), \ldots, z_{\sigma(N)}(t)) = S_\sigma T_t(z_1^{in}, \ldots, z_N^{in})$$

for all $(z_1^{in}, \ldots, z_N^{in}) \in (\mathbf{R}^d)^N$ and all $t \in \mathbf{R}$, which proves statement (1).

As for statement (2), observe that

$$\begin{aligned} S_\sigma\#F_N(t) &= S_\sigma\#(T_t\#F_N^{in}) = (S_\sigma T_t)\#F_N^{in} \\ &= (T_t S_\sigma)\#F_N^{in} = T_t\#(S_\sigma\#F_N^{in}) = T_t\#F_N^{in} = F_N(t) \end{aligned}$$

for all $t \in \mathbf{R}$ and all $\sigma \in \mathfrak{S}_N$, which is precisely the desired relation.

Finally, we discuss the growth of $F_N(t, z_1, \ldots, z_N)$ as $|z_1| + \ldots + |z_N| \to \infty$.

Lemma 1.5.1 *Under the assumptions (HK1)-(HK2) on the interaction kernel K, one has*

$$\|T_t(z_1^{in}, \ldots, z_N^{in})\|_1 \leq e^{2L|t|}\|(z_1^{in}, \ldots, z_N^{in})\|_1$$

with the notation

$$\|(z_1, \ldots, z_N)\|_1 = |z_1| + \ldots + |z_N|.$$

In particular, if $F_N^{in} \in \mathscr{P}_1((\mathbf{R}^d)^N)$, then $T_t\#F_N^{in} \in \mathscr{P}_1((\mathbf{R}^d)^N)$, and one has

$$\begin{aligned} &\int_{(\mathbf{R}^d)^N} \|(z_1, \ldots, z_N)\|_1 T_t\#F_N^{in}(dz_1 \ldots dz_N) \\ &\qquad \leq e^{2L|t|} \int_{(\mathbf{R}^d)^N} \|(z_1, \ldots, z_N)\|_1 F_N^{in}(dz_1 \ldots dz_N) \end{aligned}$$

for all $t \in \mathbf{R}$.

Proof Observe that

$$\begin{aligned}\frac{d}{dt}\|(z_1(t),\dots,z_N(t))\|_1 &= \frac{1}{N}\sum_{i,j=1}^{N} K(z_i(t),z_j(t))\cdot\frac{z_i(t)}{|z_i(t)|}\\ &\le \frac{1}{N}\sum_{i,j=1}^{N}|K(z_i(t),z_j(t))|\\ &\le \frac{1}{N}\sum_{i,j=1}^{N} L(|z_i(t)|+|z_j(t)|)\\ &= \sum_{i=1}^{N} L\left(|z_i(t)|+\frac{1}{N}\|(z_1(t),\dots,z_N(t))\|_1\right)\\ &= 2L\|(z_1(t),\dots,z_N(t))\|_1,\end{aligned}$$

and conclude by the Gronwall inequality.

Moreover, if $F_N^{in}\in\mathscr{P}_1((\mathbf{R}^d)^N)$, then

$$\begin{aligned}&\int_{(\mathbf{R}^d)^N}\|(z_1,\dots,z_N)\|_1 T_t\#F_N^{in}(dz_1\dots dz_N)\\ &\quad=\int_{(\mathbf{R}^d)^N}\|T_t(z_1,\dots,z_N)\|_1 F_N^{in}(dz_1\dots dz_N)\\ &\quad\le e^{2L|t|}\int_{(\mathbf{R}^d)^N}\|(z_1,\dots,z_N)\|_1 F_N^{in}(dz_1\dots dz_N)<\infty,\end{aligned}$$

so that $T_t\#F_N^{in}\in\mathscr{P}_1((\mathbf{R}^d)^N)$.

1.5.4 The BBGKY Hierarchy

The curious designation for this procedure finds its origin in the names of N.N. Bogoliubov, M. Born, H.S. Green, J.G. Kirkwood and J. Yvon, who introduced it in various contexts.

Before presenting the BBGKY hierarchy in detail, we first discuss the main reason for considering it in the first place.

As explained above, the N-particle distribution F_N is defined on the N-particle phase space $(\mathbf{R}^d)^N$, whose dimension increases as $N\to\infty$. In other words, the number of variables in F_N goes to infinity with N, so that the exact nature of the limiting object associated with F_N is not entirely obvious. Therefore, we seek to describe the behavior of F_N in the large N limit by considering instead its first marginal $F_{N:1}$

in that limit. By doing so, we avoid the problem of having the number of variables going to infinity with N, at the expense of losing some amount of information by reducing F_N to $F_{N:1}$.

To fulfill this program, it would be desirable to know the evolution of $F_{N:1}$, typically by means of a PDE in the 1-particle phase space to be satisfied by $F_{N:1}$.

Unfortunately, because of the interaction modeled by the kernel K, it is impossible to find a closed equation for $F_{N:1}$. Instead, the equation for $F_{N:1}$ deduced from the N-particle Liouville equation involves $F_{N:2}$. Again it is impossible to find a closed equation governing the evolution of $F_{N:2}$ when $N > 2$, as the pairwise interaction between particles will involve $F_{N:3}$ in the equation for $F_{N:2}$ deduced from the N-particle Liouville equation. By the same token, one can check that it is impossible to derive from the N-particle Liouville equation a closed system of equations for *finitely many*—i.e. $m < N$—of the marginals $F_{N:k}$.

Even though the system of PDEs governing the evolution of the marginal distributions $F_{N:k}$ cannot be put in closed form, these equations are nevertheless interesting, and we explain how to derive them below.

We start from thc N-particle Liouville equation satisfied by F_N:

$$\partial_t F_N + \frac{1}{N}\sum_{i,j=1}^{N} \operatorname{div}_{z_i}(K(z_i, z_j)F_N) = 0.$$

That $t \mapsto F_N(t)$ is a weak solution of this equation means that, for each test function $\Phi \equiv \Phi(z_1, \dots, z_N)$ in $C_b^1((\mathbf{R}^d)^N)$, one has

$$\partial_t \int_{(\mathbf{R}^d)^N} \Phi_N(z_1, \dots, z_N)F_N(t, dz_1 \dots dz_N)$$
$$= \frac{1}{N}\sum_{i,j=1}^{N} \int_{(\mathbf{R}^d)^N} K(z_i, z_j)\nabla_{z_i}\Phi(z_1, \dots, z_N)F_N(t, dz_1 \dots dz_N)$$

in the sense of distributions on $\mathbf{R}$.

We first seek an equation for the first marginal $F_{N:1}$. Let $\phi \in C_b^1(\mathbf{R}^d)$; writing the weak formulation of the N-particle Liouville equation for the test function $\Phi(z_1, \dots, z_N) = \phi(z_1)$, we see that

$$\partial_t \int_{(\mathbf{R}^d)^N} \phi(z_1)F_N(t, dz_1 \dots dz_N)$$
$$= \frac{1}{N}\sum_{j=2}^{N} \int_{(\mathbf{R}^d)^N} K(z_1, z_j) \cdot \nabla_{z_j}\phi(z_1)F_N(t, dz_1 \dots dz_N).$$

(Notice that the term associated to $j = 1$ in the summation on the right hand side of the equality above vanishes identically since K vanishes on the diagonal by assumption (HK1).)

In the term

$$\int_{(\mathbf{R}^d)^N} K(z_1, z_j) \cdot \nabla\phi(z_1) F_N(t, dz_1 \ldots dz_N)$$

we exchange the variables z_2 and z_j. Denoting by $\sigma_{2j} \in \mathfrak{S}_N$ the transposition exchanging 2 and j, one has

$$\begin{aligned}\int_{(\mathbf{R}^d)^N} & K(z_1, z_j) \cdot \nabla\phi(z_1) F_N(t, dz_1 \ldots dz_N) \\ &= \int_{(\mathbf{R}^d)^N} K(z_1, z_2) \cdot \nabla\phi(z_1) S_{\sigma_{2j}} F_N(t, dz_1 \ldots dz_N) \\ &= \int_{(\mathbf{R}^d)^N} K(z_1, z_2) \cdot \nabla\phi(z_1) F_N(t, dz_1 \ldots dz_N).\end{aligned}$$

by symmetry of $F_N(t)$.

Therefore

$$\begin{aligned}\partial_t & \int_{(\mathbf{R}^d)^N} \phi(z_1) F_N(t, dz_1 \ldots dz_N) \\ &= \frac{N-1}{N} \int_{(\mathbf{R}^d)^N} K(z_1, z_2) \cdot \nabla\phi(z_1) F_N(dz_1 \ldots dz_N).\end{aligned}$$

The integral on the left hand side of the equality above is recast as follows:

$$\int_{(\mathbf{R}^d)^N} \phi(z_1) F_N(t, dz_1 \ldots dz_N) = \int_{\mathbf{R}^d} \phi(z_1) F_{N:1}(t, dz_1).$$

By the same token, the integral on the right hand side becomes

$$\begin{aligned}\int_{(\mathbf{R}^d)^N} & K(z_1, z_2) \cdot \nabla\phi(z_1) F_N(t, dz_1 \ldots dz_N) \\ &= \int_{(\mathbf{R}^d)^2} K(z_1, z_2) \cdot \nabla\phi(z_1) F_{N:2}(t, dz_1 dz_2).\end{aligned}$$

Thus, for each $\phi \in C_b^1(\mathbf{R}^d)$, one has

$$\partial_t \int_{\mathbf{R}^d} \phi(z_1) F_{N:1}(t, dz_1) = \frac{N-1}{N} \int_{(\mathbf{R}^d)^2} K(z_1, z_2) \cdot \nabla\phi(z_1) F_{N:2}(t, dz_1 dz_2),$$

which is the weak formulation of

$$\partial_t F_{N:1} + \frac{N-1}{N} \operatorname{div}_{z_1} \int_{\mathbf{R}^d} K(z_1, z_2) F_{N:2}(\cdot, dz_2) = 0.$$

Equivalently

$$\partial_t F_{N:1} + \frac{N-1}{N} \operatorname{div}_{z_1} [K(z_1, z_2) F_{N:2}]_{:1} = 0,$$

where $K(z_1, z_2)F_{N:2}$ designates the Radon measure defined on $(\mathbf{R}^d)^2$ as the linear functional on bounded continuous functions given by the formula

$$\langle K(z_1, z_2) F_{N:2}, \psi \rangle := \int_{(\mathbf{R}^d)^2} \psi(z_1, z_2) K(z_1, z_2) F_{N:2}(dz_1 dz_2),$$

while $[K(z_1, z_2)F_{N:2}]_{:1}$ designates the Radon measure defined on $\mathbf{R}^d$ by the formula

$$\langle [K(z_1, z_2) F_{N:2}]_{:1}, \phi \rangle := \int_{(\mathbf{R}^d)^2} \phi(z_1) K(z_1, z_2) F_{N:2}(dz_1 dz_2).$$

(Equivalently,

$$[K(z_1, z_2) F_{N:2}]_{:1} := \mathbf{P}_2^1 \# [K(z_1, z_2) F_{N:2}]$$

where we recall that $\mathbf{P}_2^1$ is the orthogonal projection defined by $\mathbf{P}_2^1(z_1, z_2) := z_1$.)

In any case, as anticipated, the equation for the first marginal distribution $F_{N:1}$ involves the second marginal distribution $F_{N:2}$.

We next proceed to derive the equations satisfied by the sequence of marginal distributions $F_{N:j}$ for $j = 2, \ldots, N$; this derivation will proceed as in the case $j = 1$, except for one additional term.

For $1 < j < N$, we write the weak formulation of the N-particle Liouville equation with test function $\Phi(z_1, \ldots, z_N) = \phi(z_1, \ldots, z_j)$, where $\phi \in C_b((\mathbf{R}^d)^j)$. Thus

$$\partial_t \int\limits_{(\mathbf{R}^d)^N} \phi(z_1, \ldots, z_j) F_N(t, dz_1 \ldots dz_N)$$

$$= \frac{1}{N} \sum_{l=1}^{j} \sum_{k=j+1}^{N} \int\limits_{(\mathbf{R}^d)^N} K(z_l, z_k) \cdot \nabla_{z_l} \phi(z_1, \ldots, z_j) F_N(t, dz_1 \ldots dz_N)$$

$$+ \frac{1}{N} \sum_{l=1}^{j} \sum_{k=1}^{j} \int\limits_{(\mathbf{R}^d)^N} K(z_l, z_k) \cdot \nabla_{z_l} \phi(z_1, \ldots, z_j) F_N(t, dz_1 \ldots dz_N).$$

Notice that the range of the index l is limited to $\{1, \ldots, j\}$ since the test function ϕ does not depend on the variables $z_{j+1}, \ldots, z_N$. The range of the index k remains $\{1, \ldots, N\}$, and we have decomposed it into $\{1, \ldots, j\}$ and $\{j+1, \ldots, N\}$. This decomposition is quite natural, as the sum involving $k, l \in \{1, \ldots, j\}$ accounts for the pairwise interactions between the j particles whose state is described by $F_{N:j}$, while the sum involving $l \in \{1, \ldots, j\}$ and $k \in \{j+1, \ldots, N\}$ accounts for the pairwise interactions between each one of the j particles whose state is described by $F_{N:j}$ and the $N - j$ other particles in the system.

As in the case $j = 1$,

$$\int\limits_{(\mathbf{R}^d)^N} \phi(z_1, \ldots, z_j) F_N(t, dz_1 \ldots dz_N) = \int\limits_{(\mathbf{R}^d)^j} \phi(z_1, \ldots, z_j) F_{N:j}(t, dz_1 \ldots dz_j),$$

and by the same token, if $1 \leq k, l \leq j$

$$\int\limits_{(\mathbf{R}^d)^N} K(z_l, z_k) \cdot \nabla_{z_l} \phi(z_1, \ldots, z_j) F_N(t, dz_1 \ldots dz_N)$$

$$= \int\limits_{(\mathbf{R}^d)^j} K(z_l, z_k) \cdot \nabla_{z_l} \phi(z_1, \ldots, z_j) F_{N:j}(t, dz_1 \ldots dz_j).$$

If $1 \leq l \leq j < k \leq N$, denote by $\sigma_{j+1,k} \in \mathfrak{S}_N$ the transposition exchanging $j + 1$ and k. Then

$$\int\limits_{(\mathbf{R}^d)^N} K(z_l, z_k) \cdot \nabla_{z_l} \phi(z_1, \ldots, z_j) F_N(t, dz_1 \ldots dz_N)$$

$$= \int\limits_{(\mathbf{R}^d)^N} K(z_l, z_{j+1}) \cdot \nabla_{z_l} \phi(z_1, \ldots, z_j) S_{\sigma_{j+1,k}} F_N(t, dz_1 \ldots dz_N)$$

$$= \int\limits_{(\mathbf{R}^d)^N} K(z_l, z_{j+1}) \cdot \nabla_{z_l} \phi(z_1, \ldots, z_j) F_N(t, dz_1 \ldots dz_N)$$

by symmetry of $F_N(t)$. Then

$$\int_{(\mathbf{R}^d)^N} K(z_l, z_{j+1}) \cdot \nabla_{z_l}\phi(z_1, \dots, z_j) F_N(t, dz_1 \dots dz_N)$$
$$= \int_{(\mathbf{R}^d)^{j+1}} K(z_l, z_{j+1}) \cdot \nabla_{z_l}\phi(z_1, \dots, z_j) F_{N:j+1}(t, dz_1 \dots dz_{j+1}).$$

Finally, we obtain the equality

$$\frac{d}{dt}\int_{(\mathbf{R}^d)^j} \phi(z_1, \dots, z_j) F_{N:j}(t, dz_1 \dots dz_j)$$
$$= \frac{1}{N}\sum_{k,l=1}^{j}\int_{(\mathbf{R}^d)^j} K(z_l, z_k) \cdot \nabla_{z_l}\phi(z_1, \dots, z_j) F_{N:j}(t, dz_1 \dots dz_j)$$
$$+ \frac{N-j}{N}\sum_{l=1}^{j}\int_{(\mathbf{R}^d)^{j+1}} K(z_l, z_{j+1}) \cdot \nabla_{z_l}\phi(z_1, \dots, z_j) F_{N:j+1}(t, dz_1 \dots dz_{j+1})$$

to be verified for each $\phi \in C_b^1((\mathbf{R}^d)^j)$.

This is the weak formulation of the equation

$$\partial_t F_{N:j} + \frac{N-j}{N}\sum_{l=1}^{j} \operatorname{div}_{z_l}\int_{\mathbf{R}^d} K(z_l, z_{j+1}) F_{N:j+1}(\cdot, dz_{j+1})$$
$$+ \frac{1}{N}\sum_{k,l=1}^{j} \operatorname{div}_{z_l}(K(z_l, z_k) F_{N:j}) = 0.$$

Equivalently

$$\partial_t F_{N:j} + \frac{N-j}{N}\sum_{l=1}^{j} \operatorname{div}_{z_l}[K(z_l, z_{j+1}) F_{N:j+1}]_{:j}$$
$$+ \frac{1}{N}\sum_{k,l=1}^{j} \operatorname{div}_{z_l}(K(z_l, z_k) F_{N:j}) = 0,$$

where $K(z_l, z_{j+1})F_{N:j+1}$ designates the Radon measure defined on $(\mathbf{R}^d)^{j+1}$ as the linear functional on bounded continuous functions given by the formula

$$\langle K(z_l, z_{j+1}) F_{N:j+1}, \psi\rangle$$
$$:= \int_{(\mathbf{R}^d)^{j+1}} \psi(z_1, \dots, z_{j+1}) K(z_l, z_{j+1}) F_{N:j+1}(dz_1 \dots dz_{j+1}),$$

while $[K(z_l, z_{j+1})F_{N:j+1}]_{:j}$ designates the Radon measure defined on $(\mathbf{R}^d)^j$ by the formula

$$\langle [K(z_l, z_{j+1})F_{N:j+1}]_{:j}, \phi \rangle := \int_{(\mathbf{R}^d)^j} \phi(z_1, \ldots, z_j) K(z_l, z_{j+1}) F_{N:j+1}(dz_1 \ldots dz_{j+1}).$$

Equivalently,

$$[K(z_l, z_{j+1})F_{N:j+1}]_{:j} := \mathbf{P}^j_{j+1} \# [K(z_l, z_{j+1})F_{N:j+1}],$$

where we recall that $\mathbf{P}^j_{j+1}$ is the orthogonal projection defined by the formula

$$\mathbf{P}^j_{j+1}(z_1, \ldots, z_{j+1}) := (z_1, \ldots, z_j).$$

The equation obtained in the case $j = N$ is nothing but the N-particle Liouville equation itself since $F_{N:N} = F_N$ and $F_{N:j} = 0$ for all $j > N$: thus

$$\partial_t F_{N:N} + \frac{1}{N} \sum_{k,l=1}^{N} \operatorname{div}_{z_l}(K(z_l, z_k) F_{N:N}) = 0.$$

We summarize the above lengthy computations in the following theorem, where the notation $\mathscr{P}_{1,sym}((\mathbf{R}^d)^N)$ designates

$$\mathscr{P}_{1,sym}((\mathbf{R}^d)^N) = \mathscr{P}_1((\mathbf{R}^d)^N) \cap \mathscr{P}_{sym}((\mathbf{R}^d)^N).$$

Theorem 1.5.1 *Assume that the interaction kernel $K \in C^1(\mathbf{R}^d \times \mathbf{R}^d, \mathbf{R}^d)$ satisfies (HK1-HK2). Let $F_N^{in} \in \mathscr{P}_{1,sym}((\mathbf{R}^d)^N)$, and let $F_N(t) = T_t \# F_N^{in}$ for all $t \in \mathbf{R}$, where T_t is the flow defined on $(\mathbf{R}^d)^N$ by the N-particle ODE system as in Theorem 1.3.1. The sequence of marginal distributions $F_{N:j}$ of F_N with $j = 1, \ldots, N$ is a weak solution of the string of equations*

$$\begin{cases} \partial_t F_{N:1} + \dfrac{N-1}{N} \operatorname{div}_{z_1}[K(z_1, z_2) F_{N:2}]_{:1} = 0, \\ \partial_t F_{N:j} + \dfrac{N-j}{N} \displaystyle\sum_{l=1}^{j} \operatorname{div}_{z_l}[K(z_l, z_{j+1}) F_{N:j+1}]_{:j} \\ \qquad + \dfrac{1}{N} \displaystyle\sum_{k,l=1}^{j} \operatorname{div}_{z_l}(K(z_l, z_k) F_{N:j}) = 0, \qquad j = 2, \ldots, N-1, \\ \partial_t F_{N:N} + \dfrac{1}{N} \displaystyle\sum_{k,l=1}^{N} \operatorname{div}_{z_l}(K(z_l, z_k) F_{N:N}) = 0, \end{cases}$$

and satisfies the initial conditions

$$F_{N:j}\big|_{t=0} = F_{N:j}^{in}, \quad j = 1, \ldots, N.$$

This string of equations bears the name of *BBGKY hierarchy* for the N-particle system whose dynamics is defined by the ODE system

$$\dot{z}_i(t) = \frac{1}{N}\sum_{j=1}^{N} K(z_i(t), z_j(t)), \quad 1 \le i \le N.$$

In spite of all the (somewhat) technical computations involved in the derivation of the BBGKY hierarchy, the careful reader will notice that

(a) the BBGKY hierarchy is a consequence of the N-particle Liouville equation, but (b) it contains the N-particle Liouville equation (which is the last equation in the hierarchy).

This observation might cast some doubts as to the interest of considering the BBGKY hierarchy instead of the Liouville equation itself, since both contain exactly the same amount of information.

References for this section are Chaps. 3–4 in [26, 96, 98], Chap. 3 in [10, 17].

1.5.5 The Mean Field Hierarchy and Factorized Distributions

In the present section, our discussion of the mean field limit becomes purely formal.

Our aim is to pass to the limit in each equation in the BBGKY hierarchy as $N \to \infty$, keeping $j \ge 1$ fixed. Assume that $F_{N:j} \to F_j$ as $N \to \infty$ (in some sense to be made precise) for all $j \ge 1$. Then, in the limit as $N \to \infty$

$$\frac{N-j}{N}\int_{\mathbf{R}^d} K(z_l, z_{j+1})F_{N:j+1}(dz_{j+1}) \to \int_{\mathbf{R}^d} K(z_l, z_{j+1})F_{j+1}(dz_{j+1}),$$

while

$$\frac{1}{N}K(z_l, z_k)F_{N:j} \to 0,$$

so that

$$\partial_t F_j + \sum_{l=1}^{j} \operatorname{div}_{z_l} \int_{\mathbf{R}^d} K(z_l, z_{j+1})F_{j+1}(dz_{j+1}) = 0, \qquad j \ge 1.$$

This hierarchy of equations is henceforth referred to as *the mean field hierarchy*, or *the Vlasov hierarchy*. The similarities between this mean field hierarchy and the BBGKY hierarchy are striking. Yet there is an important difference: the mean field hierarchy is an *infinite* hierarchy of equations—unlike the BBGKY hierarchy, which contains only N equations, where N is the total number of particles. The physical meaning of this infinite hierarchy of equations will be explained in Sect. 1.7.2.

However, this infinite hierarchy is directly related to the mean field equation by the following observation.

Proposition 1.5.2 *Assume that the interaction kernel $K \in C^1(\mathbf{R}^d \times \mathbf{R}^d, \mathbf{R}^d)$ satisfies assumptions (HK1)-HK2). Let f^{in} be a smooth (at least C^1) probability density on $\mathbf{R}^d$ such that*

$$\int_{\mathbf{R}^d} |z| f^{in}(z) dz < \infty.$$

Assume that the Cauchy problem for the mean field equation

$$\begin{cases} \partial_t f(t,z) + \operatorname{div}_z \left(f(t,z) \int_{\mathbf{R}^d} K(z,z') f(t,z') dz' \right) = 0, \\ f\big|_{t=0} = f^{in}, \end{cases}$$

has a classical (at least of class C^1) solution $f \equiv f(t,z)$. Set $f_j(t,\cdot) = f(t,\cdot)^{\otimes j}$, i.e.

$$f_j(t, z_1, \ldots, z_j) = \prod_{k=1}^{j} f(t, z_k)$$

for each $t \in \mathbf{R}$ and each $z_1, \ldots, z_j \in \mathbf{R}^d$.

Then the sequence $(f_j)_{j \geq 1}$ is a solution of the infinite mean field hierarchy

$$\partial_t f_j(z_1, \ldots, z_j) + \sum_{l=1}^{j} \operatorname{div}_{z_l} \int_{\mathbf{R}^d} K(z_l, z_{j+1}) f_{j+1}(z_1, \ldots, z_{j+1}) dz_{j+1} = 0$$

for all $j \geq 1$.

Proof Since f is of class C^1 at least, one has

$$\begin{aligned} \partial_t f_j(t, z_1, \ldots, z_j) &= \sum_{k=1}^{j} \prod_{\substack{l=1 \\ l \neq k}}^{j} f(t, z_l) \partial_t f(t, z_k) \\ &= - \sum_{k=1}^{j} \prod_{\substack{l=1 \\ l \neq k}}^{j} f(t, z_l) \operatorname{div}_{z_k} \left(f(t, z_k) \int_{\mathbf{R}^d} K(z_k, z') f(t, z') dz' \right) \end{aligned}$$

$$= -\sum_{k=1}^{j} \operatorname{div}_{z_k} \left(\prod_{l=1}^{j} f(t, z_l) \int_{\mathbf{R}^d} K(z_k, z') f(t, z') dz' \right)$$

$$= -\sum_{k=1}^{j} \operatorname{div}_{z_k} \left(\int_{\mathbf{R}^d} K(z_k, z') f_{j+1}(t, z_1, \ldots, z_j, z') dz' \right),$$

which is precisely the jth equation in the mean field hierarchy.

This crucial observation suggests the following strategy to prove the mean field limit by the method of hierarchies.

Choose factorized initial data for N-particle Liouville equation: given f^{in} a probability density on $\mathbf{R}^d$ such that

$$\int_{\mathbf{R}^d} |z| f(t, z) dz < \infty,$$

define

$$F_N^{in} = (f^{in})^{\otimes N} \qquad \text{for each } N \geq 1,$$

i.e.

$$F_N^{in}(t, z_1, \ldots, z_N) = \prod_{k=1}^{N} f^{in}(t, z_k).$$

Let F_N be the solution of the Cauchy problem for the N-particle Liouville equation

$$\begin{cases} \partial_t F_N + \dfrac{1}{N} \displaystyle\sum_{k,l=1}^{N} \operatorname{div}_{z_k}(F_N K(z_k, z_l)) = 0, \\ F_N\big|_{t=0} = (f^{in})^{\otimes N}, \end{cases}$$

with that initial data.

Assume that one can prove

(a) that $F_{N:j} \to F_j$ (in some sense to be made precise) for each $j \geq 1$, where F_j is a solution of the infinite hierarchy, and
(b) that the Cauchy problem for the infinite hierarchy has a unique solution.

Let f be a solution of the mean field PDE with initial data f^{in}.

Since the sequence $f_j := f^{\otimes j}$ (for $j \geq 1$) is a solution of the infinite mean field hierarchy with initial data $(f^{in})^{\otimes j}$ by Proposition 1.5.2, statement (b) implies that it is *the* solution of the infinite hierarchy for that initial data. Therefore

$$F_{N:j} \to F_j = f^{\otimes j} \qquad \text{as } N \to \infty \text{ for all } j \geq 1.$$

In particular, for $j = 1$, one finds that the solution F_N of the Liouville equation satisfies

$$F_{N:1} \to f \qquad \text{as } N \to \infty.$$

In other words, the first marginal of the solution of the N-particle Liouville equation with factorized initial data converges to the solution of the mean field PDE in the large N limit.

This is precisely the strategy outlined by Cercignani [25] for justifying rigorously the Boltzmann equation in the case of the hard sphere gas.

Notice that, in this approach, one needs to know that the Cauchy problem for the mean field PDE is well-posed (i.e. that it has a unique solution for each initial data in some appropriate functional space). On the contrary, in the previous approach based on the notion of empirical measure, the existence of a solution of the Cauchy problem for the mean field PDE is a consequence of the existence for finitely many particles and of the mean field limit itself, and the uniqueness of that solution is a consequence of Dobrushin's estimate.

The reader might be under the impression that proving the uniqueness of the solution of the Cauchy problem for the infinite mean field hierarchy is a matter of pure routine, since the mean field hierarchy is a linear problem, at variance with the mean field PDE, which is nonlinear. This is obviously wrong, since the uniqueness of the solution of the infinite mean field hierarchy implies the uniqueness of the solution of the mean field PDE. In fact, the uniqueness property for the infinite mean field hierarchy is a very strong property and proving it is by no means obvious. See [97] and Sect. 1.7.2 for a precise discussion of this point—as well as of the physical meaning of the infinite mean field hierarchy.

However, the strategy described in the present section (starting from the N-particle Liouville equation, deriving the BBGKY hierarchy, passing to the limit to arrive at an infinite hierarchy of equations and concluding with the uniqueness of the solution of the infinite hierarchy with given initial data) has been used successfully in a greater variety of problems than the mean field limit considered in this course. For instance, the only rigorous derivation of the Boltzmann equation of the kinetic theory of gases known to this date (proposed by Lanford [70]) follows exactly these steps[12]—see also Chap. 3 of [17] and Chaps. 2–4 of [26] for an account of this fundamental result. The recent monograph [37] extends Lanford's result to short range potentials other than hard spheres, and gives a more detailed presentation of the Boltzmann-Grad limit than all previous references, even in the hard sphere case.

[12]For the case of the Boltzmann-Grad limit for a system of N hard spheres, the infinite hierarchy cannot be derived rigorously from the Liouville equation by passing to the limit in the sense of distributions in each equation of the (finite) BBGKY hierarchy: see the discussion on pp. 74–75 in [26]. The infinite Boltzmann hierarchy is derived by a different, more subtle procedure that is the core of the Lanford proof—see Sect. 4.4 in [26].

1.6 Chaotic Sequences, Empirical Measures and BBGKY Hierarchies

Our discussion of BBGKY hierarchies shows the importance of the following property of symmetric N-particle probability measures F_N:

$$F_{N:j} \to f^{\otimes j} \quad \text{weakly as } N \to \infty \text{ for all } j \geq 1 \text{ fixed.}$$

Of course, if ϕ is a probability density on $\mathbf{R}^d$, one has

$$\Phi_N = \phi^{\otimes N} \Rightarrow \Phi_{N:j} = \phi^{\otimes j}.$$

But if ψ is another probability density on $\mathbf{R}^d$, defining

$$\tilde{\Phi}_N = \frac{1}{N} \sum_{k=1}^{N} \phi^{\otimes (k-1)} \otimes \psi \otimes \phi^{\otimes N-k},$$

which is in general a non-factorized symmetric probability density on $(\mathbf{R}^d)^N$, then

$$\tilde{\Phi}_{N:j} = \frac{N-j}{N} \phi^{\otimes j} + \frac{1}{N} \sum_{k=1}^{j} \phi^{\otimes (k-1)} \otimes \psi \otimes \phi^{\otimes j-k} \to \phi^{\otimes j}$$

for all $j \geq 1$ as $N \to \infty$.

Thus, the property above can be verified, in the limit as $N \to \infty$, by sequences of N-particle probability measures that are not factorized exactly for each finite $N \geq 1$.

Definition 1.6.1 Let p be a Borel probability measure on $\mathbf{R}^d$. A sequence P_N of symmetric N-particle probability measures on $(\mathbf{R}^d)^N$ for all $N \geq 1$ is said to be chaotic, and more precisely p-chaotic, if

$$P_{N:j} \to p^{\otimes j} \quad \text{weakly in } \mathscr{P}((\mathbf{R}^d)^j)$$

as $N \to \infty$, for all $j \geq 1$ fixed.

The notion of chaotic sequences appeared in the context of the derivation of kinetic equations from particle dynamics, for the first time in [62]. Perhaps the reason for this terminology is that this property corresponds to asymptotic independence of the N-particles in the large N limit.

1.6.1 Chaotic Sequences and Empirical Measures

We begin our discussion of chaotic sequences with a characterization of chaotic sequences in terms of empirical measures.

Theorem 1.6.1 *Let $p \in \mathscr{P}(\mathbf{R}^d)$, and let $P_N \in \mathscr{P}_{sym}((\mathbf{R}^d)^N)$ for each $N \geq 1$. Then the two following properties are equivalent:*
(a) for each $j \geq 1$

$$P_{N:j} \to p^{\otimes j}$$

weakly in $\mathscr{P}((\mathbf{R}^d)^j)$ as $N \to \infty$;
(b) for each $\phi \in C_b(\mathbf{R}^d)$ and each $\varepsilon > 0$,

$$P_N(\{Z_N \in (\mathbf{R}^d)^N \text{ s.t. } |\langle \mu_{Z_N} - p, \phi\rangle| \geq \varepsilon\}) \to 0$$

as $N \to \infty$, where we recall that

$$\mu_{Z_N} := \frac{1}{N}\sum_{k=1}^{N} \delta_{z_k}$$

with $Z_N = (z_1, \ldots, z_N)$.

In very informal terms,

$$P_{N:j} \to p^{\otimes j}$$

weakly in $\mathscr{P}((\mathbf{R}^d)^j)$ as $N \to \infty$ for all $j \geq 1$ if and only if

$$P_N \to \delta_p$$

"weakly in $\mathscr{P}(\mathscr{P}(\mathbf{R}^d))$" as $N \to \infty$. In this statement, each N-tuples Z_N is identified with the corresponding empirical measure μ_{Z_N} and P_N is viewed as a probability measure on $\mathscr{P}(\mathbf{R}^d)$ that is concentrated on the set of N-particle empirical measures. This identification goes back to Grünbaum [51] and is discussed in Sect. 1.7.3.

Remark Observe that property (a) is equivalent to
(a') $P_{N:1} \to p$ and $P_{N:2} \to p \otimes p$ weakly in $\mathscr{P}(\mathbf{R}^d)$ and $\mathscr{P}((\mathbf{R}^d)^2)$ as $N \to \infty$.
In fact, as we shall see, the proof given below establishes that (a')$\Rightarrow$(b)$\Rightarrow$ *(a)*.

Proof First we prove that property (a) implies property (b).
Applying Bienaymé-Chebyshev's inequality shows that

$$P_N(\{Z_N \in (\mathbf{R}^d)^N \text{ s.t. } |\langle \mu_{Z_N} - p, \phi\rangle| \geq \varepsilon\}) \leq \frac{1}{\varepsilon^2}\mathbf{E}^{P_N}|\langle \mu_{Z_N} - p, \phi\rangle|^2.$$

Then we compute

$$
\begin{aligned}
&\mathbf{E}^{P_N}|\langle\mu_{Z_N}-p,\phi\rangle|^2\\
&=\mathbf{E}^{P_N}\left(\frac{1}{N}\sum_{j=1}^N\phi(z_j)\right)^2+\mathbf{E}^{P_N}\langle p,\phi\rangle^2-2\mathbf{E}^{P_N}\left(\langle p,\phi\rangle\frac{1}{N}\sum_{j=1}^N\phi(z_j)\right)\\
&=\mathbf{E}^{P_N}\left(\frac{1}{N}\sum_{j=1}^N\phi(z_j)\right)^2+\langle p,\phi\rangle^2-2\langle p,\phi\rangle\mathbf{E}^{P_N}\frac{1}{N}\sum_{j=1}^N\phi(z_j).
\end{aligned}
$$

Observe that, by symmetry of P_N, one has

$$
\mathbf{E}^{P_N}(\phi(z_j)\phi(z_k))=\begin{cases}\mathbf{E}^{P_N}(\phi(z_1)\phi(z_2)) & \text{if } j\neq k,\\ \mathbf{E}^{P_N}(\phi(z_1)^2) & \text{if } j=k.\end{cases}
$$

Thus

$$
\begin{aligned}
\mathbf{E}^{P_N}\left(\frac{1}{N}\sum_{j=1}^N\phi(z_j)\right)^2&=\frac{1}{N^2}\sum_{j,k=1}^N\mathbf{E}^{P_N}(\phi(z_j)\phi(z_k))\\
&=\frac{1}{N}\mathbf{E}^{P_N}(\phi(z_1)^2)+\frac{N-1}{N}\mathbf{E}^{P_N}(\phi(z_1)\phi(z_2))\\
&=\frac{1}{N}\langle P_{N:1},\phi^2\rangle+\frac{N-1}{N}\langle P_{N:2},\phi\otimes\phi\rangle\\
&\to\langle p^{\otimes 2},\phi^{\otimes 2}\rangle=\langle p,\phi\rangle^2
\end{aligned}
$$

as $N\to\infty$ by (a), while

$$
\begin{aligned}
\mathbf{E}^{P_N}\left(\frac{1}{N}\sum_{j=1}^N\phi(z_j)\right)&=\frac{1}{N}\sum_{j=1}^N\mathbf{E}^{P_N}(\phi(z_j))\\
&=\frac{1}{N}\sum_{j=1}^N\mathbf{E}^{P_N}(\phi(z_1))=\langle P_{N:1},\phi\rangle\to\langle p,\phi\rangle
\end{aligned}
$$

as $N\to\infty$, again by (a).

Therefore, property (a) implies that

$$
\mathbf{E}^{P_N}|\langle\mu_{Z_N}-p,\phi\rangle|^2\to\langle p,\phi\rangle^2+\langle p,\phi\rangle^2-2\langle p,\phi\rangle^2=0
$$

as $N\to\infty$, and therefore

$$
P_N(\{Z_N\in(\mathbf{R}^d)^N \text{ s.t. } |\langle\mu_{Z_N}-p,\phi\rangle|\geq\varepsilon\})\leq\frac{1}{\varepsilon^2}\cdot o(1)
$$

by Bienaymé-Chebyshev's inequality, which is precisely property (b).

Next we prove that, conversely, property (b) implies property (a).

Step 1: let us prove first that property (b) implies that $P_{N:1} \to p$ weakly as $N \to \infty$.

Let $\phi \in C_b(\mathbf{R}^d)$. Denote

$$U_N^\varepsilon(\phi) := \{Z_N \in (\mathbf{R}^d)^N \text{ s.t. } |\langle \mu_{Z_N} - p, \phi\rangle| > \varepsilon\},$$
$$V_N^\varepsilon(\phi) := (\mathbf{R}^d)^N \setminus U_N^\varepsilon(\phi).$$

Then

$$|\mathbf{E}^{P_N}\langle \mu_{Z_N}, \phi\rangle - \langle p, \phi\rangle| = |\mathbf{E}^{P_N}\langle \mu_{Z_N} - p, \phi\rangle|$$
$$\leq \mathbf{E}^{P_N}(|\langle \mu_{Z_N} - p, \phi\rangle| \mathbf{1}_{U_N^\varepsilon(\phi)}) + \mathbf{E}^{P_N}(|\langle \mu_{Z_N} - p, \phi\rangle| \mathbf{1}_{V_N^\varepsilon(\phi)}).$$

Obviously

$$\mathbf{E}^{P_N}(|\langle \mu_{Z_N} - p, \phi\rangle| \mathbf{1}_{U_N^\varepsilon(\phi)}) \leq 2\|\phi\|_{L^\infty} P_N(U_N^\varepsilon(\phi)),$$

and

$$\mathbf{E}^{P_N}(|\langle \mu_{Z_N} - p, \phi\rangle| \mathbf{1}_{V_N^\varepsilon(\phi)}) \leq \varepsilon.$$

By (b), there exists $N(\varepsilon, \phi)$ such that

$$N > N(\varepsilon, \phi) \Rightarrow P_N(U_N^\varepsilon(\phi)) < \varepsilon,$$

so that

$$|\mathbf{E}^{P_N}\langle \mu_{Z_N}, \phi\rangle - \langle p, \phi\rangle| \leq (2\|\phi\|_{L^\infty} + 1)\varepsilon.$$

On the other hand

$$\mathbf{E}^{P_N}(\langle \mu_{Z_N}, \phi\rangle) = \mathbf{E}^{P_N}\left(\frac{1}{N}\sum_{j=1}^N \phi(z_j)\right) = \mathbf{E}^{P_N}(\phi(z_1)) = \langle P_{N:1}, \phi\rangle,$$

so that

$$N > N(\varepsilon, \phi) \Rightarrow |\langle P_{N:1}, \phi\rangle - \langle p, \phi\rangle| \leq (2\|\phi\|_{L^\infty} + 1)\varepsilon.$$

which concludes step 1.

Step 2: next we prove that property (b) implies that $P_{N:j} \to p^{\otimes j}$ weakly as $N \to \infty$, for all $j > 1$.

Let $\phi \in C_b(\mathbf{R}^d) \setminus \{0\}$. Let $\mathscr{E}_N^j := \{1, \ldots, N\}^{\{1,\ldots,j\}}$ (the set of maps from $\{1, \ldots, j\}$ to $\{1, \ldots, N\}$) and let $\mathscr{A}_N^j$ be the set of one-to-one elements of $\mathscr{E}_N^j$. Then

$$\begin{aligned}
\mathbf{E}^{P_N}(\langle \mu_{Z_N}^{\otimes j}, \phi^{\otimes j}\rangle) &= \mathbf{E}^{P_N}\left(\frac{1}{N^j}\sum_{s\in\mathscr{E}_N^j}\phi(z_{s(1)})\dots\phi(z_{s(j)})\right)\\
&= \frac{1}{N^j}\sum_{s\in\mathscr{A}_N^j}\mathbf{E}^{P_N}(\phi(z_{s(1)})\dots\phi(z_{s(j)}))\\
&+ \frac{1}{N^j}\sum_{s\in\mathscr{E}_N^j\setminus\mathscr{A}_N^j}\mathbf{E}^{P_N}(\phi(z_{s(1)})\dots\phi(z_{s(j)})).
\end{aligned}$$

For $s \in \mathscr{A}_N^j$, one has

$$\mathbf{E}^{P_N}(\phi(z_{s(1)})\dots\phi(z_{s(j)})) = \mathbf{E}^{P_N}(\phi(z_1)\dots\phi(z_j)) = \langle P_{N:j}, \phi^{\otimes j}\rangle$$

by symmetry of P_N, while, for all $s \in \mathscr{E}_N^j$

$$|\mathbf{E}^{P_N}(\phi(z_{s(1)})\dots\phi(z_{s(j)}))| \leq \|\phi\|_{L^\infty}^j.$$

Now, for all j fixed

$$\begin{cases} \#\mathscr{A}_N^j = N(N-1)\dots(N-j+1) \sim N^j \text{ as } N \to \infty,\\ \#(\mathscr{E}_N^j \setminus \mathscr{A}_N^j) = N^j - N(N-1)\dots(N-j+1) = o(N^j), \end{cases}$$

so that

$$\left|\mathbf{E}^{P_N}\langle \mu_{Z_N}^{\otimes j}, \phi^{\otimes j}\rangle - \langle P_{N:j}, \phi^{\otimes j}\rangle\right| \leq 2\frac{N^j - \#\mathscr{A}_N^j}{N^j}\|\phi\|_{L^\infty}^j.$$

Introduce

$$\begin{cases} X_N^\varepsilon(j,\phi) = \{Z_N \in (\mathbf{R}^d)^N \text{ s.t. } |\langle \mu_{Z_N}^{\otimes j} - p^{\otimes j}, \phi^{\otimes j}\rangle| > \varepsilon\},\\ Y_N^\varepsilon(j,\phi) = (\mathbf{R}^d)^N \setminus X_N^\varepsilon(j,\phi). \end{cases}$$

Observe that

$$\langle \mu_{Z_N}^{\otimes j} - p^{\otimes j}, \phi^{\otimes j}\rangle = \sum_{k=1}^{j}\langle \mu_{Z_N}, \phi\rangle^{k-1}\langle \mu_{Z_N} - p, \phi\rangle\langle p, \phi\rangle^{j-k},$$

so that

$$|\langle \mu_{Z_N}^{\otimes j} - p^{\otimes j}, \phi^{\otimes j}\rangle| \leq j\|\phi\|_{L^\infty}^{j-1}|\langle \mu_{Z_N} - p, \phi\rangle|.$$

Therefore, property (b) implies that

$$P_N(X_N^\varepsilon(j,\phi)) \leq P_N(U_N^{\varepsilon/j\|\phi\|_{L^\infty}^{j-1}}(\phi)) \to 0$$

as $N \to \infty$ for all $j > 1$, all $\varepsilon > 0$ and all $\phi \in C_b(\mathbf{R}^d) \setminus \{0\}$.

In particular, for each $j \geq 1$, there exists $N_j(\varepsilon, \phi) \geq j$ such that

$$N > N_j(\varepsilon, \phi) \Rightarrow P_N(X_N^\varepsilon(j, \phi)) < \varepsilon.$$

Thus

$$\begin{aligned}|\mathbf{E}^{P_N}\langle \mu_{Z_N}^{\otimes j} - p^{\otimes j}, \phi^{\otimes j}\rangle| &\leq \mathbf{E}^{P_N}(|\langle \mu_{Z_N}^{\otimes j} - p^{\otimes j}, \phi^{\otimes j}\rangle|\mathbf{1}_{X_N^\varepsilon(j,\phi)}) \\ &\quad + \mathbf{E}^{P_N}(|\langle \mu_{Z_N}^{\otimes j} - p^{\otimes j}, \phi^{\otimes j}\rangle|\mathbf{1}_{Y_N^\varepsilon(j,\phi)}) \\ &\leq 2\|\phi\|_{L^\infty}^j P_N(X_N^\varepsilon(j, \phi)) + \varepsilon\end{aligned}$$

so that

$$N > N_j(\varepsilon, \phi) \Rightarrow |\mathbf{E}^{P_N}\langle \mu_{Z_N}^{\otimes j} - p^{\otimes j}, \phi^{\otimes j}\rangle| < (2\|\phi\|_{L^\infty}^j + 1)\varepsilon.$$

Therefore, for each $\varepsilon > 0$ and each $j > 1$, and for all $N > N_j(\varepsilon, \phi)$, one has

$$\begin{aligned}\left|\langle p^{\otimes j} - P_{N:j}, \phi^{\otimes j}\rangle\right| &\leq |\mathbf{E}^{P_N}(\langle p^{\otimes j} - \mu_{Z_N}^{\otimes j}, \phi^{\otimes j}\rangle)| \\ &\quad + |\mathbf{E}^{P_N}(\mu_{Z_N}^{\otimes j}, \phi^{\otimes j}\rangle) - \langle P_{N:j}, \phi^{\otimes j}\rangle| \\ &\leq (2\|\phi\|_{L^\infty}^j + 1)\varepsilon + 2\frac{N^j - \#\mathscr{A}_N^j}{N^j}\|\phi\|_{L^\infty}^j,\end{aligned}$$

and since $N^j - \#\mathscr{A}_N^j = o(N^j)$ as $N \to \infty$,

$$\overline{\lim_{N\to\infty}} \left|\langle p^{\otimes j} - P_{N:j}, \phi^{\otimes j}\rangle\right| \leq (2\|\phi\|_{L^\infty}^j + 1)\varepsilon.$$

Since this holds for each $\varepsilon > 0$, we conclude that

$$\langle P_{N:j}, \phi^{\otimes j}\rangle \to \langle p^{\otimes j}, \phi^{\otimes j}\rangle \quad \text{as } N \to \infty.$$

This property holds for each $\phi \in C_b(\mathbf{R}^d) \setminus \{0\}$ so that, by a classical density argument, we conclude that the sequence P_N is p-chaotic.

Exercise The purpose of this exercise is to complete the "classical density argument" used at the end of the proof of Theorem 1.6.1. For each $\Phi \in C((\mathbf{R}^d)^j)$, denote by $S_j\Phi$ the element of $C((\mathbf{R}^d)^j)$ defined by the formula

$$S_j\Phi(z_1, \ldots, z_j) := \frac{1}{j!}\sum_{\sigma\in\mathfrak{S}_j} \Phi(z_{\sigma(1)}, \ldots, z_{\sigma(j)}).$$

Finally, for each $\phi_1, \dots, \phi_j \in C(\mathbf{R}^d)$, we denote by $\phi_1 \otimes \dots \otimes \phi_j$ the function

$$\phi_1 \otimes \dots \otimes \phi_j : (z_1, \dots, z_j) \mapsto \phi_1(z_1) \dots \phi_j(z_j),$$

and by $\phi^{\otimes j}$ the j-fold tensor product of ϕ with itself.
(a) Prove that

$$S_j(\phi_1 \otimes \dots \otimes \phi_j) = \frac{1}{j!} \frac{\partial^j}{\partial X_1 \dots \partial X_j} (X_1\phi_1 + \dots + X_j\phi_j)^{\otimes j}.$$

(b) Prove that, for each $Q \in \mathscr{P}_{sym}((\mathbf{R}^d)^j)$ and each $\Phi \in C_b((\mathbf{R}^d)^j)$, one has

$$\langle Q, \Phi \rangle = \langle Q, S_j \Phi \rangle.$$

(c) Let[13] $\Phi \in C_0((\mathbf{R}^d)^j)$ and $\varepsilon > 0$. Prove that there exists $\alpha_1, \dots, \alpha_n \in \mathbf{R}$ and a jn-tuple $(\phi_{i,m})_{1 \le i \le j, 1 \le m \le n}$ of elements of $C_0(\mathbf{R}^d)$ such that

$$\left\| \Phi - \sum_{m=1}^{n} \alpha_m \phi_{1,m} \otimes \dots \otimes \phi_{j,m} \right\| < \varepsilon,$$

where

$$\|\Psi\| = \sup_{z_1, \dots, z_j \in \mathbf{R}^d} |\Psi(z_1, \dots, z_j)|.$$

(Hint: apply the Stone-Weierstrass theorem.)
(d) Let $Q \in \mathscr{P}_{sym}((\mathbf{R}^d)^j)$ and $(Q_N)_{N \ge 1}$ be a sequence of elements of $\mathscr{P}_{sym}((\mathbf{R}^d)^j)$ such that

$$\langle Q_N, \phi^{\otimes j} \rangle \to \langle Q, \phi^{\otimes j} \rangle$$

as $N \to \infty$ for each $\phi \in C_0(\mathbf{R}^d)$. Prove that

$$\langle Q_N, \Phi \rangle \to \langle Q, \Phi \rangle$$

as $N \to \infty$ for each $\Phi \in C_0((\mathbf{R}^d)^j)$.
(e) Conclude that

$$\langle Q_N, \Phi \rangle \to \langle Q, \Phi \rangle$$

as $N \to \infty$ for each $\Phi \in C_b((\mathbf{R}^d)^j)$. (Hint: apply Theorem 6.8 in Chap. II of [75].)

References for this section are Chap. 1, Sect. 2 in [100], and Chap. 4, Sect. 6 in [26].

---

[13] For each locally compact topological space X and each finite dimensional vector space E on $\mathbf{R}$, we denote by $C_0(X, E)$ the set of continuous functions on X with values in E that converge to 0 at infinity. We set $C_0(X) := C_0(X, E)$.

1.6.2 From Dobrushin's Theorem to the BBGKY Hierarchy

In the proof of the mean-field limit based on the notion of empirical measure, one proves the stability of the limiting, mean field PDE in the weak topology of probability measures on the single-particle phase space. The method based on the BBGKY hierarchy involves the N-particle phase space, and the need for considering marginals of the N-particle distribution in the limit as $N \to \infty$. Perhaps the best way to understand the relation between these two approaches of the same problem is to think of the nature of the N-particle empirical measure. Indeed

$$\mu_{Z_N}(dz)$$

is a symmetric *function* of the N variables $Z_N = (z_1, \ldots, z_N)$, defined on the *N-particle phase space* $(\mathbf{R}^d)^N$, and with values in the set of *probability measures on the single-particle phase space* $\mathbf{R}^d$—in the notation above, z is the variable in the single-particle phase space.

Already the characterization of chaotic sequences in terms of empirical measures obtained in the previous section clarifies the respective roles of the single particle phase space and of the N-particle phase space in this limit. The N-particle symmetric distribution function F_N can be viewed as a probability on $\mathscr{P}(\mathbf{R}^d)$ concentrated on the set of empirical measures—in other words, the N-tuple Z_N is, up to permutations of its N components, identified with the empirical measure μ_{Z_N}. (This idea can be found in [51] and will be discussed in more detail in Sect. 1.7.3.)

Although this is a static picture, it provides the right point of view in order to unify the two approaches of the mean field limit presented above, i.e. the approach based on empirical measures and Dobrushin's estimate, and the one based on the BBGKY hierarchy. More precisely, we shall prove that the sequence $F_N(t)$ of solutions of the N-particle Liouville equation with factorized initial data is $f(t)\mathscr{L}^d$-chaotic for each $t \in \mathbf{R}$, where f is the solution of the mean field PDE.

The proof of this result goes as follows.

Since F_N^{in} is factorized, it is of course chaotic (factorized probability measures being the first example of chaotic measures). But in fact, F_N^{in} satisfies a stronger estimate than property (b) in Theorem 1.6.1, in terms of the Monge-Kantorovich distance $\text{dist}_{MK,2}$ (see Theorem 1.6.3). This estimate is propagated for all $t \in \mathbf{R}$ by Dobrushin's inequality, involving the weaker distance $\text{dist}_{MK,1}$. The resulting bound implies that property (b) in Theorem 1.6.1 holds for $F_N(t)$ for all t with $p := f(t)\mathscr{L}^d$. Applying Theorem 1.6.1, we conclude that $F_N(t)$ is $f(t)\mathscr{L}^d$-chaotic.

Our main result in this section is summarized in the following statement.

Theorem 1.6.2 *Assume that the interaction kernel $K \in C^1(\mathbf{R}^d \times \mathbf{R}^d, \mathbf{R}^d$ satisfies assumptions (HK1)-(HK2). Let f^{in} be a probability density on $\mathbf{R}^d$ such that*

$$\int_{\mathbf{R}^d} |z|^{d+5} f(z)dz < \infty.$$

Let $F_N^{in} = (f^{in}\mathscr{L}^d)^{\otimes N}$, and let $F_N(t) = T_t \# F_N^{in}$ be the solution of the N-particle Liouville equation

$$\begin{cases} \partial_t F_N + \frac{1}{N}\sum_{j=1}^{N} \operatorname{div}_{z_i}(K(z_i, z_j)F_N) = 0, \\ F_N\big|_{t=0} = F_N^{in}. \end{cases}$$

Then, for each $j \geq 1$

$$F_{N:j}(t) \to (f(t,\cdot)\mathscr{L}^d)^{\otimes j}$$

weakly in $\mathscr{P}((\mathbf{R}^d)^j)$ as $N \to \infty$, where the probability density $f(t,\cdot)$ is the solution of the mean field PDE

$$\begin{cases} \partial_t f(t,z) + \operatorname{div}_z \left(f(t,z) \int\limits_{\mathbf{R}^d} K(z,z')f(t,z')dz' \right) = 0, \\ f\big|_{t=0} = f^{in}. \end{cases}$$

The rather stringent moment condition on the initial single-particle density f^{in} comes from the following important result from statistics, that can be viewed as a quantitative variant of the law of large numbers.

Theorem 1.6.3 (Horowitz-Karandikar) *For all $p \in \mathscr{P}(\mathbf{R}^d)$ such that*

$$a := \langle p, |z|^{d+5} \rangle < \infty,$$

one has

$$\int\limits_{(\mathbf{R}^d)^N} \operatorname{dist}_{MK,2}(\mu_{Z_N}, p)^2 p^{\otimes N}(dZ_N) \leq \frac{C(a,d)^2}{N^{2/(d+4)}},$$

where $C(a,d)$ is a positive constant that depends only on a and the space dimension d.

References for this result are the original article [61], and the monograph [89].

Taking this estimate for granted, we give the proof of Theorem 1.6.2.

Proof (Proof of Theorem 1.6.2) By the Cauchy-Schwarz inequality, for each pair of Borel probability measures $\mu, \nu \in \mathscr{P}_2(\mathbf{R}^d)$ and all $\pi \in \Pi(\mu,\nu)$, one has

$$\iint\limits_{\mathbf{R}^d} |x-y|\pi(dxdy) \leq \left(\iint\limits_{\mathbf{R}^d} |x-y|^2 \pi(dxdy) \right)^{1/2}.$$

Taking the infimum of both sides of the inequality above as π runs through $\Pi(\mu, \nu)$ shows that

$$\mathrm{dist}_{MK,1}(\mu, \nu) \leq \mathrm{dist}_{MK,2}(\mu, \nu).$$

Hence

$$\int_{(\mathbf{R}^d)^N} \mathrm{dist}_{MK,1}(\mu_{Z_N}, f^{in}\mathscr{L}^d)^2 (f^{in})^{\otimes N}(Z_N)dZ_N \leq \frac{C(a,d)}{N^{2/(d+4)}}.$$

Since $F_N(t) = T_t \# F_N^{in}$, one has

$$\begin{aligned} &F_N(t)(\{Z_N \in (\mathbf{R}^d)^N \text{ s.t. } |\langle \mu_{Z_N} - f(t)\mathscr{L}^d, \phi\rangle| \geq \varepsilon\}) \\ &= F_N^{in}(\{Z_N^{in} \in (\mathbf{R}^d)^N \text{ s.t. } |\langle \mu_{T_t Z_N^{in}} - f(t)\mathscr{L}^d, \phi\rangle| \geq \varepsilon\}). \end{aligned}$$

For each $\pi \in \Pi(\mu_{T_t Z_N^{in}}, f(t)\mathscr{L}^d)$,

$$\begin{aligned} |\langle \mu_{T_t Z_N^{in}} - f(t)\mathscr{L}^d, \phi\rangle| &= \left| \iint_{\mathbf{R}^d \times \mathbf{R}^d} (\phi(x) - \phi(y))\pi(dxdy) \right| \\ &\leq \mathrm{Lip}(\phi) \iint_{\mathbf{R}^d \times \mathbf{R}^d} |x-y|\pi(dxdy). \end{aligned}$$

Taking the inf of both sides of this inequality as π runs through the set of couplings $\Pi(\mu_{T_t Z_N^{in}}, f(t)\mathscr{L}^d)$ shows that

$$|\langle \mu_{T_t Z_N^{in}} - f(t)\mathscr{L}^d, \phi\rangle| \leq \mathrm{Lip}(\phi)\, \mathrm{dist}_{MK,1}(\mu_{T_t Z_N^{in}}, f(t)\mathscr{L}^d).$$

By Proposition 1.4.1 and Dobrushin's inequality (Theorem 1.4.1),

$$\begin{aligned} |\langle \mu_{T_t Z_N^{in}} - f(t)\mathscr{L}^d, \phi\rangle| &\leq \mathrm{Lip}(\phi)\, \mathrm{dist}_{MK,1}(\mu_{T_t Z_N^{in}}, f(t)\mathscr{L}^d) \\ &\leq \mathrm{Lip}(\phi) e^{2L|t|}\, \mathrm{dist}_{MK,1}(\mu_{Z_N^{in}}, f^{in}\mathscr{L}^d). \end{aligned}$$

Therefore

$$\begin{aligned} &F_N(t)(\{Z_N \in (\mathbf{R}^d)^N \text{ s.t. } |\langle \mu_{Z_N} - f(t)\mathscr{L}^d, \phi\rangle| \geq \varepsilon\}) \\ &\leq F_N^{in}(\{Z_N^{in} \in (\mathbf{R}^d)^N \text{ s.t. } \mathrm{dist}_{MK,1}(\mu_{Z_N^{in}}, f^{in}\mathscr{L}^d) \geq e^{-2L|t|}\varepsilon/\mathrm{Lip}(\phi)\}). \end{aligned}$$

Applying the Horowitz-Karandikar theorem recalled above and the Bienaymé-Chebyshev inequality shows that

$$
\begin{aligned}
F_N^{in}(\{Z_N^{in} \in (\mathbf{R}^d)^N \text{ s.t. } \mathrm{dist}_{MK,1}(\mu_{Z_N^{in}}, f^{in}\mathscr{L}^d) \geq e^{-2L|t|}\varepsilon / \mathrm{Lip}(\phi)\}) \\
\leq \frac{e^{4L|t|}\,\mathrm{Lip}(\phi)^2}{\varepsilon^2} \int_{(\mathbf{R}^d)^N} \mathrm{dist}_{MK,1}(\mu_{Z_N^{in}}, f^{in}\mathscr{L}^d)^2 (f^{in})^{\otimes N}(Z_N) dZ_N \\
\leq \frac{e^{4L|t|}\,\mathrm{Lip}(\phi)^2}{\varepsilon^2} \frac{C(a,d)^2}{N^{2/(d+4)}}.
\end{aligned}
$$

Hence

$$
F_N(t)(\{Z_N \in (\mathbf{R}^d)^N \text{ s.t. } |\langle \mu_{Z_N} - f(t)\mathscr{L}^d, \phi\rangle| \geq \varepsilon\}) \to 0
$$

as $N \to \infty$ for each $\phi \in L^\infty \cap \mathrm{Lip}(\mathbf{R}^d)$. By density of $L^\infty \cap \mathrm{Lip}(\mathbf{R}^d)$ in $C_b(\mathbf{R}^d)$ and the implication (b) $\Rightarrow$ (a) in Theorem 1.6.1, we conclude that

$$
F_{N:j}(t) \to (f(t)\mathscr{L}^d)^{\otimes j}
$$

weakly in $\mathscr{P}((\mathbf{R}^d)^j)$ as $N \to \infty$ for all $j \geq 1$.

In particular

$$
F_{N:1}(t) \to f(t)\mathscr{L}^d
$$

weakly in $\mathscr{P}(\mathbf{R}^d)$ as $N \to \infty$, where $f(t,\cdot)$ is the solution of the mean field PDE, and this concludes the proof.

1.7 Further Results on Mean Field Limits in Classical Mechanics

1.7.1 Propagation of Chaos and Quantitative Estimates

Following carefully our proof in the previous section, the convergence result stated in Theorem 1.6.2 could obviously have been formulated as a quantitative estimate of the form

$$
\mathrm{dist}_{MK,1}(P_{N:j}(t), (f(t)\mathscr{L}^d)^{\otimes j}) \leq \varepsilon_j(N),
$$

where $\varepsilon_j(N) \to 0$ for each $j \geq 1$ in the limit as $N \to \infty$. Such an estimate would obviously imply Theorem 1.6.2 since the Monge-Kantorovich distance $\mathrm{dist}_{MK,1}$ metricizes the weak topology on $\mathscr{P}_1(\mathbf{R}^n)$ for all $n \geq 1$—see Theorem 6.9 in [104], although what is needed here is a consequence of Proposition 1.4.1.

In fact, there is another approach of Theorem 1.6.2 which is perhaps worth mentioning, since it provides additional information on the relation between the approach with the empirical measure and the approach based on BBGKY hierarchies.

Specifically, one can prove that, if $F_N^{in} \in \mathscr{P}_{1,sym}((\mathbf{R}^d)^N)$ and if $t \mapsto F_N(t)$ is the solution of the Cauchy problem for the N-particle Liouville equation with initial data F_N^{in}, assuming that the interaction kernel K satisfies the assumptions (HK1-HK2), then

$$\int_{(\mathbf{R}^d)^N} \mu_{T_t Z_N^{in}}^{\otimes m} F_N^{in}(dZ_N^{in}) = \frac{N!}{(N-m)!N^m} F_{N:m}(t) + R_{N,m}(t)$$

where T_t is the flow defined by the N-particle ODE system, while $R_{N,m}(t)$ is a positive Radon measure on $(\mathbf{R}^d)^m$ whose total mass satisfies

$$\langle R_{N,m}(t), 1\rangle = 1 - \frac{N!}{(N-m)!N^m} \leq \frac{m(m-1)}{2N}.$$

This explicit formula, which can be found in [46] expresses the mth marginal of the N-particle distribution $F_N(t)$ in terms of the empirical measure of the N-particle system at time t, i.e. $\mu_{T_t Z_N^{in}}$, up to an error that vanishes as $N \to \infty$.

With this formula, one easily arrives at the following quantitative estimate for the propagation of chaos in the mean field problem. Assume that the initial data is factorized, i.e.

$$P_N^{in} = (f^{in}\mathscr{L}^d)^{\otimes N}$$

and that

$$a := \int_{\mathbf{R}^d} |z|^{d+5} f^{in}(z)dz < \infty.$$

Then

$$\|P_{N:m}(t) - (f(t)\mathscr{L}^d)^{\otimes m}\|_{W^{-1,1}((\mathbf{R}^d)^m)} \leq m\left(\frac{m-1}{N} + e^{2L|t|}\frac{C(a,d)}{N^{1/(d+4)}}\right).$$

for all $t \in \mathbf{R}$ and all $N \geq m \geq 1$. In particular, for $m = 1$, one has

$$\text{dist}_{MK,1}(P_{N:1}(t), f(t)\mathscr{L}^d) \leq C(a,d)e^{2L|t|}/N^{1/(d+4)}.$$

This estimate can be found in [46]; it is established independently and in a slightly different manner in [79].

In fact, the idea of obtaining quantitative estimates for the propagation of chaos in various situations of non equilibrium classical statistical mechanics has been systematically pursued in the reference [79], which bears on a more general class of models than the one considered so far in the present notes. In particular, the discussion in [79] applies to situations where the empirical measure of the N-particle system is *not an exact weak solution* of the limiting mean field equation.

1.7.2 Infinite Hierarchies and Statistical Solutions

While the physical content of the BBGKY hierarchy is transparent (it is the string of equations satisfied by the marginals of the N-particle distribution), the physical meaning of the infinite mean field hierarchy is somewhat less obvious. We discuss this point in the present section.

Consider the Cauchy problem for the mean field PDE:

$$\begin{cases} \partial_t f + \operatorname{div}_z(f\mathscr{K}f) = 0, \\ f\big|_{t=0} = f^{in}, \end{cases}$$

where we recall that

$$\mathscr{K}f(t,z) := \int_{\mathbf{R}^d} K(z,z')f(t,dz'),$$

and where the interaction kernel $K \in C^1(\mathbf{R}^d \times \mathbf{R}^d, \mathbf{R}^d)$ is assumed to satisfy conditions (HK1-HK2).

Denote by $G_t : \mathscr{P}_1(\mathbf{R}^d) \mapsto \mathscr{P}_1(\mathbf{R}^d)$ the nonlinear 1-parameter group defined by

$$G_t f^{in} := f(t,\cdot)$$

where f is the unique solution of the Cauchy problem above.

The notion of statistical solutions of the mean field PDE is defined by analogy with the case of an ODE. Consider the Cauchy problem for an ODE with unknown $t \mapsto x(t) \in \mathbf{R}^n$ in the form

$$\begin{cases} \dot{x}(t) = v(x(t)), \\ x(0) = x_0. \end{cases}$$

Assuming that $v \in \text{Lip}(\mathbf{R}^n, \mathbf{R}^n)$, the Cauchy-Lipschitz theorem provides the existence of a global solution flow $X : \mathbf{R} \times \mathbf{R}^n \to \mathbf{R}^n$, so that the map

$$t \mapsto X(t, x_0)$$

is the solution of the ODE above satisfying $X(0,x_0) = x_0$. The flow X corresponds with a completely deterministic notion of solution: knowing the initial condition x_0 exactly determines the solution $t \mapsto X(t,x_0)$ for all times. Suppose that, instead of knowing exactly the initial condition x_0, one is given a probability distribution μ_0 on the set $\mathbf{R}^n$ of initial data x_0. In other words, μ_0 can be regarded as a "statistical initial condition" for the Cauchy problem above. With the flow X and the statistical initial condition μ_0, one defines

$$\mu(t) := X(t,\cdot)\#\mu_0, \qquad t \in \mathbf{R}.$$

Applying the method of characteristics (see the exercise following Theorem 1.3.1) shows that

$$\begin{cases} \partial_t \mu(t) + \operatorname{div}(\mu(t)v) = 0, \\ \mu(0) = \mu_0. \end{cases}$$

In other words, we recover the well-known fact that solutions of the transport equation (a 1st order PDE) can be viewed as statistical solutions of the ODE defined by the characteristic field of the transport operator. The idea of considering the time-dependent probability $\mu(t)$ (or distribution function, when $\mu(t)$ is absolutely continuous with respect to the Lebesgue measure on the single-particle phase space) instead of the deterministic solution $X(t, x_0)$ for each initial data x_0 lies at the core of the kinetic theory of gases.

Let us return to the problem of defining a notion of statistical solution of the mean field PDE recalled above. First we need a probability measure ν_0 on the space of initial conditions f^{in}, in this case on $\mathscr{P}_1(\mathbf{R}^d)$: this probability ν_0 will be the statistical initial condition for the mean field PDE. By analogy with the case of the simple ODE presented above, we define

$$\nu(t) := G_t \# \nu_0, \qquad t \in \mathbf{R}.$$

The next question is to find the analogue of the transport equation satisfied by $\mu(t)$.

First we need to have a better grasp on $\nu(t)$. One way to understand the formula above expressing $\nu(t)$ as the push-forward of ν_0 under the map G_t defined by the mean field evolution is to write

$$\int_{\mathscr{P}_1(\mathbf{R}^d)} \mathscr{F}(p)\nu(t, dp) = \int_{\mathscr{P}_1(\mathbf{R}^d)} \mathscr{F}(G_t p)\nu_0(dp)$$

for some appropriate class of continuous (in some sense to be defined) functions $\mathscr{F}$ on $\mathscr{P}_1(\mathbf{R}^d)$. Certainly this class of functions should contain "polynomials" on $\mathscr{P}_1(\mathbf{R}^d)$, i.e. linear combinations of "monomials". A monomial of degree k on $\mathscr{P}(\mathbf{R}^d)$ is a expression of the form

$$M_{k,\phi}(p) := \int_{(\mathbf{R}^d)^k} \phi(z_1, \ldots, z_k) p(dx_1) \ldots p(dx_k) = \langle p^{\otimes k}, \phi \rangle$$

where $\phi \in C_b((\mathbf{R}^d)^k)$—without loss of generality one can assume that ϕ is symmetric in the variables $z_1, \ldots, z_k$.

Specializing the formula above to the case $\mathscr{F} = M_{j,\phi}$ as ϕ runs through $C_b((\mathbf{R}^d)^j)$ results in the equality

$$\int_{\mathscr{P}_1(\mathbf{R}^d)} p^{\otimes j} \nu(t, dp) = \int_{\mathscr{P}_1(\mathbf{R}^d)} (G_t p)^{\otimes j} \nu_0(dp).$$

Defining

$$F_j(t) := \int_{\mathscr{P}_1(\mathbf{R}^d)} (G_t p)^{\otimes j} \nu_0(dp), \quad j \geq 1,$$

we claim that the sequence $(F_j)_{j\geq 1}$ is a solution of the infinite mean field hierarchy. Indeed, for each initial single particle probability distribution $p \in \mathscr{P}_1(\mathbf{R}^d)$, the sequence $((G_t p)^{\otimes j})_{j\geq 1}$ is a solution of the infinite mean field hierarchy, which is a sequence of linear equations. Therefore $(F_j)_{j\geq 1}$ is also a solution of the infinite mean field hierarchy, being an average under ν_0 of solutions of this hierarchy.

There is another formulation of this observation. Set $\Omega := (\mathbf{R}^d)^{\mathbf{N}^*}$, equipped with its product topology and the associated Borel σ-algebra $\mathscr{B}(\Omega)$. For each Borel probability measure $p \in \mathscr{P}_1(\mathbf{R}^d)$, we denote by $p^{\otimes\infty}$ the unique Borel probability measure defined on Ω by the formula

$$p^{\otimes\infty}\left(\prod_{k\geq 1} E_k\right) = \prod_{k\geq 1} p(E_k)$$

for each sequence $(E_k)_{k\geq 1}$ of Borel subsets of $\mathbf{R}^d$ such that $E_k = \mathbf{R}^d$ for all but finitely many ks. Notice that

$$\left(p^{\otimes\infty}\right)_{:j} = p^{\otimes j}, \qquad j \geq 1.$$

Define

$$\mathbf{F}(t) := \int_{\mathscr{P}_1(\mathbf{R}^d)} p^{\otimes\infty} \nu(t, dp) = \int_{\mathscr{P}_1(\mathbf{R}^d)} (G_t p)^{\otimes\infty} \nu_0(dp)\,;$$

then, for each $j \geq 1$ and each $t \in \mathbf{R}$,

$$\mathbf{F}(t)_{:j} = F_j(t)$$

so that $(\mathbf{F}(t)_{:j})_{j\geq 1}$ is a solution of the infinite mean field hierarchy. Let us write the jth equation of the infinite mean field hierarchy in terms of $\mathbf{F}(t)$:

$$\partial_t \mathbf{F}(t)_{:j} + \sum_{i=1}^{j} \operatorname{div}_{z_i}(\mathbf{F}(t)_{:j+1} K(z_i, z_{j+1}))_{:j} = 0.$$

This equality is equivalent to the following weak formulation: for each test function $\psi_j \in C_c^1((\mathbf{R}^d)^j)$

$$\partial_t \langle \mathbf{F}(t), \psi_j \rangle = \partial_t \langle \mathbf{F}(t)_{:j}, \psi_j \rangle = \sum_{i=1}^{j} \langle \mathbf{F}(t)_{:j+1}, K(z_i, z_{j+1}) \cdot \nabla_{z_i} \psi_j \rangle$$
$$= \left\langle \mathbf{F}(t), \sum_{i=1}^{j} K(z_i, z_{j+1}) \cdot \nabla_{z_i} \psi_j \right\rangle$$

This suggests the following definition. We denote by $\mathscr{P}_{sym}(\Omega)$ the set of Borel probability measures μ on Ω such that

$$U_\sigma \# \mu = \mu$$

for each $N > 1$ and each $\sigma \in \mathfrak{S}_N$, where

$$U_\sigma(z_1, z_2, \ldots) := (z_{\sigma(1)}, \ldots, z_{\sigma(N)}, z_{N+1}, z_{N+2}, \ldots).$$

Definition 1.7.1 (*Spohn* [97]) Let $I \subset \mathbf{R}$ be an interval. A map

$$I \ni t \mapsto \mathbf{P}(t) \in \mathscr{P}_{sym}(\Omega)$$

is a measure-valued solution of the mean field hierarchy if and only if, for each $j \in \mathbf{N}^*$ and each[14] $\psi_j \in C_0^1((\mathbf{R}^d)^j)$, the map $t \mapsto \langle \mathbf{P}(t), \psi_j \rangle$ is of class C^1 on the interval I and

$$\partial_t \langle \mathbf{P}(t), \psi_j \rangle = \partial_t \langle \mathbf{P}(t)_{:j}, \psi_j \rangle = \left\langle \mathbf{P}(t), \sum_{i=1}^{j} K(z_i, z_{j+1}) \cdot \nabla_{z_i} \psi_j \right\rangle$$

for all $t \in I$.

With this definition, the map $\mathbf{R} \ni t \mapsto \mathbf{F}(t) \in \mathscr{P}_{sym}(\Omega)$ defined above is a measure-valued solution of the mean-field hierarchy satisfying the initial condition

$$\mathbf{F}(0) = \int_{\mathscr{P}_1(\mathbf{R}^d)} p^{\otimes \infty} \nu_0(dp).$$

This observation raises two natural questions:

(a) are measure-valued solutions of the mean-field hierarchy uniquely determined by their initial data?

---

[14] For each $n, k \geq 1$ and each finite dimensional vector space E on $\mathbf{R}$, we denote by $C_0^k(\mathbf{R}^n, E)$ the set of functions of class C^k defined on $\mathbf{R}^n$ with values in E all of whose partial derivatives converge to 0 at infinity. In other words,

$$C_0^k(\mathbf{R}^n, E) := \{f \in C^k(\mathbf{R}^n, E) \text{ s.t. } \partial^\alpha f(x) \to 0 \text{ as } |x| \to \infty \text{ for each } \alpha \in \mathbf{N}^n\}.$$

We denote $C_0^k(\mathbf{R}^n) := C_0^k(\mathbf{R}^n, \mathbf{R})$.

(b) are all the measure-valued solutions of the mean-field hierarchy defined by statistical solutions of the mean field PDE?

A first useful tool in answering these questions is the following result.

Theorem 1.7.1 (Hewitt-Savage [56]) *For each $\mathbf{P} \in \mathcal{P}_{sym}(\Omega)$, there exists a unique Borel probability measure π on $\mathscr{P}(\mathbf{R}^d)$ such that*

$$\mathbf{P} = \int_{\mathscr{P}(\mathbf{R}^d)} p^{\otimes\infty}\pi(dp).$$

The Hewitt-Savage theorem is often quoted and used as follows. Given a sequence of probability measures $(P_j)_{j\geq 1}$ such that $P_j \in \mathscr{P}_{sym}((\mathbf{R}^d)^j)$ satisfying the compatibility condition

$$(CC) \qquad P_{k:j} = P_j \quad \text{for each } k > j \geq 1,$$

by the Kolmogorov extension theorem [99], there exists a unique probability measure $\mathbf{P} \in \mathcal{P}_{sym}(\Omega)$ such that

$$\mathbf{P}_{:j} = P_j \quad \text{for each } j \in \mathbf{N}^*.$$

Therefore, by the Hewitt-Savage theorem, there exists a unique Borel probability measure π on $\mathscr{P}(\mathbf{R}^d)$ such that

$$P_j = \int_{\mathscr{P}(\mathbf{R}^d)} p^{\otimes j}\pi(dp).$$

With this, we can answer the questions (a) and (b) above.

First we consider the problem of admissible initial data. In view of the discussion above, the initial data for the mean field hierarchy can be equivalently either a sequence $(F_j^{in})_{j\geq 1}$ such that $F_j^{in} \in \mathscr{P}_{1,sym}((\mathbf{R}^d)^j)$ for $j \geq 1$ and satisfying the compatibility conditions (CC), or a unique element $\mathbf{F}^{in} \in \mathscr{P}(\Omega)$ such that

$$\mathbf{F}^{in}_{:j} = F_j, \quad j \geq 1,$$

or a unique Borel probability measure ν^{in} on $\mathscr{P}_1(\mathbf{R}^d)$ such that

$$F_j^{in} = \int_{\mathscr{P}_1(\mathbf{R}^d)} p^{\otimes j}\nu^{in}(dp), \quad j \geq 1,$$

or equivalently

$$\mathbf{F}^{in} = \int_{\mathscr{P}_1(\mathbf{R}^d)} p^{\otimes\infty}\nu^{in}(dp).$$

The same reasoning applies for each instant of time $t \neq 0$. A solution at time t of the infinite mean field hierarchy is a sequence $(P_j(t))_{j\geq 1}$ that satisfies in particular $P_j(t) \in \mathscr{P}_{1,sym}((\mathbf{R}^d)^j)$ for $j \geq 1$ together with the compatibility condition (CC). Equivalently, this defines a unique probability measure $\mathbf{P}(t) \in \mathscr{P}_{sym}(\Omega)$ such that $\mathbf{P}(t)_{:j} = P_j(t)$ for each $j \geq 1$, or a unique Borel probability measure $\pi(t)$ on $\mathscr{P}(\mathbf{R}^d)$ such that

$$P_j(t) = \int_{\mathscr{P}(\mathbf{R}^d)} p^{\otimes j} \pi(t, dp), \qquad j \geq 1,$$

or equivalently

$$\mathbf{P}(t) = \int_{\mathscr{P}(\mathbf{R}^d)} p^{\otimes\infty} \pi(t, dp).$$

The following important result was proved by Spohn [97]. It answers questions (a) and (b) above.

Theorem 1.7.2 *Under the assumptions (HK1-HK2) on the interaction kernel* $K \in C^1(\mathbf{R}^d \times \mathbf{R}^d, \mathbf{R}^d)$, *for each Borel probability measure* ν^{in} *on* $\mathscr{P}_1(\mathbf{R}^d)$, *the only measure-valued solution of the infinite mean field hierarchy with initial data*

$$\mathbf{F}^{in} = \int_{\mathscr{P}_1(\mathbf{R}^d)} p^{\otimes\infty} \nu^{in}(dp)$$

is

$$\mathbf{F}(t) = \int_{\mathscr{P}_1(\mathbf{R}^d)} (G_t p)^{\otimes\infty} \nu^{in}(dp), \quad t \in \mathbf{R}.$$

In other words, Spohn's theorem proves that the only measure valued solution of the infinite mean field hierarchy is the statistical solution of the mean field PDE

$$\nu(t) := G_t \# \nu^{in}, \quad t \in \mathbf{R},$$

where ν^{in} is the initial probability measure on the space $\mathscr{P}_1(\mathbf{R}^d)$ of initial data for the mean field PDE such that the initial condition for the mean field hierarchy is the element of $\mathscr{P}_{sym}(\Omega)$ given by

$$\mathbf{F}^{in} = \int_{\mathscr{P}_1(\mathbf{R}^d)} p^{\otimes\infty} \nu^{in}(dp),$$

or equivalently the sequence

$$F_j^{in} = \int\limits_{\mathscr{P}_1(\mathbf{R}^d)} p^{\otimes j} \nu^{in}(dp) \in \mathscr{P}_{sym}((\mathbf{R}^d)^j), \qquad j \geq 1.$$

Notice that, in the case where $\nu^{in} = \delta_{f^{in}}$, one has

$$\mathbf{F}^{in} = (f^{in})^{\otimes\infty}, \text{ or equivalently } F_j^{in} = (f^{in})^{\otimes j} \text{ for all } j \geq 1,$$

and

$$\mathbf{F}(t) = (G_t f^{in})^{\otimes\infty}, \text{ or equivalently } F_j(t) = (G_t f^{in})^{\otimes j} \text{ for all } j \geq 1.$$

In other words the statistical solution $\nu(t)$ of the mean field PDE at time t is given by the formula

$$\nu(t) = \delta_{f(t)} = \delta_{G_t f^{in}} = G_t \# \delta_{f^{in}}.$$

That the statistical solution $\nu(t)$ is a Dirac measure for each $t \in \mathbf{R}$, knowing that $\nu(0) = \nu^{in}$ is a Dirac measure, is equivalent to the propagation of chaos in the context of the mean field limit.

Therefore, Spohn's theorem contains as a particular case the uniqueness theorem with factorized initial data which justifies the validity of the mean field limit in the approach with the BBGKY hierarchy.

But more generally, Spohn's theorem shows that solutions of the infinite mean field hierarchy coincide with the notion of statistical solutions of the mean field PDE. This important piece of information clarifies the meaning of the infinite hierarchy.

Spohn's uniqueness theorem can also be combined with the following uniform stability result on the BBGKY hierarchy to produce a quantitative stability estimate on statistical solutions of the mean field PDE. The following quantitative stability estimate uses the following variant of Monge-Kantorovich distance. For each pair of symmetric Borel probability measures $P \in \mathscr{P}_{1,sym}((\mathbf{R}^d)^M)$ and $Q \in \mathscr{P}_{1,sym}((\mathbf{R}^d)^N)$, consider

$$\mathrm{Dist}_{MK,1}(P, Q) = \inf_{\rho \in \Pi(P,Q)} \iint\limits_{\mathbf{R}^{dM} \times \mathbf{R}^{dN}} \mathrm{dist}_{MK,1}(\mu_{X_M}, \mu_{Y_N}) \rho(dX_M, dY_N).$$

In this formula, P and Q are viewed as Borel probability measures on $\mathscr{P}_1(\mathbf{R}^d)$ concentrated on the set of M- and N-particle empirical measures respectively.

Theorem 1.7.3 *Let $M, N \geq 1$, and consider two probability measures on Euclidean spaces, $P_M^{in} \in \mathscr{P}_{1,sym}((\mathbf{R}^d)^M)$ and $Q_N^{in} \in \mathscr{P}_{1,sym}((\mathbf{R}^d)^N)$, possibly with different dimensions. Assume that the interaction kernel K satisfies the conditions (HK1-HK2). Let $t \mapsto P_M(t)$ and $t \mapsto Q_N(t)$ be respectively the solutions of the M-particle and the N-particle Liouville equations with initial data P_M^{in} and Q_N^{in} respectively. Then*

(a) for each $t \in \mathbf{R}$, one has

$$\mathrm{Dist}_{MK,1}(P_M(t), Q_N(t)) \le e^{2L|t|}\,\mathrm{Dist}_{MK,1}(P_M^{in}, Q_N^{in}).$$

(b) for each $t \in \mathbf{R}$, each $m, M, N \in \mathbf{N}^*$ such that $M, N \ge m$, and for each bounded and Lipschitz continuous function ϕ_m defined on $(\mathbf{R}^d)^m$, one has

$$\begin{aligned}&|\langle P_{N:m}(t) - Q_{N:m}(t), \phi_m\rangle| \\ &\le m\left(e^{2L|t|}\,\mathrm{Lip}(\phi_m)\,\mathrm{Dist}_{MK,1}(P_M^{in}, Q_N^{in}) + (m-1)\|\phi_m\|_{L^\infty}\left(\frac{1}{M}+\frac{1}{N}\right)\right).\end{aligned}$$

For a proof of this result, see [46].

Notice however that Spohn's uniqueness theorem for the infinite mean field hierarchy, even in the particular case of factorized initial data, is more than what is needed to justify the mean field limit. It would be enough to prove the uniqueness of those solutions of the infinite hierarchy that are limits of the sequence of marginals of N-particle distributions as $N \to \infty$. This weaker uniqueness property follows from Theorem 1.7.3, without using Spohn's uniqueness theorem.

1.7.3 Symmetric Functions of Infinitely Many Variables

In various places in this notes—and especially in the last theorem—we encountered the idea of viewing elements of $\mathscr{P}_{sym}((\mathbf{R}^d)^N)$ as Borel probability measures on $\mathscr{P}(\mathbf{R}^d)$ concentrated on the set of empirical measures.

In fact, the identification

$$(\mathbf{R}^d)^N/\mathfrak{S}_N \ni (z_1, \ldots, z_N) \mapsto \mu_{Z_N} := \frac{1}{N}\sum_{k=1}^N \delta_{z_k} \in \mathscr{P}_{sym}(\mathbf{R}^d)$$

can be found in [51] (see especially p. 330 there).

This point of view was pushed much further by P.-L. Lions. He constructed a complete mathematical framework for handling continuous symmetric functions of infinitely many variables that depend weakly on each variable, and used it systematically in his 2007–2008 lectures at the Collège de France on mean field games [72]. This remarkable circle of ideas also appears in the recent work of Mischler, Mouhot and Wennberg (see [78, 79]) on the mean field limit and on quantitative estimates on the propagation of chaos in classical statistical mechanics.

We introduce, for each $X, Y \in Q^N$, the notation

$$d_{LP}(X, Y) := \inf\{\varepsilon > 0 \text{ s.t. } \#\{k = 1, \ldots, N \mid |x_k - y_k| > \varepsilon\} < N\varepsilon\}.$$

This quantity is related to the Levy-Prokhorov distance dist_{LP} on Borel probability measures on Q in the following manner:

$$\text{dist}_{LP}(\mu_X, \mu_Y) = \inf_{\sigma \in \mathfrak{S}_N} d_{LP}(X, Y_\sigma),$$

where $Y_\sigma := (y_{\sigma(1)}, \ldots, y_{\sigma(N)})$. We recall the definition of the Levy-Prokhorov distance dist_{LP} on $\mathscr{P}(Q)$:

$$\text{dist}_{LP}(P_1, P_2) := \inf \left\{ \varepsilon > 0 \mid \inf_{\pi \in \Pi(P_1, P_2)} \iint_{Q \times Q} \mathbf{1}_{|x-y|>\varepsilon} \pi(dxdy) < \varepsilon \right\}.$$

We also recall that the Levy-Prokhorov distance metricizes the weak topology on $\mathscr{P}(Q)$, so that $(\mathscr{P}(Q), \text{dist}_{LP})$ is a compact metric space—see for instance [14].

Lions' key observation is summarized in the following theorem.

Theorem 1.7.4 (Lions [72]) *Let Q be a compact metric space and consider, for each $N \geq 1$, a symmetric function $u_N \in C(Q^N)$. Assume that*

$$\sup_{N \geq 1} \sup_{(z_1, \ldots, z_N) \in Q^N} |u_N(z_1, \ldots, z_N)| < \infty,$$

and that

$$\sup\{|u_N(X) - u_N(Y)| \text{ s.t. } X, Y \in Q^N \text{ and } d_{LP}(X, Y) < \varepsilon\} \to 0$$

uniformly in $N \geq 1$ as $\varepsilon \to 0$. Then there exists $U \in C(\mathscr{P}(Q))$ and a subsequence u_{N_k} of u_N such that

$$\sup_{Z \in Q^{N_k}} |u_{N_k}(Z) - U(\mu_Z)| \to 0 \text{ as } N_k \to \infty.$$

This point of view is obviously dual of the Hewitt-Savage theorem quoted above. In fact, as noticed by Lions, it leads to a very simple proof of the Hewitt-Savage theorem. Lions' argument [72] is sketched below.

Consider a sequence $(P_j)_{j \geq 1}$ of probability measures satisfying the compatibility condition (CC) above. Lions' idea is to consider

$$C(\mathscr{P}(Q)) \ni U \mapsto L(U) := \lim_{N \to \infty} \int_{Q^N} U(\mu_{Z_N}) P_N(dZ_N) \in \mathbf{R}.$$

That the limit on the right hand side of the equality above exists follows from considering the case of a monomial. Indeed, when $U = M_{k,\phi}$ with the notation used in the previous section, i.e.

$$U(\mu) = M_{k,\phi}(\mu) := \int_{Q^k} \phi(x_1, \ldots, x_k)\mu^{\otimes k}(dx_1 \ldots dx_k)$$

where $\phi \in C(Q^k)$, a straightforward computation shows that

$$\int_{Q^N} M_{k,\phi}(\mu_{Z_N})P_N(dZ_N) = \int_{Q^k} \phi(z_1, \ldots, z_k)P_k(dz_1 \ldots dz_k) + O(1/N).$$

Hence

$$L(M_{k,\phi}) = \int_{Q^k} \phi(z_1, \ldots, z_k)P_k(dz_1 \ldots dz_k)$$

for all $k \geq 1$ and all $\phi \in C(Q^k)$.

The set of polynomials, i.e. of linear combinations of monomials, is a subalgebra of $C(\mathscr{P}(Q))$ since

$$M_{k,\phi}(\mu)M_{l,\psi}(\mu) = M_{k+l,\phi\otimes\psi}(\mu)$$

with

$$\phi \otimes \psi(x_1, \ldots, x_{k+l}) := \phi(x_1, \ldots, x_k)\psi(x_{k+1}, \ldots, x_{k+l}).$$

We also use the convention

$$M_{0,1} = 1.$$

This subalgebra is dense in $C(\mathscr{P}(Q))$ for the topology of uniform convergence by the Stone-Weierstrass theorem.

Since

$$\left|\int_{Q^N} U(\mu_{Z_N})P_N(dZ_N)\right| \leq \sup_{\mu\in\mathscr{P}(Q)} |U(\mu)|,$$

for each[15] $U \in C(\mathscr{P}(Q))$, and since the limit

$$L(M_{k,\phi}) := \lim_{N\to\infty} \int_{Q^N} M_{k,\phi}(\mu_{Z_N})P_N(dZ_N) = \int_{Q^k} \phi(z_1, \ldots, z_k)P_k(dz_1 \ldots dz_k)$$

exists for each $k \geq 1$ and each $\phi \in C(Q^k)$, we conclude that this limit exists for each $U \ni C(\mathscr{P}(Q))$.

[15]Since $(\mathscr{P}(Q), \mathrm{dist}_{LP})$ is compact, any element U of $C(\mathscr{P}(Q))$ is bounded on $\mathscr{P}(Q)$, so that

$$\sup_{\mu\in\mathscr{P}(Q)} |U(\mu)| < \infty.$$

Since obviously $L \geq 0$ (being the limit of linear functionals defined by P_N which is a positive measure), and since

$$L(1) = \int_{Q^N} P_N(dZ_N) = 1 \quad \text{for each } N \geq 1$$

we conclude that the linear functional L is represented by a unique probability measure $\pi \in \mathscr{P}(\mathscr{P}(Q))$, i.e.

$$L(U) = \int_{\mathscr{P}(Q)} U(\mu)\pi(d\mu).$$

Specializing to monomials

$$\begin{aligned} L(M_{k,\phi}) &= \int_{Q^k} \phi(z_1, \ldots, z_k) P_k(dz_1 \ldots dz_k) = \int_{\mathscr{P}(Q)} M_{k,\phi}(\mu)\pi(d\mu) \\ &= \int_{\mathscr{P}(Q)} \left(\int_{Q^k} \phi(z_1, \ldots, z_k)\mu^{\otimes k}(dz_1 \ldots dz_k) \right) \pi(d\mu) \\ &= \int_{Q^k} \phi(z_1, \ldots, z_k) \left(\int_{\mathscr{P}(Q)} \mu^{\otimes k}(dz_1 \ldots dz_k)\pi(d\mu) \right) \end{aligned}$$

and this means that

$$P_k = \int_{\mathscr{P}(Q)} \mu^{\otimes k}\pi(d\mu)$$

with the unique probability measure π defined above. This is precisely the representation formula in the Hewitt-Savage theorem.

1.7.4 The Case of Singular Interaction Kernels

The method for proving the mean field limit presented above is based on Dobrushin's estimate and, as such, is limited to cases where the interaction kernel k is Lipschitz continuous in both its arguments. This is most annoying since many interaction kernels of physical interest are singular on the diagonal. All the examples presented in the first section (i.e. the Vlasov-Poisson system and the vorticity formulation of the Euler equation for incompressible fluids in two space dimensions) involve the fundamental solution of the Laplacian, leading to interaction kernels that become singular as the distance between the two interacting particles vanishes. Obtaining rigorous derivations of both models as the mean field limit of large particle systems remains an important open problem.

However, some remarkable results have been obtained in this direction.

In the case of the vorticity formulation of the two-dimensional Euler equation for incompressible fluids, the mean field limit for the dynamics of a large number of vortex centers is analogous to the convergence problem for vortex methods in the numerical analysis of the Euler equation. These methods approximate the vorticity field $\omega \equiv \omega(t,x)$ as follows:

$$\omega(t,\cdot) \simeq \sum_{k=1}^{N} \omega_k \delta_{x_k(t)},$$

where ω_k is the (constant) intensity of the vortex centered at $x_k(t)$. (Notice the slight difference with the mean-field limit discussed above, where each vortex would have the same intensity $\omega_k = 1/N$.) The motion of the vortex centers $x_k(t)$ is governed by the following ODE system:

$$\dot{x}_k(t) = \sum_{\substack{l=1\\ l\neq k}}^{N} \omega_l K_\varepsilon(x_k(t) - x_l(t)), \quad k = 1, \dots, N,$$

where K_ε is an approximation of the vortex interaction kernel

$$K(x) = -\tfrac{1}{2\pi}\frac{Jx}{|x|^2},$$

(with J standing for the rotation of an angle $-\frac{\pi}{2}$).

In one variant of these methods, called "the vortex blob method", the interaction potential is truncated near the singularity at the origin so as to remain smooth as the distance between interacting vortices vanishes. The method based on Dobrushin's estimate presented above for proving the mean field limit applies without modification to the convergence of the vortex blob method. See Chap. 5 in [76] for more details on this topic.

In another variant of these methods, called "the vortex point method", there is no regularization of the interaction kernel, i.e. one takes $K_\varepsilon = K$ in the system of ODEs above governing the motion of vortices. The vorticity field $\omega \equiv \omega(t,x)$ is approximated as follows

$$\omega(t,\cdot) \simeq \omega^h(t,\cdot) := \sum_{k\in\mathbf{Z}^2} \omega_k^h \delta_{x_k^h(t)},$$

where $h > 0$, and the vortex centers satisfy

$$\begin{cases} \dot{x}_k^h(t) = \sum_{\substack{l \in \mathbf{Z}^2 \\ l \neq k}}^{N} \omega_l^h K(x_l^h(t) - x_k^h(t)), \\ x_k^h(0) = hk, \quad k \in \mathbf{Z}^2, \end{cases}$$

while the vortex intensities are chosen so that

$$\omega_k^h = \omega(0, hk), \qquad k \in \mathbf{Z}^2.$$

The convergence of the vortex point method for initial data in the Schwartz class $\mathscr{S}(\mathbf{R}^2)$ of infinitely differentiable functions with rapidly decaying derivatives of all orders has been proved in [48]—see also [49, 52, 94]. The key argument is a control of the minimal distance between vortex centers in Proposition 1 of [52].

One should also mention recent attempts to justify the derivation of the Vlasov-Poisson system as the mean field limit of the Liouville equation for a large number of identical point particles with (unmollified) Coulomb interaction: see [53, 54]. These papers prove the mean field limit and the propagation of chaos for large systems of point particles with singular interaction force field of order $O(d^{-\alpha})$ for $\alpha \leq 1$, where d designates the distance between the two interacting particles.

1.7.5 From Particle Systems to the Vlasov-Maxwell System

There are several difficulties in adapting the method for proving the mean field limit presented above to the case of the Vlasov-Maxwell system; see [44] for a detailed discussion of this problem.

First, the source term in the system of Maxwell's equations is not a probability distribution—as in the case of the Vlasov-Poisson system, where the electric field $E \equiv E(t, x)$ is given by

$$E(t, x) = -\nabla_x \phi_f(t, x), \quad \text{with } -\Delta_x \phi_f(t, x) = \rho_f(t, x).$$

(We recall that

$$\rho_f(t, x) := \int_{\mathbf{R}^3} f(t, x, v) dv$$

so that $\rho_f(t, \cdot)$ is a probability distribution on $\mathbf{R}^3$ if $f(t, \cdot, \cdot)$ is a probability distribution on $\mathbf{R}^3 \times \mathbf{R}^3$, which can be assumed without loss of generality since the total particle number is invariant under the Vlasov-Poisson dynamics.)

In the case of the Vlasov-Maxwell system, the source term in Maxwell's equations is the 4-vector (ρ_f, j_f), defined as follows:

$$\rho_f(t,x) := \int_{\mathbf{R}^3} f(t,x,\xi)d\xi, \qquad j_f(t,x) := \int_{\mathbf{R}^3} v(\xi)f(t,x,\xi)d\xi.$$

This difficulty is handled by an idea introduced in earlier works on Vlasov-Maxwell type systems [15, 16]. The idea is to represent the solution of Maxwell's equations for the electromagnetic field in terms of a *single momentum distribution* of electromagnetic potential, as follows. Consider the Cauchy problem for the wave equation

$$\begin{cases} \Box_{t,x} u_f(t,x,\xi) = f(t,x,\xi), \\ u_f\big|_{t=0} = \partial_t u_f\big|_{t=0} = 0, \end{cases}$$

where $\Box_{t,x} := \partial_t^2 - \Delta_x$ is the d'Alembert operator in the variables t and x. In other words, the momentum variable is a simple parameter in the wave equation above. The self-consistent electromagnetic field in the Vlasov-Maxwell system is represented as

$$\begin{aligned} E(t,x) = &- \partial_t \int_{\mathbf{R}^d} v(\xi)u_f(t,x,\xi)d\xi - \nabla_x \int_{\mathbf{R}^d} u_f(t,x,\xi)d\xi \\ &- \partial_t A_0(t,x) - \nabla_x \phi_0(t,x), \\ B(t,x) = &\operatorname{rot}_x \int_{\mathbf{R}^d} v(\xi)u_f(t,x,\xi)d\xi + \operatorname{rot}_x A_0(t,x), \end{aligned}$$

where $\phi_0 \equiv \phi_0(t,x) \in \mathbf{R}$ and $A_0 \equiv A_0(t,x) \in \mathbf{R}^3$ are respectively a scalar and a vector potential satisfying

$$\Box_{t,x}\phi_0 = 0, \qquad \text{and } \Box_{t,x} A_0 = 0,$$

together with appropriate initial conditions so that the formulas above for E and B match the prescribed initial conditions in the Cauchy problem for the Vlasov-Maxwell system.

With this representation for the electromagnetic field in the Vlasov-Maxwell system, the source term in the field equation is now f itself, a probability distribution in the single particle phase space $\mathbf{R}^3_x \times \mathbf{R}^3_\xi$ whenever $f\big|_{t=0}$ is a probability distribution, since the integral

$$\iint_{\mathbf{R}^3 \times \mathbf{R}^3} f(t,x,\xi)dxd\xi$$

is an invariant of the dynamics defined by the Vlasov-Maxwell equations.

A second difficulty is that the solution of the Cauchy problem for the wave equation defining u_f involves the space-time convolution with Kirchoff's kernel (the forward fundamental solution of the d'Alembert operator): see for instance [95]. This formula is equivalent to the retarded potential formula for solving the system of Maxwell's equations. At variance with the solution of the Poisson equation

$$-\Delta_x \phi_f(t,x) = \rho_f(t,x),$$

the formula giving the solution u_f of the wave equation above in terms of f is not local in the time variable t. Physically, this is due to the fact that the electromagnetic field is propagated at the speed of light $c > 0$ in the Vlasov-Maxwell system, while the electrostatic field in the Vlasov-Poisson system is propagated instantaneously—in other words, the speed of light is considered as infinite in the Vlasov-Poisson system. It remains to check that this difference in structure between the Vlasov-Poisson and the Vlasov-Maxwell systems does not rule out the possibility of an estimate *à la* Dobrushin.

This second difficulty was partially handled in an earlier work [31] for the simpler Vlasov-d'Alembert system

$$\begin{cases} (\partial_t + v(\xi)\cdot\nabla_x)f(t,x,\xi) - \nabla_x\phi_f(t,x)\cdot\nabla_\xi f(t,x,\xi) = 0, \\ \square_{t,x,}\phi_f(t,x) = \rho_f(t,x), \end{cases}$$

where ρ_f is defined in terms of f as above, and $v(\xi) = \nabla_\xi\sqrt{1+|\xi|^2}$ as in the Vlasov-Maxwell system.

A third difficulty—albeit a less essential one—is to choose a regularization procedure for the interaction potential that does not destroy the delicate invariance properties of the Vlasov-Maxwell system. Let $\chi_\varepsilon \equiv \chi_\varepsilon(x)$ be a regularizing sequence in $\mathbf{R}^3$, chosen so that the function χ_ε is even for each $\varepsilon > 0$. The Vlasov-Maxwell system is regularized by replacing the momentum distribution of electromagnetic potential u_f with the solution of the Cauchy problem

$$\begin{cases} \square_{t,x}u_f^\varepsilon(t,x,\xi) = \chi_\varepsilon \star_x \chi_\varepsilon \star_x f(t,x,\xi), \\ u_f\big|_{t=0} = \partial_t u_f\big|_{t=0} = 0, \end{cases}$$

where $\star_x$ designates the convolution product in the x variable. This regularization procedure, originally due to E. Horst, is such that both the conservation of the total particle number and the conservation of some variant of the total energy are satisfied by the mollified system (see [90]).

The interested reader is referred to [44] for a complete discussion of the material presented in this section.

1.8 The Mean Field Problem in Quantum Mechanics

1.8.1 The N-Body Problem in Quantum Mechanics

At variance with classical mechanics, the N-body problem in quantum mechanics is a PDE, and not a system of ODEs. The state at time t of a system of N identical point particles is defined by its N-body wave function

$$\Psi_N \equiv \Psi_N(t, x_1, \ldots, x_N) \in \mathbf{C}, \qquad x_1, \ldots, x_N \in \mathbf{R}^d.$$

We assume that the reader is more or less familiar with the formalism of quantum mechanics, and we shall not attempt to recall more than a few basic facts. An excellent introduction to quantum mechanics can be found in [12].

The meaning of the wave function is such that $|\Psi(x_1, \ldots, x_N)|^2$ is the (joint) probability density of having particle 1 at the position x_1, particle 2 at the position x_2,...,and particle N at the position x_N at time t. This implies the normalization[16]

---

[16]This normalization condition is not satisfied by "generalized eigenfunctions" of an operator with continuous spectrum. Consider the two following examples, where $\mathfrak{H} = L^2(\mathbf{R})$.

(a) Let $\mathcal{H} = -\frac{1}{2}\frac{d^2}{dx^2} + \frac{1}{2}x^2$ (the quantum harmonic oscillator), which has discrete spectrum only. The sequence of eigenvalues of $\mathcal{H}$ is $n + \frac{1}{2}$ with $n \in \mathbf{N}$. Besides $\mathrm{Ker}(\mathcal{H} - (n + \frac{1}{2})I) = \mathbf{C}h_n$ for each $n \in \mathbf{N}$, with

$$h_n(x) := \frac{1}{\sqrt{2^n n!}\pi^{1/4}} e^{-x^2/2} H_n(x), \quad \text{where } H_n(x) = (-1)^n e^{x^2} \frac{d^n}{dx^n} e^{-x^2}.$$

The function H_n is the nth Hermite polynomial, and the sequence $(h_n)_{n\geq 0}$ is a Hilbert basis of $\mathfrak{H}$. In particular one has the orthonormality condition

$$\int_{\mathbf{R}} h_n(x) h_m(x) dx = \delta_{mn}, \quad \text{for all } m, n \geq 0$$

where δ_{mn} is the Kronecker symbol (i.e. $\delta_{mn} = 0$ if $m \neq n$ and $\delta_{mn} = 1$ if $m = n$).

(b) Let $P = -i\frac{d}{dx}$ (the momentum operator), which has continuous spectrum only. The spectrum of P is the real line $\mathbf{R}$, and the generalized eigenfunctions of P are the functions $e_k : x \mapsto e_k(x) := e^{i2\pi kx}$. For each $k \in \mathbf{R}$, one has $Pe_k = 2\pi k e_k$ but $e_k \notin \mathfrak{H}$. However one has the "formula" analogous to the orthonormality condition in case (a):

$$\int_{\mathbf{R}} \overline{e_k(x)} e_l(x) dx = \delta_0(k - l)$$

where δ_0 is the Dirac mass at the origin. The integrand on the left hand side of the equality above is not an element of $L^1(\mathbf{R})$, and the integral is not a Lebesgue integral. The identity above should be understood as the Fourier inversion formula on the class $\mathscr{S}'(\mathbf{R})$ of tempered distributions on the real line $\mathbf{R}$.

$$\int_{(\mathbf{R}^d)^N} |\Psi_N(t, x_1, \ldots, x_N)|^2 dx_1 \ldots dx_N = 1,$$

that is satisfied for all $t \in \mathbf{R}$.

We assume that the interaction between a particle at position x and a particle at position y is given by a 2-body potential $V(x - y) \in \mathbf{R}$. Henceforth, we assume that V is even, so that the force exerted by the particle at position y on the particle at position x, i.e. $-\nabla V(x - y)$, exactly balances the force exerted by the particle at position x on the particle at position y, i.e. $-\nabla V(y - x)$. (Indeed, ∇V is odd since V is even.)

With these data, we can write the Schrödinger equation governing the N-body wave function for a system of N identical particles of mass m with 2-body interaction given by the potential V:

$$i\hbar\partial_t \Psi_N = -\tfrac{\hbar^2}{2m} \sum_{k=1}^{N} \Delta_{x_k} \Psi_N + \sum_{1 \le k < l \le N} V(x_k - x_l)\Psi_N,$$

where $\hbar$ is the reduced Planck constant.

The N-body Schrödinger equation above is a PDE which is the analogue in quantum mechanics of Newton's second law

$$\begin{cases} m\dot{x}_k = \xi_k, \\ \dot{\xi}_k = -\displaystyle\sum_{\substack{l=1 \\ l \neq k}}^{N} \nabla V(x_k - x_l), \qquad k = 1, \ldots, N \end{cases}$$

in classical mechanics.

The question of existence and uniqueness of a solution of the Schrödinger equation of N-particle quantum dynamics is settled by the following remarkable result. (Without loss of generality, we assume that $\hbar^2/m = 1$.)

Theorem 1.8.1 (Kato) *Assume that the space dimension is $d = 3$. If, for some $R > 0$,*

$$V\big|_{B(0,R)} \in L^2(B(0, R)) \quad \text{and} \quad V\big|_{\mathbf{R}^3 \setminus B(0,R)} \in L^\infty(\mathbf{R}^3 \setminus B(0, R)),$$

(Footnote 16 continued)
In the case (a), if the system is in an eigenstate corresponding with the eigenvalue $n + \frac{1}{2}$ of the operator $\mathcal{H}$, its wave function is of the form $\psi = \omega h_n$ with $|\omega| = 1$, so that $\|\psi\|_{\mathfrak{H}} = 1$.

In the case (b), if the system is in an eigenstate corresponding with the element k of the spectrum of the operator P, it cannot be described by any wave function in $\mathfrak{H}$, but only by a generalized eigenfunction of P, that does not belong to $\mathfrak{H}$.

In the discussion below, we shall never consider quantum states described by generalized eigenfunctions as in (b), but only quantum states corresponding with normalized wave functions.

then, for all $N \geq 1$, the unbounded operator

$$\mathfrak{H}_N := -\tfrac{1}{2}\sum_{k=1}^{N} \Delta_{x_k} + \sum_{1\leq k<l\leq N} V(x_k - x_l)$$

has a self-adjoint extension on $\mathfrak{H}_N := L^2((\mathbf{R}^3)^N)$ and generates a unitary group $e^{-it\mathcal{H}_N}$ on $\mathfrak{H}_N$.

See Chap. V, Sect. 5.3 in [63, 64] for a proof of this result.

This result is the analogue in quantum mechanics of the existence and uniqueness of the solution of the Cauchy problem for the motion equations in classical mechanics that follow from the Cauchy-Lipschitz theorem. Observe that applying the Cauchy-Lipschitz theorem to the system of ODEs resulting from Newton's second law applied to each particle would require V to have Lipschitz continuous first order derivatives. The assumptions on the regularity of the potential in Kato's result are obviously much less stringent—for instance, the Coulomb potential $V(z) = C/|z|$ obviously satisfies these assumptions. (That the Coulomb potential satisfies the assumptions in Kato's theorem is of course very satisfying since most computations in atomic physics involve the Coulomb interaction between the electrons and the nuclei in atoms and molecules.)

As in the case of classical mechanics, our goal is to study the behavior of the wave function $\Psi_N(t) = e^{-it\mathcal{H}_N}\Psi_N^{in}$ in the limit as $N \to \infty$, under appropriate scaling assumptions involving the interaction potential V and the particle number N.

First we need to define the analogue of the mean field scaling used in the case of classical mechanics. Our argument for choosing this scaling is based on considering the energy of the particle system.

In quantum mechanics, the energy of a system of N identical point particles of mass m with pairwise interaction described in terms of the real-valued 2-body potential V is

$$\sum_{k=1}^{N} \int_{(\mathbf{R}^d)^N} \tfrac{\hbar^2}{2m} |\nabla_{x_k} \Psi_N(x_1, \ldots, x_N)|^2 dx_1 \ldots, dx_N$$
$$+ \sum_{1\leq k<l\leq N} \int_{(\mathbf{R}^d)^N} V(x_k - x_l)|\Psi_N(x_1, \ldots, x_N)|^2 dx_1 \ldots, dx_N,$$

where the first term is the kinetic energy, while the second term is the potential energy.

Pick a typical length scale L and energy scale E in the system of N particles under consideration. We introduce the dimensionless space variables and potential as follows:

$$\hat{x}_k = \frac{x_k}{L}, \qquad \hat{V}(\hat{x}_k - \hat{x}_l) = \frac{1}{E} V(x_k - x_l).$$

Likewise, the N-particle wave function is scaled as

$$\hat{\Psi}_N(\hat{t}, \hat{x}_1, \dots, \hat{x}_N) = L^{dN/2}\Psi_N(t, x_1, \dots, x_N),$$

where T is a time scale to be defined later, and where

$$\hat{t} = \frac{t}{T}$$

is the dimensionless time variable.

Observe that

$$\int |\hat{\Psi}_N(\hat{t}, \hat{x}_1, \dots, \hat{x}_N)|^2 d\hat{x}_1 \dots d\hat{x}_N = \int |\Psi_N(t, x_1, \dots, x_N)|^2 dx_1 \dots dx_N = 1$$

with this scaling.

With these dimensionless quantities, the kinetic and potential energies become respectively

$$\text{Kinetic energy} = \sum_{k=1}^{N} \int_{(\mathbf{R}^d)^N} \tfrac{\hbar^2}{2mL^2} |\nabla_{\hat{x}_k} \hat{\Psi}_N|^2 d\hat{x}_1 \dots d\hat{x}_N,$$

$$\text{Potential energy} = \sum_{1 \le k < l \le N} \int_{(\mathbf{R}^d)^N} E\hat{V}(\hat{x}_k - \hat{x}_l)|\hat{\Psi}_N|^2 d\hat{x}_1 \dots d\hat{x}_N.$$

Observe that there are N terms in the kinetic energy, while the potential energy involves $\frac{1}{2}N(N-1)$ terms. Therefore, we scale the system so that the kinetic energy and the potential energy are of same order of magnitude, by choosing E and L so that

$$\frac{\hbar^2}{mL^2} = NE.$$

Thus

$$\begin{aligned}
&-\tfrac{\hbar^2}{2m} \sum_{k=1}^{N} \Delta_{x_k} \Psi_N + \sum_{1 \le k < l \le N} V(x_k - x_l)\Psi_N \\
&= L^{-dN/2} \frac{\hbar^2}{mL^2} \left(-\tfrac{1}{2} \sum_{k=1}^{N} \Delta_{\hat{x}_k} \hat{\Psi}_N + \frac{1}{N} \sum_{1 \le k < l \le N} \hat{V}(\hat{x}_k - \hat{x}_l)\hat{\Psi}_N \right).
\end{aligned}$$

At this point, we set

$$\hat{H}_N \hat{\Psi}_N := -\tfrac{1}{2} \sum_{k=1}^{N} \Delta_{\hat{x}_k} \hat{\Psi}_N + \frac{1}{N} \sum_{1 \le k < l \le N} \hat{V}(\hat{x}_k - \hat{x}_l)\hat{\Psi}_N,$$

and define the time scale T as follows:

$$T = \frac{mL^2}{\hbar}.$$

With the rescaled time variable $\hat{t} = t/T$, the N-body Schrödinger equation in mean-field scaling becomes

$$i\partial_{\hat{t}}\hat{\Psi}_N = -\tfrac{1}{2}\sum_{k=1}^{N}\Delta_{\hat{x}_k}\hat{\Psi}_N + \frac{1}{N}\sum_{1\le k<l\le N}\hat{V}(\hat{x}_k - \hat{x}_l)\hat{\Psi}_N.$$

Henceforth we drop all hats on variables, and consider as our starting point the scaled equation

$$i\partial_t\Psi_N = -\tfrac{1}{2}\sum_{k=1}^{N}\Delta_{x_k}\Psi_N + \frac{1}{N}\sum_{1\le k<l\le N}V(x_k - x_l)\Psi_N.$$

A formal computation based on the fact that V is real-valued shows that

$$\frac{d}{dt}\int_{(\mathbf{R}^d)^N}|\Psi_N(t,x_1,\dots,x_N)|^2dx_1\dots dx_N = 0,$$

so that

$$\int_{(\mathbf{R}^d)^N}|\Psi_N(t,x_1,\dots,x_N)|^2dx_1\dots dx_N = 1.$$

for all $t \in \mathbf{R}$.

The rigorous argument is based on Kato's theorem (Theorem 1.8.1) stated above in space dimension $d = 3$: with the notation

$$\mathcal{H}_N := -\tfrac{1}{2}\sum_{k=1}^{N}\Delta_{x_k} + \frac{1}{N}\sum_{1\le k<l\le N}V(x_k - x_l)$$

the one-parameter group $e^{-it\mathcal{H}_N}$ is unitary on $\mathfrak{H}_N = L^2((\mathbf{R}^3)^N)$, so that

$$\int_{(\mathbf{R}^3)^N}|\Psi_N(t,x_1,\dots,x_N)|^2dx_1\dots dx_N = \|e^{-it\mathcal{H}_N}\Psi_N^{in}\|_{\mathfrak{H}_N}^2 = 1$$

for all $t \in \mathbf{R}$.

1.8.2 The Target Mean Field Equation

The target quantum mean field equation is the quantum analogue of the Vlasov-Poisson equation.

Its unknown is the single-particle wave function $\psi \equiv \psi(t,x) \in \mathbf{C}$, with $x \in \mathbf{R}^d$

The basic idea in the mean field approximation is the same as in the classical case. Since the probability of finding a particle in an infinitesimal volume element dx at time t is $|\psi(t,x)|^2 dx$, one expects that the action on the kth particle located at x_k is defined in terms of the potential V by the formula

$$\frac{1}{N}\sum_{\substack{l=1\\ k\neq l}}^{N} V(x_k - x_l) \sim \int_{\mathbf{R}^d} V(x_k - z)|\psi(t,z)|^2 dz.$$

This suggests that the target mean field equation for single-particle wave function obtained as the limit of the N-body Schrödinger equation is

$$i\partial_t\psi(t,x) = -\tfrac{1}{2}\Delta_x\psi(t,x) + \psi(t,x)\int_{\mathbf{R}^d} V(x-y)|\psi(t,y)|^2 dy.$$

This equation is known as the Hartree equation—and also sometimes referred to as the Schrödinger-Poisson equation when $V(z)$ is the Coulomb potential in space dimension 3. Indeed, if

$$V(z) = \frac{1}{4\pi|z|}, \qquad z \in \mathbf{R}^3 \setminus \{0\},$$

the equation above is equivalent to the system

$$\begin{cases} i\partial_t\psi(t,x) = -\tfrac{1}{2}\Delta_x\psi(t,x) + U(t,x)\psi(t,x), \\ -\Delta_x U(t,x) = |\psi(t,x)|^2. \end{cases}$$

This system is the quantum analogue of the Vlasov-Poisson system.

The Hartree equation has been studied extensively by various authors: see [18, 39]. The following statement is a consequence of Theorem 3.1 in [39].

Proposition 1.8.1 *Assume that V is a real-valued, even element of $L^\infty(\mathbf{R}^d)$. For each $\psi^{in} \in H^k(\mathbf{R}^d)$, there exists a unique mild solution*[17] *of the Cauchy problem for Hartree's equation*

---

[17] A mild solution of the Cauchy problem

$$\begin{cases} \dot{u}(t) = Au(t) + F[u(t)], \\ u\big|_{t=0} = u^{in}, \end{cases}$$

$$\begin{cases} i\partial_t\psi(t,x) = -\frac{1}{2}\Delta_x\psi(t,x) + (V \star_x |\psi|^2)\psi(t,x), & x \in \mathbf{R}^d \\ \psi\big|_{t=0} = \psi^{in} \end{cases}$$

such that $\psi \in C_b(\mathbf{R}; H^1(\mathbf{R}^d)) \cap C(\mathbf{R}; H^k(\mathbf{R}^2))$. *Besides, this solution satisfies the conservation laws of mass and energy, viz.*

$$\|\psi(t,\cdot)\|_{L^2(\mathbf{R}^d)} = Const.$$

and

$$\tfrac{1}{2}\|\nabla_x\psi(t,\cdot)\|^2_{L^2(\mathbf{R}^d)} + \tfrac{1}{2}\iint\limits_{\mathbf{R}^d\times\mathbf{R}^d} V(x-y)|\psi(t,x)|^2|\psi(t,y)|^2 dxdy = Const.$$

for all $t \in \mathbf{R}$.

1.8.3 The Formalism of Density Matrices

Before studying the N-particle Schrödinger equation in the large N limit, we need another formulation of the quantum N-body problem.

Henceforth we denote by $\langle\cdot|\cdot\rangle$ (or by $\langle\cdot|\cdot\rangle_{\mathfrak{H}_N}$ to avoid ambiguity if needed) the inner product of the Hilbert space $\mathfrak{H}_N := L^2((\mathbf{R}^d)^N)$, and by $|\cdot|$ or $|\cdot|_{\mathfrak{H}_N}$ the associated norm, defined by the formula

$$|\Phi|_{\mathfrak{H}_N} := \langle\Phi|\Phi\rangle^{1/2}_{\mathfrak{H}_N}.$$

To the N-particle wave function $\Psi_N \equiv \Psi_N(x_1,\ldots,x_N)$ that is an element of the Hilbert space $\mathfrak{H}_N := L^2((\mathbf{R}^d)^N)$, one associates the bounded linear operator D_N on $\mathfrak{H}_N$ defined as follows:

$$D_N := \text{ orthogonal projection on } \mathbf{C}\Psi_N \text{ in } \mathfrak{H}_N,$$

i.e.

(Footnote 17 continued)
where A generates a strongly continuous contraction semigroup on a Banach space X and $F : X \to X$ is a continuous map, is an X-valued continuous function $I \ni t \mapsto u(t) \in X$ defined on $I = [0,\tau]$ with $\tau \in [0,+\infty]$ that is a solution of the integral equation

$$u(t) = e^{tA}u^{in} + \int_0^t e^{(t-s)A}F[u(s)]ds$$

for each $t \in I$.

$$D_N\Phi_N = \Psi_N \int_{(\mathbf{R}^d)^N} \overline{\Psi_N}\Phi_N(x_1,\ldots,x_N)dx_1\ldots dx_N = \langle\Psi_N|\Phi_N\rangle\Psi_N.$$

This linear operator is called the "density matrix" of the N-particle system.

Whenever convenient, we use the notation common in the physics literature

$$D_N := |\Psi_N\rangle\langle\Psi_N|$$

which is the equivalent of

$$D_N := \Psi_N \otimes L_{\Psi_N}, \qquad \text{with } L_{\Psi_N} := \langle\Psi_N|\cdot\rangle \in \mathfrak{H}_N^*.$$

Obviously, D_N is an integral operator, whose integral kernel is

$$D_N(x_1,\ldots,x_N,y_1,\ldots,y_N) = \Psi_N(x_1,\ldots,x_N)\overline{\Psi_N(y_1,\ldots,y_N)}.$$

In the sequel, we shall sometimes abuse the notation and use the same letter to designate an integral operator and its integral kernel.

Being an orthogonal projection in $\mathfrak{H}_N$, the operator D_N is self-adjoint and non-negative:

$$D_N = D_N^* = D_N^2 \geq 0.$$

This property is the quantum analogue of the positivity of the N-particle distribution function in classical statistical mechanics.

The density matrix D_N for a system of identical, indistinguishable particles satisfies the following symmetry: for all $\sigma \in \mathfrak{S}_N$, one has

$$S_\sigma D_N = D_N,$$

where S_σ is the transformation on integral kernels defined by the formula

$$S_\sigma D_N(x_1,\ldots,x_N,y_1,\ldots,y_N) := D_N(x_{\sigma^{-1}(1)},\ldots,x_{\sigma^{-1}(N)},y_{\sigma^{-1}(1)},\ldots,y_{\sigma^{-1}(N)}).$$

This assumption is not to be confused with the condition

$$U_\sigma\Psi_N = \Psi_N,$$

where U_σ is the unitary operator defined on $\mathfrak{H}_N$ by the formula

$$U_\sigma\Psi_N(x_1,\ldots,x_N) := \Psi_N(x_{\sigma^{-1}(1)},\ldots,x_{\sigma^{-1}(N)}).$$

This assumption is an assumption on the *statistics* of particles, i.e. on whether the particles under consideration follow the Bose-Einstein or the Fermi-Dirac statistics. Particles following the Bose-Einstein statistics are called bosons, and the wave function of any system of N identical bosons satisfies the condition

$$U_\sigma \Psi_N = \Psi_N .$$

Particles following the Fermi-Dirac statistics are called fermions, and the wave function of any system of N identical fermions satisfies the condition

$$U_\sigma \Psi_N = (-1)^{\mathrm{sign}(\sigma)} \Psi_N .$$

Exercise Express $U_{\sigma\sigma'}$ and U_σ^* in terms of U_σ and $U_{\sigma'}$ for all $\sigma, \sigma' \in \mathfrak{S}_N$, and check that U_σ is a unitary operator on $\mathfrak{H}_N$. Express the transformation S_σ in terms of U_σ for all $\sigma \in \mathfrak{S}_N$.

In fact, whenever

$$U_\sigma \Psi_N = \omega(\sigma)\Psi_N \quad \text{with } \omega(\sigma) \in \mathbf{C} \text{ and } |\omega(\sigma)| = 1$$

one has

$$S_\sigma D_N = D_N .$$

In particular, the density matrix of a system of N identical bosons ($\omega(\sigma) = 1$) or N identical fermions ($\omega(\sigma) = (-1)^{\mathrm{sign}(\sigma)}$) satisfies

$$S_\sigma D_N = D_N .$$

Exercise Assume that, for some $\Psi_N \in \mathfrak{H}_N$ such that $|\Psi_N| = 1$ and for all $\sigma \in \mathfrak{S}_N$, one has

$$U_\sigma \Psi_N = \omega(\sigma)\Psi_N \quad \text{with } \omega(\sigma) \in \mathbf{C} \text{ and } |\omega(\sigma)| = 1.$$

Compute $\omega(\sigma\sigma')$ in terms of $\omega(\sigma)$ and $\omega(\sigma')$ for all $\sigma, \sigma' \in \mathfrak{S}_N$. Prove that one has $\omega(\sigma) = \pm 1$ for all $\sigma \in \mathfrak{S}_N$.

Assuming that the time-dependent wave function $\Psi_N(t, \cdot)$ of a system of N identical particles satisfies the N-body Schrödinger equation with initial data Ψ_N^{in}, i.e.

$$\begin{cases} i\partial_t \Psi_N = \mathcal{H}_N \Psi_N := -\frac{1}{2}\sum_{k=1}^{N} \Delta_{x_k}\Psi_N + \frac{1}{N}\sum_{1\le k<l\le N} V(x_k - x_l)\Psi_N, \\ \Psi_N\big|_{t=0} = \Psi_N^{in}, \end{cases}$$

our next task is to find the equation satisfied by $D_N := |\Psi_N\rangle\langle\Psi_N|$.

Since $e^{-it\mathcal{H}_N}$ is a unitary operator on $\mathfrak{H}_N$ provided that the interaction potential V satisfies the assumptions of Kato's theorem above, one has

$$\begin{aligned}D_N(t) &= |e^{-it\mathcal{H}_N}\Psi_N^{in}\rangle\langle e^{-it\mathcal{H}_N}\Psi_N^{in}| \\ &= e^{-it\mathcal{H}_N}D_N^{in}(e^{-it\mathcal{H}_N})^* = e^{-it\mathcal{H}_N}D_N^{in}e^{it\mathcal{H}_N},\end{aligned}$$

where

$$D_N^{in} := |\Psi_N^{in}\rangle\langle\Psi_N^{in}|.$$

It is instructive to do the analogous computation in finite dimension (i.e. with matrices). Assume that $A, B \in M_n(\mathbf{C})$, and set

$$M(t) := e^{tA}Be^{-tA}.$$

Then the function $\mathbf{R} \ni t \mapsto M(t) \in M_n(\mathbf{C})$ is obviously of class C^∞ (and even real analytic) and satisfies[18]

$$\begin{aligned}\frac{d}{dt}M(t) = \frac{d}{dt}(e^{tA}Be^{-tA}) &= Ae^{tA}Be^{-tA} - e^{tA}BAe^{-tA} \\ &= Ae^{tA}Be^{-tA} - e^{tA}Be^{-tA}A \\ &= AM(t) - M(t)A = [A, M(t)].\end{aligned}$$

since A commutes with e^{-tA}.

In view of this elementary computation, we leave it to the reader to verify the following statement.

Proposition 1.8.2 *Assume that the real-valued potential V satisfies the assumptions in Kato's theorem (Theorem 1.8.1), and let $\Psi_N^{in} \in \mathfrak{H}_N = L^2((\mathbf{R}^d)^N)$.*

Let $\Psi_N \equiv \Psi_N(t, x_1, \ldots, x_N)$ be the solution of the N-particle Schrödinger equation

$$\begin{cases} i\partial_t\Psi_N = -\frac{1}{2}\sum_{k=1}^N \Delta_{x_k}\Psi_N + \sum_{1\le k<l\le N} V(x_k - x_l)\Psi_N, & x_1, \ldots, x_N \in \mathbf{R}^d, \ t \in \mathbf{R}, \\ \Psi_N\big|_{t=0} = \Psi_N^{in}, \end{cases}$$

i.e.

$$\Psi_N(t, \cdot) = e^{-it\mathcal{H}_N}\Psi_N^{in},$$

with scaled Hamiltonian

---

[18]For linear operators A, B on the Hilbert space $\mathfrak{H}$, we designate by $[A, B]$ their commutator in other words $[A, B] := AB - BA$. There is obviously a difficulty with the domain of $[A, B]$ viewed as an unbounded operator on $\mathfrak{H}$ if A and B are unbounded operators on $\mathfrak{H}$. This difficulty will be deliberately left aside, as we shall mostly consider $B \mapsto [A, B]$ as an unbounded operator on $\mathcal{L}(\mathfrak{H})$, which is different matter.

$$\mathcal{H}_N := -\tfrac{1}{2}\sum_{k=1}^{N}\Delta_k + \frac{1}{N}\sum_{1\le k<l\le N} V_{kl},$$

where Δ_k designates the Laplacian in the kth variable, while V_{kl} is the operator defined as the multiplication by $V(x_k - x_l)$.

Then the density matrix

$$D_N(t) := |\Psi_N(t,\cdot)\rangle\langle\Psi_N(t,\cdot)| = e^{-it\mathcal{H}_N} D_N(0) e^{it\mathcal{H}_N}$$

satisfies the von Neumann equation

$$i\dot{D}_N(t) = [\mathcal{H}_N, D_N(t)] = -\tfrac{1}{2}\sum_{k=1}^{N}[\Delta_k, D_N(t)] + \frac{1}{N}\sum_{1\le k<l\le N}[V_{kl}, D_N].$$

In terms of integral kernels

$$\begin{aligned} & i\partial_t D_N(t, x_1, \ldots, x_N, y_1, \ldots, y_N) \\ &= -\tfrac{1}{2}\sum_{k=1}^{N}(\Delta_{x_k} - \Delta_{y_k}) D_N(t, x_1, \ldots, x_N, y_1, \ldots, y_N) \\ &+ \frac{1}{N}\sum_{1\le k<l\le N}(V(x_k - x_l) - V(y_k - y_l)) D_N(t, x_1, \ldots, x_N, y_1, \ldots, y_N). \end{aligned}$$

The fundamental properties of the density matrix are propagated under the flow associated to the von Neumann equation.

Proposition 1.8.3 *Assume that the real-valued potential V satisfies the assumptions in Kato's theorem, and let*

$$D_N^{in} = (D_N^{in})^* = (D_N^{in})^2 \ge 0.$$

Then for each $t \in \mathbf{R}$, one has

$$D_N(t) := e^{-it\mathcal{H}_N} D_N^{in} e^{it\mathcal{H}_N} = D_N(t)^* = D_N(t)^2 \ge 0,$$

with scaled Hamiltonian

$$\mathcal{H}_N := -\tfrac{1}{2}\sum_{k=1}^{N}\Delta_k + \frac{1}{N}\sum_{1\le k<l\le N} V_{kl}$$

where Δ_k designates the Laplacian in the kth variable, while V_{kl} is the operator defined as the multiplication by $V(x_k - x_l)$.

Likewise, if particles are indistinguishable initially, i.e. if

$$S_\sigma D_N^{in} = D_N^{in},$$

then

$$S_\sigma D_N(t) = D_N(t)$$

for all $t \in \mathbf{R}$.

Proof First

$$D_N(t)^* = (e^{-it\mathcal{H}_N} D_N^{in} e^{it\mathcal{H}_N})^* = e^{-it\mathcal{H}_N}(D_N^{in})^* e^{it\mathcal{H}_N}$$

since $\mathcal{H}_N$ is self adjoint, so that $(e^{it\mathcal{H}_N})^* = e^{-it\mathcal{H}_N}$. Then

$$\begin{aligned} D_N(t)^2 &= (e^{-it\mathcal{H}_N} D_N^{in} e^{it\mathcal{H}_N})^2 \\ &= e^{-it\mathcal{H}_N}(D_N^{in})^2 e^{it\mathcal{H}_N} = (e^{-it\mathcal{H}_N} D_N^{in} e^{it\mathcal{H}_N})^2 = D_N(t). \end{aligned}$$

That $S_\sigma D_N(t) = D_N(t)$ for all $t \in \mathbf{R}$ if $S_\sigma D_N^{in} = D_N^{in}$ is a straightforward consequence of the following lemma.

Lemma 1.8.1 *Let U_σ be the unitary operator defined on $\mathfrak{H}_N$ for each $\sigma \in \mathfrak{S}_N$ by*

$$U_\sigma \Psi_N(x_1, \ldots, x_N) := \Psi_N(x_{\sigma^{-1}(1)}, \ldots, x_{\sigma^{-1}(N)}).$$

Then

$$U_\sigma e^{it\mathcal{H}_N} = e^{it\mathcal{H}_N} U_\sigma$$

for each $t \in \mathbf{R}$ and each $\sigma \in \mathfrak{S}_N$.

Thus, for each $t \in \mathbf{R}$ and each $\sigma \in \mathfrak{S}_N$, one has

$$\begin{aligned} S_\sigma D_N(t) &= U_\sigma D_N(t) U_\sigma^* = U_\sigma e^{-it\mathcal{H}_N} D_N^{in} e^{it\mathcal{H}_N} U_\sigma^* \\ &= e^{-it\mathcal{H}_N} U_\sigma D_N^{in} U_\sigma^* e^{it\mathcal{H}_N} = e^{-it\mathcal{H}_N} D_N^{in} e^{it\mathcal{H}_N} = D_N(t), \end{aligned}$$

provided that

$$S_\sigma D_N^{in} = U_\sigma D_N^{in} U_\sigma^* = D_N^{in}.$$

Proof ([Proof of Lemma 1.8.1) If $\Phi \in C_c^\infty((\mathbf{R}^d)^N)$, one has

$$\Delta_k U_\sigma \Phi = U_\sigma \Delta_{\sigma(k)} \Phi$$

and

$$V_{kl} U_\sigma \Phi = U_\sigma V_{\sigma(k),\sigma(l)} \Phi,$$

so that

$$\mathcal{H}_N U_\sigma \Phi = U_\sigma \mathcal{H}_N \Phi.$$

Therefore

$$U_\sigma e^{itH_N} = e^{itH_N} U_\sigma$$

for each $t \in \mathbf{R}$ and each $\sigma \in \mathfrak{S}_N$.

1.9 Elements of Operator Theory

We recall that $\mathfrak{H}$ designates a complex, separable Hilbert space $\mathfrak{H}$, with inner product denoted by $\langle\cdot|\cdot\rangle$ and norm denoted by $|\cdot|$. (To avoid ambiguity, we also use the notation $\langle\cdot|\cdot\rangle_{\mathfrak{H}}$ and $|\cdot|_{\mathfrak{H}}$ whenever needed.) We always assume that the inner product is antilinear in its first argument and linear in its second argument. The set of Hilbert basis of $\mathfrak{H}$—i.e. the set of orthonormal and total families in $\mathfrak{H}$—is denoted by $HB(\mathfrak{H})$.

The set of bounded operators on $\mathfrak{H}$ is denoted by $\mathscr{L}(\mathfrak{H})$, with operator norm

$$\|A\| = \sup_{|x|=1} |Ax|.$$

We recall that $\mathscr{L}(\mathfrak{H})$ endowed with the operator norm $\|\cdot\|$ is a Banach algebra—and even a C^*-algebra, since

$$\|A^*A\| = \|A\|^2 \quad \text{for each } A \in \mathscr{L}(\mathfrak{H}).$$

A bounded operator A on $\mathfrak{H}$ is said to be compact if $A(\overline{B(0,1)})$ is relatively compact in $\mathfrak{H}$. The set of compact operators on $\mathfrak{H}$ is denoted by $\mathscr{K}(\mathfrak{H})$; one easily checks that $\mathscr{K}(\mathfrak{H})$ is a closed (for the operator norm) two-sided ideal in $\mathscr{L}(\mathfrak{H})$.

We recall the following fundamental facts about compact operators:

- if A is a compact operator on $\mathfrak{H}$, its adjoint A^* is also a compact operator;
- $\mathscr{K}(\mathfrak{H})$ is the operator-norm closure of the set of finite rank operators on $\mathfrak{H}$.

Next we introduce the trace norm: for each $A \in \mathscr{L}(\mathfrak{H})$

$$\|A\|_1 := \sup_{e,f \in HB(\mathfrak{H})} \sum_{k\geq 0} |\langle Ae_k|f_k\rangle| \in [0,\infty].$$

A bounded operator A on $\mathfrak{H}$ is said to be a trace-class operator if and only if it has finite trace norm. The set of trace-class operators on $\mathfrak{H}$ is denoted by $\mathscr{L}^1(\mathfrak{H})$; in other words,

$$\mathscr{L}^1(\mathfrak{H}) := \{A \in \mathscr{L}(\mathfrak{H}) \text{ s.t. } \|A\|_1 < \infty\}.$$

One easily checks that $\mathscr{L}^1(\mathfrak{H})$ is a two-sided ideal in $\mathscr{L}(\mathfrak{H})$, that $A \mapsto \|A\|_1$ defines a norm on $\mathscr{L}^1(\mathfrak{H})$, and that $\mathscr{L}^1(\mathfrak{H})$ equipped with the trace norm $\|\cdot\|_1$ is a separable Banach space. Besides, the map

$$\mathcal{L}^1(\mathfrak{H}) \ni A \mapsto A^* \in \mathcal{L}^1(\mathfrak{H})$$

is an isometry for the trace norm. The following properties can be checked easily:
(a) $\mathcal{L}^1(\mathfrak{H}) \subset \mathcal{K}(\mathfrak{H}) \subset \mathcal{L}(\mathfrak{H})$ with $\|A\| \leq \|A\|_1$ for all $A \in \mathcal{L}^1(\mathfrak{H})$, and
(b) for all $A \in \mathcal{L}^1(\mathfrak{H})$ and $B \in \mathcal{L}(\mathfrak{H})$

$$\|AB\|_1 \text{ and } \|BA\|_1 \leq \|A\|_1\|B\|, \quad A \in \mathcal{L}^1(\mathfrak{H}) \text{ and } B \in \mathcal{L}(\mathfrak{H}).$$

There is a natural notion of trace for operators in infinite dimensional spaces, that can be defined for each trace-class operator. For each $A \in \mathcal{L}^1(\mathfrak{H})$,

$$\operatorname{trace}(A) := \sum_{k\geq 0}\langle Ae_k|e_k\rangle,$$

for each $(e_k)_{k\geq 0} \in HB(\mathfrak{H})$. We leave it to the reader to check that the expression on the right hand side of the previous equality is independent of the choice of the Hilbert basis $(e_k)_{k\geq 0} \in HB(\mathfrak{H})$.

Perhaps the most important property of the trace is the following identity: for all $A \in \mathcal{L}^1(\mathfrak{H})$ and $B \in \mathcal{L}(\mathfrak{H})$, the operators AB and BA are both trace-class operators on $\mathfrak{H}$, and one has

$$\operatorname{trace}(AB) = \operatorname{trace}(BA).$$

Let us recall the notion of polar decomposition of an operator. For each $A \in \mathcal{L}(\mathfrak{H})$, there exists $|A|, U \in \mathcal{L}(\mathfrak{H})$ such that

$$UU^* = U^*U = I, \quad |A| = |A|^* \geq 0 \text{ and } A = |A|U.$$

In this decomposition

$$|A| = \sqrt{AA^*}.$$

Then

$$A \in \mathcal{L}^1(\mathfrak{H}) \Leftrightarrow |A| \in \mathcal{L}^1(\mathfrak{H}) \text{ and } \|A\|_1 = \operatorname{trace}(|A|).$$

The polar decomposition is the analogue of polar coordinates in operator theory: $|A|$ is the analogue of $|z|$ for $z \in \mathbf{C}$, while U is the analogue of $z/|z|$ for $z \in \mathbf{C}^*$.

Exercise Let $A \in \mathcal{L}^1(\mathfrak{H})$. Prove that $\operatorname{trace}(\sqrt{AA^*}) = \operatorname{trace}(\sqrt{A^*A}) = \|A\|_1$.

The two following facts are important:

- $\mathcal{L}^1(\mathfrak{H})$ is the (topological) dual of $\mathcal{K}(\mathfrak{H})$, with duality defined by the trace as follows

$$\mathcal{L}^1(\mathfrak{H}) \times \mathcal{K}(\mathfrak{H}) \ni (A, K) \mapsto \operatorname{trace}(AK) \in \mathbf{C};$$

- $\mathcal{L}(\mathfrak{H})$ is the (topological) dual of $\mathcal{L}^1(\mathfrak{H})$, with duality defined by the trace as follows

$$\mathcal{L}(\mathfrak{H}) \times \mathcal{L}^1(\mathfrak{H}) \ni (B, A) \mapsto \operatorname{trace}(BA) \in \mathbf{C}.$$

Another important class of bounded operators on $\mathfrak{H}$ is the class of Hilbert-Schmidt operators, denoted by $\mathscr{L}^2(\mathfrak{H})$, defined as follows:

$$A \in \mathscr{L}^2(\mathfrak{H}) \Leftrightarrow A^*A \in \mathscr{L}^1(\mathfrak{H}) \Leftrightarrow AA^* \in \mathscr{L}^1(\mathfrak{H}).$$

The class $\mathscr{L}^2(\mathfrak{H})$ is a closed two-sided ideal in $\mathscr{L}(\mathfrak{H})$. It is a Banach space for the Hilbert-Schmidt norm defined as follows:

$$\|A\|_2 := \|A^*A\|_1^{1/2} = \|AA^*\|_1^{1/2}.$$

One easily checks that

$$\mathscr{L}^1(\mathfrak{H}) \subset \mathscr{L}^2(\mathfrak{H}) \subset \mathscr{K}(\mathfrak{H}).$$

An important particular case is $\mathfrak{H} = L^2(\mathscr{X})$, where $\mathscr{X}$ is a measurable subset of $\mathbf{R}^d$, equipped with the Lebesgue measure. In that case, a bounded operator A on $L^2(\mathscr{X})$ is a Hilbert-Schmidt operator if and only if it is an integral operator of the form

$$Af(x) = \int_{\mathscr{X}} a(x, y)f(y)dy \quad \text{for all } f \in L^2(\mathscr{X}),$$

with

$$a \in L^2(\mathscr{X} \times \mathscr{X}).$$

In that case

$$\|A\|_2 = \left(\iint_{\mathscr{X} \times \mathscr{X}} |a(x, y)|^2 dxdy \right)^{1/2} = \|a\|_{L^2(\mathscr{X} \times \mathscr{X})}.$$

References for the material presented so far in this section are Chap. VI of [22] and chapter XIX, section 1 of [58].

According to the discussion above, each $A \in \mathscr{L}^1(L^2(\mathbf{R}^n))$ is an integral operator with integral kernel $a \in L^2(\mathbf{R}^n \times \mathbf{R}^n)$. The integral kernel of a trace-class operator on $L^2(\mathbf{R}^n)$ has an additional interesting property, recalled below.

Lemma 1.9.1 *If $A \in \mathscr{L}^1(L^2(\mathbf{R}^n))$, its integral kernel $a \equiv a(x, y)$ is such that the map*[19]

$$z \mapsto a(x, x + z) \textit{ belongs to } C_b(\mathbf{R}^n_z; L^1(\mathbf{R}^n_x)).$$

[19]For each topological space X and each vector space E on $\mathbf{R}$ equipped with the norm $\| \cdot \|_E$, we designate by $C(X, E)$ the set of continuous maps from X to E, and we denote

$$C_b(X, E) := \{f \in C(X, E) \text{ s.t. } \sup_{x \in X} \|f(x)\|_E < \infty\}.$$

In other words, $C_b(X, E)$ is the set of bounded continuous functions defined on X with values in E.

Exercise (Outline of the proof of Lemma 1.9.1). Let $A \in \mathscr{L}^1(\mathfrak{H})$ with $\mathfrak{H} := L^2(\mathbf{R}^n)$.
(1) Prove that the integral kernel a of A can be represented as

$$a(x, y) = \sum_{n\geq 1} \lambda_n e_n(x)\overline{f_n(y)}$$

with $e_n, f_n \in \mathfrak{H}$ such that $|e_n|_{\mathfrak{H}} = |f_n|_{\mathfrak{H}} = 1$ and with

$$\sum_{n\geq 1} |\lambda_n| < \infty.$$

(2) Prove that, for each $\phi \in \mathfrak{H}$, one has

$$\int_{\mathbf{R}^n} |\phi(x+h) - \phi(x)|^2 dx \to 0$$

as $|h| \to 0$.
(3) Prove that

$$\int_{\mathbf{R}^n} |a(x, x+z) - a(x, x+z')|dx \to 0 \quad \text{as } |z - z'| \to 0,$$

and conclude.

In particular, if $A \in \mathscr{L}^1(L^2(\mathbf{R}^n))$, then its integral kernel $a \equiv a(x, y)$ is such that the map $x \mapsto a(x, x)$ is well-defined as an element of $L^1(\mathbf{R}^n)$, and

$$\operatorname{trace}(A) = \int_{\mathbf{R}^n} a(x, x)dx.$$

The converse of this last statement is obviously false: if $A \in \mathscr{L}(\mathfrak{H})$ is an integral operator with integral kernel $a(x, y)$, that the map $x \mapsto a(x, x)$ belongs to $L^1(\mathbf{R}^n)$ does not imply in general that $A \in \mathscr{L}^1(L^2(\mathbf{R}^n))$.

Exercise Construct an example of operator that belongs to $\mathscr{L}^2(L^2(\mathbf{R}^n))$ but not to $\mathscr{L}^1(L^2(\mathbf{R}^n))$.

In the sequel, we shall need the notion of "partial trace" for a trace-class operator defined on the tensor product of two Hilbert spaces. Let $\mathfrak{H}_1$ and $\mathfrak{H}_2$ be two complex, separable Hilbert spaces.

Definition 1.9.1 For each $A \in \mathscr{L}^1(\mathfrak{H}_1 \otimes \mathfrak{H}_2)$, the first partial trace of A, denoted

$$A_{:1} = \operatorname{trace}_1(A) \in \mathscr{L}^1(\mathfrak{H}_1),$$

is the unique element of $\mathscr{L}^1(\mathfrak{H}_1)$ such that

$$\text{trace}_{\mathfrak{H}_1}(A_{:1}B) = \text{trace}_{\mathfrak{H}_1\otimes\mathfrak{H}_2}(A(B\otimes I_{\mathfrak{H}_2}))$$

holds for each $B \in \mathcal{L}(\mathfrak{H}_1)$, where $I_{\mathfrak{H}_2}$ is the identity on $\mathfrak{H}_2$.

The second partial trace of A is the element of $\mathcal{L}^1(\mathfrak{H}_2)$ defined analogously.

Let A be an integral operator with integral kernel $a \equiv a(x_1, x_2, y_1, y_2)$ in the case where $\mathfrak{H}_j = L^2(\mathbf{R}^{n_j})$ for $j = 1, 2$. In that case, $\mathfrak{H}_1 \otimes \mathfrak{H}_2 = L^2(\mathbf{R}^{n_1+n_2})$, and one easily checks that $A_{:1}$ is the integral operator with integral kernel

$$a_{:1}(x_1, y_1) = \int_{\mathbf{R}^{n_2}} a(x_1, z, y_1, z)dz$$

on $\mathfrak{H}_1 = L^2(\mathbf{R}^{n_1})$.

There is an easy generalization of the notion of partial trace to the case of trace-class operators defined on N-fold tensor products of separable Hilbert spaces. Given N complex, separable Hilbert spaces $\mathfrak{H}_1, \dots, \mathfrak{H}_N$, then for each trace-class operator $A \in \mathcal{L}^1(\mathfrak{H}_1 \otimes \dots \otimes \mathfrak{H}_N)$, for each $1 \leq k \leq N$, we define

$$A_{:k} = \text{trace}_k(A)$$

to be the first partial trace $\text{trace}_1(A)$ in the decomposition

$$\mathfrak{H}_1 \otimes \dots \otimes \mathfrak{H}_N \simeq (\mathfrak{H}_1 \otimes \dots \otimes \mathfrak{H}_k) \otimes (\mathfrak{H}_{k+1} \otimes \dots \otimes \mathfrak{H}_N).$$

If one views $A \in \mathcal{L}^1(\mathfrak{H}_1 \otimes \dots \otimes \mathfrak{H}_N)$ such that $A = A^* \geq 0$ and $\text{trace}(A) = 1$ as the operator analogue of an N-particle probability measure, then $A_{:k}$ is the analogue of its (first) k-particle marginal considered above in our description of the BBGKY hierarchy for the mean field limit in classical statistical mechanics.

More details on partial traces can be found in [7, 8].

Remark All the quantum states of the N-particle system considered so far are defined by an N-particle wave function $\Psi_N(t, \cdot)$ that belongs to $L^2((\mathbf{R}^d)^N)$ and satisfies the normalization condition $\|\Psi_N(t, \cdot)\|_{L^2((\mathbf{R}^d)^N)} = 1$ for each $t \in \mathbf{R}$. An N-particle density matrix $D_N(t)$ is associated to this N-particle wave function by the formula $D_N(t) := |\Psi_N(t, \cdot)\rangle\langle\Psi_N(t, \cdot)|$, and we have seen that this N-particle density matrix $D_N(t)$ satisfies

$$D_N(t) = D_N(t)^* = D_N(t)^2, \quad \text{and } \text{trace}(D_N(t)) = 1.$$

Conversely, if $D_N(t) \in \mathcal{L}^1(L^2((\mathbf{R}^d)^N))$ satisfies the conditions above, it is an orthogonal projection with rank 1, and therefore there exists $\Phi_N(t, \cdot) \in L^2((\mathbf{R}^d)^N)$ such that $\|\Phi_N(t, \cdot)\|_{L^2((\mathbf{R}^d)^N)} = 1$ and $D_N(t) := |\Phi_N(t, \cdot)\rangle\langle\Phi_N(t, \cdot)|$. Besides, the function $\Phi(t, \cdot)$ is defined uniquely up to multiplication by a complex number of modulus 1.

Such quantum states are referred to as "pure states".

However, there also exist more general quantum states, that are described by an N-particle density matrix $D_N(t) \in \mathcal{L}^1(L^2((\mathbf{R}^d)^N))$ satisfying

$$D_N(t) = D_N(t)^* \geq 0, \quad \text{and trace}(D_N(t)) = 1.$$

The condition $D_N(t) = D_N(t)^* \geq 0$ is equivalent to the fact that

$$\langle \Phi_N | D_N(t) \Phi_N \rangle_{L^2((\mathbf{R}^d)^N)} \geq 0 \quad \text{for all } \Phi_N \in L^2((\mathbf{R}^d)^N).$$

Equivalently, the spectrum of $D_N(t)$ consists of nonnegative eigenvalues (since $D_N(t) \in \mathcal{L}^1(L^2((\mathbf{R}^d)^N))$, we already know that $D_N(t)$ is compact, so that its spectrum consists of eigenvalues only).

With the normalization condition trace$(D_N(t)) = 1$, one sees that $D_N(t)$ is of the form

$$D_N(t) = \sum_{n \geq 1} \lambda_n(t) |\psi_n(t, \cdot)\rangle \langle \psi_n(t, \cdot)|$$

where $(\psi_n(t, \cdot))_{n \geq 1}$ is a complete orthonormal system in $L^2(\mathbf{R}^d)^N)$ and

$$\lambda_n(t) \geq 0, \quad \sum_{n \geq 1} \lambda_n(t) = 1.$$

In other words, $D_N(t)$ is a (possibly infinite) convex combination of the density matrices $|\psi_n(t, \cdot)\rangle \langle \psi_n(t, \cdot)|$ corresponding with pure states.

While the Schrödinger equation governs the evolution of quantum states described by wave functions, the von Neumann equation governs the evolution of mixed as well as pure states in quantum mechanics. If $D_N(t)$ satisfies the von Neumann equation

$$i\dot{D}_N(t) = [\mathcal{H}_N, D_N(t)]$$

where

$$\mathcal{H}_N := -\tfrac{1}{2} \sum_{k=1}^{N} \Delta_{x_k} + \frac{1}{N} \sum_{1 \leq k < l \leq N} V(x_k - x_l)$$

and V is such that $i\mathcal{H}_N$ is the generator of a unitary group on $L^2(\mathbf{R}^d)^N)$ (as in Kato's theorem for instance), then the eigenvalues of $D_N(t)$ satisfy

$$\lambda_n(t) = \lambda_n(0), \quad \text{for each } n \geq 1.$$

The eigenfunctions of $D_N(t)$ are transformed as

$$\psi_n(t, \cdot) = e^{-it\mathcal{H}_N} \psi_n(0, \cdot), \quad \text{for each } n \geq 1.$$

Throughout the following sections, we shall be dealing with pure states, although most of our arguments could easily be extended to mixed states.

The notion of partial trace discussed above leads to a very natural way of constructing mixed states, discussed in the exercise below.

Exercise Let $\Psi_M \in L^2((\mathbf{R}^d)^M)$ be an M-particle wave function such that one has $\|\Psi_M\|_{L^2((\mathbf{R}^d)^M)} = 1$, and let $N \geq 1$ be such that $N < M$. Set $D_M := |\Psi_M\rangle\langle\Psi_M|$ and consider the operator $D_N := D_{M:N}$.
(1) Check that $D_N = D_N^* \geq 0$, and that $\operatorname{trace}(D_N) = 1$.
(2) Assume that $D_N = |\psi\rangle\langle\psi|$ with $\psi \in L^2((\mathbf{R}^d)^N)$ such that $\|\psi\|_{L^2((\mathbf{R}^d)^N)} = 1$ and compute $\operatorname{trace}(D_N|\psi\rangle\langle\psi|)$ in terms of Ψ_M.
(3) Under which condition on Ψ_M does one have $D_N = D_N^2$? (Hint: use the result of question 2 and the equality case in the Cauchy-Schwarz inequality to find that Ψ_M is of the form $\Psi_M(x_1, \ldots, x_M) = \psi(x_1, \ldots, x_N)\phi(x_{N+1}, \ldots, x_M)$.)

This exercise is inspired from the second paragraph in Sect. 14 of [69], where it is explained that mixed states arise for instance as quantum states for subsystems of a larger system in some pure state. The result obtained in the exercise above suggests that the state of such a subsystem (particles 1 to N) is pure if and only if the subsystem considered (the N first particles are independent from the M-N remaining particles.)

1.10 The BBGKY Hierarchy in Quantum Mechanics

In this section, we outline the BBGKY method presented above in the case of classical statistical mechanics, and explain how it leads to a rigorous derivation of the mean-field Hartree equation from the N-particle Schrödinger equation.

First we introduce some elements of notation.

For all $N \geq 1$, we set $X_N := (x_1, \ldots, x_N)$ and $Y_N := (y_1, \ldots, y_N)$; likewise for all $N \geq 1$ and $1 \leq k < N$, we adopt the notation $Z_N^k = (z_{k+1}, \ldots, z_N)$.

Given an integral operator D_N acting on $L^2((\mathbf{R}^d)^N)$ with integral kernel $D_N(X_N, Y_N)$ we denote by $D_{N:k}$ the operator with integral kernel

$$D_{N:k}(X_k, Y_k) := \int_{(\mathbf{R}^d)^{N-k}} D_N(X_k, Z_N^k; Y_k, Z_N^k)dZ_N^k.$$

In terms of operators,

$$D_{N:k} = \operatorname{trace}_k(D_N).$$

Here again, we use the same letter to designate an integral operator and its integral kernel. This obvious abuse of notation is deliberately chosen to avoid the unnecessary multiplication of mathematical symbols.

1.10.1 The Quantum BBGKY Hierarchy

First we explain how to deduce the BBGKY hierarchy from the N-particle Schrödinger equation.

Theorem 1.10.1 *Assume that V is a real-valued, even function in $L^\infty(\mathbf{R}^d)$. Define $D_N^{in} = (D_N^{in})^* \geq 0$ be a trace-class operator with* trace$(D_N^{in}) = 1$ *on the Hilbert space $\mathfrak{H}_N = \mathfrak{H}^{\otimes N} =$
$L^2((\mathbf{R}^d)^N)$ with $\mathfrak{H} = L^2(\mathbf{R}^d)$, satisfying the indistinguishability symmetry, i.e.*

$$S_\sigma D_N^{in} = D_N^{in} \quad \textit{for each } \sigma \in \mathfrak{S}_N.$$

Let

$$D_N(t) := e^{-it\mathcal{H}_N} D_N^{in} e^{it\mathcal{H}_N}, \qquad t \in \mathbf{R},$$

where

$$\mathcal{H}_N = -\tfrac{1}{2}\sum_{k=1}^N \Delta_k + \frac{1}{N}\sum_{1\leq k<l\leq N} V_{kl}$$

is the N-particle Schrödinger operator. Here Δ_k designates the Laplacian acting on the kth variable x_k, while V_{kl} designates the multiplication by $V(x_k - x_l)$.

Then the sequence of partial traces of D_N satisfies the BBGKY hierarchy of differential equations

$$\begin{aligned} i\dot{D}_{N:k} = &-\tfrac{1}{2}\sum_{j=1}^k [\Delta_j, D_{N:k}] + \frac{N-k}{N}\sum_{j=1}^k [V_{j,k+1}, D_{N:k+1}]_{:k} \\ &+ \frac{1}{N}\sum_{1\leq l<m\leq k} [V_{lm}, D_{N:k}] \end{aligned}$$

for all $k \geq 1$, with the convention

$$D_{N:N} = D_N, \quad \textit{and } D_{N:k} = 0 \textit{ whenever } k > N.$$

In terms of integral kernels, the BBGKY hierarchy reads

$$\begin{aligned} i\partial_t D_{N:k}(t, X_k, Y_k) = &-\tfrac{1}{2}\sum_{j=1}^k (\Delta_{x_j} - \Delta_{y_j}) D_{N:k}(t, X_k, Y_k) \\ &+ \frac{N-k}{N}\sum_{j=1}^k \int_{\mathbf{R}^d} (V(x_j - z) - V(y_j - z)) D_{N:k+1}(t, X_k, z, Y_k, z)dz \\ &+ \frac{1}{N}\sum_{1\leq l<m\leq k} (V(x_l - x_m) - V(y_l - y_m)) D_{N:k}(t, X_k, Y_k) \end{aligned}$$

for all $k \geq 1$.

Proof Consider the free N-body Schrödinger group

$$W_N(t) := \exp\left(-it\tfrac{1}{2}(\Delta_1 + \ldots + \Delta_N)\right),$$

and set

$$F_N(t) := W_N(t) D_N(t) W_N(-t) = W_N(t) D_N(t) W_N(t)^*.$$

Starting from the von Neumann equation, one deduces that

$$i\dot{F}_N(t) = \frac{1}{N} \sum_{1 \leq l < m \leq N} [W_N(t) V_{lm} W_N(t)^*, F_N(t)].$$

Since $W_N(t)$ is a unitary operator on $L^2((\mathbf{R}^d)^N)$, the operator $F_N(t)$ is trace-class on $L^2((\mathbf{R}^d)^N)$. Since $V \in L^\infty(\mathbf{R}^d)$, all the operators V_{lm} are bounded on $L^2((\mathbf{R}^d)^N)$, so that $[W_N(t) V_{lm} W_N(t)^*, F_N(t)]$ is trace-class on $L^2((\mathbf{R}^d)^N)$.Taking partial traces of both sides of the equality above leads to

$$i\dot{F}_{N:k}(t) = \frac{1}{N} \sum_{1 \leq l < m \leq N} [W_N(t) V_{lm} W_N(t)^*, F_N(t)]_{:k}$$

for all $k \geq 1$.

Then we must distinguish three cases.

(a) If $k < l < m \leq N$, then

$$[W_N(t) V_{lm} W_N(t)^*, F_N(t)]_{:k} = [\tilde{V}_{lm}(t), F_{N:m}(t)]_{:k} = 0,$$

with the notation

$$\tilde{V}_{lm}(t) := W_N(t) V_{lm} W_N(t)^* = e^{-it\frac{1}{2}(\Delta_l + \Delta_m)} V_{lm} e^{it\frac{1}{2}(\Delta_l + \Delta_m)}.$$

Indeed, let $A \in \mathscr{L}(L^2((\mathbf{R}^d)^k))$, and denote by I_p the identity in $\mathscr{L}(L^2((\mathbf{R}^d)^p))$. Then

$$[A \otimes I_{m-k}, \tilde{V}_{lm}(t)] = 0,$$

so that

$$\begin{aligned} \operatorname{trace}_{\mathfrak{H}_k}([\tilde{V}_{lm}(t), F_{N:m}(t)]_{:k} A) &= \operatorname{trace}_{\mathfrak{H}_m}([\tilde{V}_{lm}(t), F_{N:m}(t)] A \otimes I_{m-k}) \\ &= \operatorname{trace}_{\mathfrak{H}_k}([A \otimes I_{m-k}, \tilde{V}_{lm}(t)] F_{N:m}(t)) = 0, \end{aligned}$$

with the notation $\mathfrak{H}_k := L^2((\mathbf{R}^d)^k)$.

(b) If $1 \leq l < m \leq k \leq N$, then

$$[W_N(t)V_{lm}W_N(t)^*, F_N(t)]_{:k} = [W_k(t)V_{lm}W_k(t)^*, F_{N:k}(t)].$$

Indeed, with the same notation as above, for each $A \in \mathscr{L}(L^2((\mathbf{R}^d)^k))$,

$$\begin{aligned}
&\operatorname{trace}_{\mathfrak{H}_N}([W_N(t)V_{lm}W_N(t)^*, F_N(t)]A \otimes I_{N-k}) \\
&\quad = \operatorname{trace}_{\mathfrak{H}_N}([\tilde{V}_{lm}(t), F_N(t)]A \otimes I_{N-k}) \\
&\quad = \operatorname{trace}_{\mathfrak{H}_N}(\tilde{V}_{lm}(t)F_N(t)A \otimes I_{N-k}) \\
&\quad\quad - \operatorname{trace}_{\mathfrak{H}_N}(F_N(t)\tilde{V}_{lm}(t)A \otimes I_{N-k}) \\
&\quad = \operatorname{trace}_{\mathfrak{H}_N}(F_N(t)A \otimes I_{N-k}\tilde{V}_{lm}(t)) \\
&\quad\quad - \operatorname{trace}_{\mathfrak{H}_N}(F_N(t)\tilde{V}_{lm}(t)A \otimes I_{N-k}) \\
&\quad = \operatorname{trace}_{\mathfrak{H}_N}(F_N(t)(A\tilde{V}_{lm}(t)) \otimes I_{N-k}) \\
&\quad\quad - \operatorname{trace}_{\mathfrak{H}_N}(F_N(t)(\tilde{V}_{lm}(t)A) \otimes I_{N-k}) \\
&\quad = \operatorname{trace}_{\mathfrak{H}_k}(F_{N:k}(t)[A, \tilde{V}_{lm}(t)]) \\
&\quad = \operatorname{trace}_{\mathfrak{H}_k}([\tilde{V}_{lm}(t), F_{N:k}(t)]A) \\
&\quad = \operatorname{trace}_{\mathfrak{H}_k}([W_k(t)V_{lm}W_k(t)^*, F_{N:k}(t)]A).
\end{aligned}$$

(c) It remains to treat the case $1 \le l \le k < m \le N$. Denote by $\sigma \in \mathfrak{S}_N$ the transposition exchanging $k+1$ and m. At this point we use the symmetry of both operators D_N and W_N, i.e. the fact that

$$S_\sigma D_N(t) = D_N(t) \quad \text{and } S_\sigma W_N(t) = W_N(t)$$

for each $t \in \mathbf{R}$. First, one can check by direct inspection that

$$S_\sigma[\tilde{V}_{l,k+1}(t), F_N(t)] = [\tilde{V}_{l,m}(t), S_\sigma F_N(t)] = [\tilde{V}_{l,m}(t), F_N(t)].$$

Then, for each $A \in \mathscr{L}(L^2((\mathbf{R}^d)^k))$ and with the same notation as above,

$$\begin{aligned}
\operatorname{trace}_{\mathfrak{H}_k}([\tilde{V}_{l,m}(t), F_N(t)]_{:k}A) &= \operatorname{trace}_{\mathfrak{H}_m}([\tilde{V}_{l,m}(t), F_N(t)]A \otimes I_{m-k}) \\
&= \operatorname{trace}_{\mathfrak{H}_m}([\tilde{V}_{l,m}(t), F_N(t)]S_\sigma(A \otimes I_{m-k})) \\
&= \operatorname{trace}_{\mathfrak{H}_m}(S_\sigma[\tilde{V}_{l,m}(t), F_N(t)]S_\sigma^2(A \otimes I_{m-k})) \\
&= \operatorname{trace}_{\mathfrak{H}_m}(S_\sigma[\tilde{V}_{l,m}(t), F_N(t)](A \otimes I_{m-k})) \\
&= \operatorname{trace}_{\mathfrak{H}_m}([\tilde{V}_{l,k+1}(t), F_N(t)](A \otimes I_{m-k})) \\
&= \operatorname{trace}_{\mathfrak{H}_k}([\tilde{V}_{l,k+1}(t), F_N(t)]_{:k}A) \quad .
\end{aligned}$$

(The second equality above follows from the fact that $S_\sigma(A \otimes I_{m-k}) = A \otimes I_{m-k}$ since σ acts as the identity on $\{1, \dots, k\}$; the third equality follows from the identities

$$\operatorname{trace}(S_\sigma T) = \operatorname{trace}(U_\sigma T U_\sigma^*) = \operatorname{trace}(T)$$

and

$$S_\sigma(BA) = U_\sigma B U_\sigma^* U_\sigma A U_\sigma = S_\sigma(B) S_\sigma(A),$$

while the fourth equality is based on the relation $S_\sigma^2 = \mathrm{Id}$).

Thus

$$[\tilde{V}_{l,m}(t), F_N(t)]_{:k} = [\tilde{V}_{l,k+1}(t), F_N(t)]_{:k} = [\tilde{V}_{l,k+1}(t), F_{N:k+1}(t)]_{:k}$$

whenever $1 \le l \le k < m \le N$—the last equality being obvious.

Therefore

$$\begin{aligned} i\dot{F}_{N:k}(t) &= \frac{1}{N} \sum_{l=1}^{k} \sum_{m=k+1}^{N} [\tilde{V}_{l,m}(t), F_N(t)]_{:k} + \frac{1}{N} \sum_{1 \le l < m \le k} [\tilde{V}_{lm}(t), F_{N:k}(t)] \\ &= \frac{N-k}{N} \sum_{l=1}^{k} [\tilde{V}_{l,k+1}(t), F_{N:k+1}(t)]_{:k} + \frac{1}{N} \sum_{1 \le l < m \le k} [\tilde{V}_{lm}, F_{N:k}(t)], \end{aligned}$$

Undoing the conjugation with $W_k(t)$, we next arrive at the equality

$$\begin{aligned} i\dot{D}_{N:k} = &- \tfrac{1}{2} \sum_{j=1}^{k} [\Delta_j, D_{N:k}] + \frac{1}{N} \sum_{1 \le l < m \le k} [V_{lm}, D_{N:k}] \\ &+ \frac{N-k}{N} \sum_{j=1}^{k} \left(e^{-it\frac{1}{2}\Delta_{k+1}} [V_{j,k+1}, D_{N:k+1}] e^{it\frac{1}{2}\Delta_{k+1}} \right)_{:k}. \end{aligned}$$

It remains to check that

$$\left(e^{-it\frac{1}{2}\Delta_{k+1}} [V_{j,k+1}, D_{N:k+1}] e^{it\frac{1}{2}\Delta_{k+1}} \right)_{:k} = [V_{j,k+1}, D_{N:k+1}]_{:k}.$$

For each $A \in \mathscr{L}(L^2((\mathbf{R}^d)^k))$ and with the same notation as above, one has

$$\begin{aligned} &\operatorname{trace}\left(\left(e^{-it\frac{1}{2}\Delta_{k+1}} [V_{j,k+1}, D_{N:k+1}] e^{it\frac{1}{2}\Delta_{k+1}} \right)_{:k} A \right) \\ &= \operatorname{trace}\left(\left(e^{-it\frac{1}{2}\Delta_{k+1}} [V_{j,k+1}, D_{N:k+1}] e^{it\frac{1}{2}\Delta_{k+1}} \right) A \otimes I_1 \right) \\ &= \operatorname{trace}\left([V_{j,k+1}, D_{N:k+1}] e^{it\frac{1}{2}\Delta_{k+1}} (A \otimes I_1) e^{-it\frac{1}{2}\Delta_{k+1}} \right) \\ &= \operatorname{trace}([V_{j,k+1}, D_{N:k+1}] A \otimes I_1) \\ &= \operatorname{trace}([V_{j,k+1}, D_{N:k+1}]_{:k} A) \end{aligned}$$

since

$$e^{it\frac{1}{2}\Delta_{k+1}}(A \otimes I_1)e^{-it\frac{1}{2}\Delta_{k+1}} = A \otimes I_1,$$

and this concludes the proof.

References for this section are [7, 10, 96].

1.10.2 The Infinite Quantum Mean Field Hierarchy

From the quantum BBGKY hierarchy, we formally deduce an infinite hierarchy that is the quantum analogue of the infinite mean field hierarchy in classical statistical mechanics. The formal argument is as follows: assume that

$$D_{N:k}(t) \to D_k(t)$$

for each fixed $k \geq 1$ as $N \to \infty$ in some sense to be made precise, so that

$$[V_{lm}, D_{N:k}(t)] \to [V_{lm}, D_k(t)].$$

Then we pass to the limit as $N \to \infty$ in both terms

$$\frac{1}{N} \sum_{1 \leq l < m \leq k} [V_{lm}, D_{N:k}]$$

and

$$\frac{N-k}{N} \sum_{j=1}^{k} [V_{j,k+1}, D_{N:k+1}]_{:k}.$$

The first term vanishes as $N \to \infty$ since

$$\frac{1}{N} \sum_{1 \leq l < m \leq k} [V_{lm}, D_{N:k}] = O\left(\frac{k^2}{N}\right) \to 0,$$

while the second term satisfies

$$\frac{N-k}{N} \sum_{j=1}^{k} [V_{j,k+1}, D_{N:k+1}]_{:k} \to \sum_{j=1}^{k} [V_{j,k+1}, D_{k+1}]_{:k}$$

for each $k \geq 1$ as $N \to \infty$.

Therefore, we expect that the sequence of operators $(D_k(t))_{k \geq 1}$ should satisfy the infinite hierarchy of differential equations

$$i\dot{D}_k = -\tfrac{1}{2}\sum_{j=1}^{k}[\Delta_j, D_k] + \sum_{j=1}^{k}[V_{j,k+1}, D_{k+1}]_{:k}, \qquad k \geq 1.$$

In terms of integral kernels

$$D_{N:k}(t, X_k, Y_k) \to D_k(t, X_k, Y_k),$$

for each $k \geq 1$ as $N \to \infty$, where $D_k(t, X_k, Y_k)$ is the integral kernel of the operator $D_k(t)$. Then, the sequence of functions $(D_k(t, X_k, Y_k))_{k\geq 1}$ satisfies the infinite hierarchy of integro-differential equations indexed by $k \geq 1$:

$$\begin{aligned} i\partial_t D_k(t, X_k, Y_k) = &-\tfrac{1}{2}\sum_{j=1}^{k}(\Delta_{x_j} - \Delta_{y_j})D_k(t, X_k, Y_k) \\ &+ \sum_{j=1}^{k}\int_{\mathbf{R}^d}(V(x_j - z) - V(y_j - z))D_{k+1}(t, X_k, z, Y_k, z)dz. \end{aligned}$$

Of course, the argument above is purely formal and remains to be justified rigorously. This is achieved by the following theorem.

Theorem 1.10.2 *Assume that the potential V is an even, real-valued function such that*

$$V \in C(\mathbf{R}^d \setminus \{0\}) \cap L^2_{loc}(\mathbf{R}^d), \quad V \to 0 \text{ at } \infty, \quad \text{and } V \geq -V_m$$

for some $V_m > 0$. Assume that the initial density matrix

$$D_N^{in} = |\Psi_N^{in}\rangle\langle\Psi_N^{in}| \quad \text{with} \int_{(\mathbf{R}^d)^N} |\Psi_N^{in}(X_N)|^2 dX_N = 1$$

satisfies the indistinguishability symmetry, i.e.

$$S_\sigma D_N^{in} = D_N^{in} \quad \text{for each } \sigma \in \mathfrak{S}_N,$$

and the energy bound

$$\begin{aligned} \mathscr{E}_N^{in} :=& \tfrac{1}{2}\sum_{j=1}^{N}\int_{(\mathbf{R}^d)^N} |\nabla_{x_j}\Psi_N^{in}(X_N)|^2 dX_N \\ &+ \frac{1}{N}\sum_{1\leq l<m\leq N}\int_{(\mathbf{R}^d)^N} V(x_l - x_m)|\Psi_N^{in}(X_N)|^2 dX_N = O(N). \end{aligned}$$

Assume that, for each $k \geq 1$ and in the limit $N \to \infty$,

$$D_{N:k}^{in} \to D_k^{in} \text{ in } \mathcal{L}^1(L^2((\mathbf{R}^d)^k)) \text{ weak-*}.$$

Then

(a) the sequence $(D_{N:k})_{k\geq 1}$ *indexed by* N *is relatively compact in the infinite product space* $\prod_{k\geq 1} L^\infty(\mathbf{R}; \mathcal{L}^1(L^2((\mathbf{R}^d)^k)))$, *with the product of weak-* topologies—viewing each space* $L^\infty(\mathbf{R}; \mathcal{L}^1(L^2((\mathbf{R}^d)^k)))$ *as the dual of the Banach space* $L^1(\mathbf{R}; \mathcal{K}(L^2((\mathbf{R}^d)^k)))$ *equipped with the norm*

$$|||K||| := \int_{\mathbf{R}} \|K(t)\|_1 dt\,;$$

(b) each limit point $(D_k)_{k\geq 1}$ *of this sequence as* $N \to \infty$ *solves the infinite quantum mean field hierarchy written above in the sense of distributions, with initial data*

$$D_k(0) = D_k^{in}, \qquad k \geq 1.$$

Although the proof of this result relies on relatively soft functional analytic techniques, we shall not give it in detail. We just sketch the key arguments and refer the interested reader to [7] for the complete proof.

The general strategy is to use the uniform bounds on the operator $D_N(t)$ deduced from the condition $\text{trace}(D_N(t)) = 1$ and the conservation of energy to obtain the compactness property in statement (a) of the theorem above.

Next, one writes each equation in the BBGKY hierarchy[20] in the sense of distributions and passes to the limit as $N \to \infty$ for each $k \geq 1$ fixed. This strategy uses in particular various observations listed below.

A first important ingredient in the proof is the Cauchy-Schwarz inequality

$$|D_{N:k}(t, X_k, Y_k)|^2 \leq D_{N:k}(t, X_k, X_k) D_{N:k}(t, Y_k, Y_k)$$

since $D_N(t)$ is a nonnegative operator. It implies in particular the bound

$$\iint_{(\mathbf{R}^d)^k} |D_{N:k}(t, X_k, Y_k)|^2 dX_k dY_k \leq 1$$

that is uniform in $t \in \mathbf{R}$ and $N, k \geq 1$. We recall the convention $D_{N:k} = 0$ whenever $k > N$ and the equality

$$\int_{(\mathbf{R}^d)^k} D_{N:k}(t, Z_k, Z_k) dZ_k = 1,$$

[20] The BBGKY hierarchy has been derived from the N-particle Schrödinger equation in the case where the potential V belongs to $L^\infty(\mathbf{R}^d)$—see Theorem 1.10.1. However its validity in the sense of distributions under the assumptions of Theorem 1.10.2 results from the same arguments as those used in the derivation of the infinite hierarchy and presented below.

since

$$\int_{(\mathbf{R}^d)^N} D_N(t, Z_N, Z_N)dZ_N = 1.$$

A second crucial ingredient in the proof is the H^1 bound

$$\tfrac{1}{2} \int_{(\mathbf{R}^d)^N} |\nabla_{x_l} \Psi_N(t, X_N)|^2 dX_N \leq \frac{\mathscr{E}_N^{in}}{N} + \frac{N-1}{2N} V_m$$

for each $l = 1, \ldots, N$. This estimate is a consequence of the initial energy bound postulated on Ψ_N^{in}, together with the conservation of energy under the Schrödinger group, implying that

$$\begin{aligned} &\tfrac{1}{2} \sum_{l=1}^{N} \int_{(\mathbf{R}^d)^N} |\nabla_{x_l} \Psi_N(t, X_N)|^2 dX_N \\ &= \mathscr{E}_N - \frac{1}{N} \sum_{1 \leq k < l \leq N} \int_{(\mathbf{R}^d)^N} |V(x_k - x_l)||\Psi_N(t, X_N)|^2 dX_N \\ &\leq \mathscr{E}_N + \tfrac{1}{2}(N-1)V_m \int_{(\mathbf{R}^d)^N} |\Psi_N(t, X_N)|^2 dX_N = \mathscr{E}_N + \tfrac{1}{2}(N-1)V_m, \end{aligned}$$

and of the assumption of indistinguishable particles, so that one has, for each pair k, l such that $1 \leq k \neq l \leq N$,

$$\int_{(\mathbf{R}^d)^N} |\nabla_{x_k} \Psi_N(t, X_N)|^2 dX_N = \int_{(\mathbf{R}^d)^N} |\nabla_{x_l} \Psi_N(t, X_N)|^2 dX_N.$$

A third key step in the proof is based on the following result bearing on trace-class operators.

Lemma 1.10.1 *Consider a sequence of integral operators $A_n \in \mathscr{L}^1(L^2(\mathbf{R}^d))$ with integral kernels $a_n \equiv a_n(x, y)$ satisfying*

$$A_n \to 0 \text{ in } \mathscr{L}^1(L^2(\mathbf{R}^d)) \text{ weak-}*$$

as $n \to \infty$, and

$$\sup_{n \geq 1} \int_{\mathbf{R}^d} |a_n(z, z+h) - a_n(z, z)| dz \to 0 \text{ as } |h| \to 0.$$

Then, for each $\chi \in C_c(\mathbf{R}^d)$

$$\int_{\mathbf{R}^d} a_n(z,z)\chi(z)dz \to 0 \quad as\ n \to \infty.$$

See Sect. 2 of [7] for the proof of this lemma.

This result is applied in the following context. For each $\Theta \in C_c((\mathbf{R}^d)^k)$, consider

$$a_{N,k}(t,z,w) := \iint_{(\mathbf{R}^d)^k \times (\mathbf{R}^d)^k} \Theta(X_k)\Theta(Y_k)D_{N:k+1}(t,X_n,z,Y_k,w)dX_k dY_k.$$

Then by the Sobolev estimate above, one has

$$\int_{\mathbf{R}^d} |a_{N,k}(t,z,z+h) - a_{N,k}(t,z,z)|dz \le |h| \|\Theta\|_{L^\infty}^2 \left(\frac{2\mathscr{E}_N^{in}}{N} + \frac{N-1}{N}V_m\right).$$

Thus

$$\int_{\mathbf{R}^d} a_{N,k}(z,z)\chi(z)dz \to \int_{\mathbf{R}^d} a_k(z,z)\chi(z)dz$$

as $N \to \infty$, where

$$a_k(t,z,w) := \iint_{(\mathbf{R}^d)^k \times (\mathbf{R}^d)^k} \Theta(X_k)\Theta(Y_k)D_{k+1}(t,X_n,z,Y_k,w)dX_k dY_k,$$

and where $\chi \in C^\infty(\mathbf{R})$ is such that $0 \le \chi \le 1$, with $\chi(z) = 1$ for $1/R \le z \le R$ and $\text{supp}(\chi) \subset [1/2R, R+1]$. This is the key step in obtaining the limit of the interaction term

$$\int_{\mathbf{R}^d} (V(x_k - z) - V(y_k - z))D_{N:k+1}(t,X_k,z,Y_k,z)dz.$$

Some remarks are in order before going further.

Observe first that the assumptions on the potential V include the (repulsive) Coulomb potential in space dimension $d = 3$

$$V(z) = \tfrac{1}{4\pi}\frac{1}{|z|}$$

between identical charged particles. This case is obviously of fundamental importance in view of the (many) applications to atomic physics and quantum chemistry, in which the interacting particles are electrons.

Our second observation is of a more mathematical nature. Under the assumptions on the potential used in the theorem, the interaction integrand

$$(V(x_j - z) - V(y_j - z))D_{k+1}(t, X_k, z, Y_k, z)$$

may fail to belong to $L^1(\mathbf{R}^d_z)$ for each (or almost every) X_k, Y_k. Yet our analysis defines the interaction integral

$$\int_{\mathbf{R}^d} (V(x_j - z) - V(y_j - z))D_{k+1}(t, X_k, z, Y_k, z)dz$$

as a Radon *measure* in (X_n, Y_n)—instead of a *function* of these same variables.

A general reference for the material presented in this section is [7].

1.11 The Mean Field Limit in Quantum Mechanics and Hartree's Equation

1.11.1 Mathematical Statement of the Mean Field Limit

The quantum analogue of the mean field limit obtained in classical statistical mechanics is the following result.

Theorem 1.11.1 *Let V be a real-valued, even, bounded measurable function defined on $\mathbf{R}^d$, such that $V \to 0$ at infinity. Let the initial single-particle wave function $\psi^{in} \in H^1(\mathbf{R}^d)$ satisfy $\|\psi^{in}\|_{L^2} = 1$, and let $\psi \equiv \psi(t, x)$ be the solution of the Cauchy problem for the Hartree equation*

$$\begin{cases} (i\partial_t + \frac{1}{2}\Delta_x)\psi(t,x) + \psi(t,x) \displaystyle\int_{\mathbf{R}^d} V(x-z)|\psi(t,z)|^2 dz = 0, \\ \psi(0,x) = \psi^{in}(x). \end{cases}$$

Let $\Psi_N^{in} = (\psi^{in})^{\otimes N}$ and let $\Psi_N(t) = e^{-it\mathcal{H}_N}\Psi_N^{in}$ be the solution of N-particle Schrödinger equation in the mean field scaling, i.e. with

$$\mathcal{H}_N = -\tfrac{1}{2}\sum_{j=1}^{N} \Delta_{x_j} + \frac{1}{N}\sum_{1\le k<l\le N} V(x_k - x_l).$$

Then, for each $k \geq 1$, one has

$$D_{N:k}(t) := |\Psi_N(t)\rangle\langle\Psi_N(t)|_{:k} \to D_k(t)$$

in $L^\infty(\mathbf{R};\mathcal{L}^1(L^2((\mathbf{R}^d)^k)))$ *weak-* as* $N\to\infty$, *where*

$$D_k(t)=|\psi(t)^{\otimes k}\rangle\langle\psi(t)^{\otimes k}|\,;$$

—in other words

$$D_k(t,X_k,Y_k)=\prod_{j=1}^k\psi(t,x_j)\overline{\psi(t,y_j)}.$$

As already observed in the case of the mean field limit in classical mechanics, using the hierarchy approach requires knowing that the mean field PDE is well-posed: see Proposition 1.8.1.

Here again, we only sketch the key arguments in the proof, and refer the interested reader to [96] or [7] for a complete account.

By Theorem 1.11.1, we know that the sequence $(D_{N:k})_{k\geq 1}$ indexed by N is relatively compact in

$$\prod_{k\geq 1}L^\infty(\mathbf{R};\mathcal{L}^1(L^2((\mathbf{R}^d)^k)))$$

equipped with the product topology, where $L^\infty(\mathbf{R};\mathcal{L}^1(L^2((\mathbf{R}^d)^k)))$ is endowed with the weak-* topology for each $k\geq 1$.

Besides, Theorem 1.11.1 implies that any limit point $(D_k)_{k\geq 1}$ of that sequence is a solution of the infinite quantum mean field hierarchy with initial data

$$D_k^{in}:=|(\psi^{in})^{\otimes k}\rangle\langle(\psi^{in})^{\otimes k}|,\qquad k\geq 1.$$

Proposition 1.11.1 *Let* V *be a real-valued, even, bounded measurable function defined on* $\mathbf{R}^d$, *such that* $V\to 0$. *Let the initial single-particle wave function* $\psi^{in}\in H^1(\mathbf{R}^d)$ *satisfy* $\|\psi^{in}\|_{L^2}=1$, *and let* $\psi\in C(\mathbf{R};H^1(\mathbf{R}^d))\cap C^1(\mathbf{R};L^2(\mathbf{R}^d))$ *be the solution of the Cauchy problem for the Hartree equation*

$$\begin{cases}(i\partial_t+\tfrac12\Delta_x)\psi(t,x)+\psi(t,x)\displaystyle\int_{\mathbf{R}^d}V(x-z)|\psi(t,z)|^2dz=0,\\[2ex] \psi(0,x)=\psi^{in}(x).\end{cases}$$

Then the sequence

$$\rho_k(t)=|\psi(t,\cdot)^{\otimes k}\rangle\langle\psi(t,\cdot)^{\otimes k}|$$

is a particular solution of the Cauchy problem for the infinite quantum mean field hierarchy with initial data

$$\rho_k^{in}:=|(\psi^{in})^{\otimes k}\rangle\langle(\psi^{in})^{\otimes k}|,\qquad k\geq 1.$$

Proof Since $\psi \in C(\mathbf{R}; H^1(\mathbf{R}^d)) \cap C^1(\mathbf{R}; L^2(\mathbf{R}^d))$ satisfies the Hartree equation, one has

$$i\partial_t(\psi(t,x)\overline{\psi(t,y)}) = -\tfrac{1}{2}(\Delta_x - \Delta_y)\psi(t,x)\overline{\psi(t,y)} + \psi(t,x)\overline{\psi(t,y)} \int_{\mathbf{R}^d} (V(x-z) - V(y-z))|\psi(t,z)|^2 dz$$

in the sense of distributions on $\mathbf{R}_t \times \mathbf{R}^d_x \times \mathbf{R}^d_y$. Next, an elementary computation shows that, for each $k > 1$,

$$\begin{aligned}
i\partial_t \rho_k(t, X_k, Y_k) &= i \sum_{j=1}^{k} \prod_{\substack{l=1\\ l\neq j}}^{k} \psi(t,x_l)\overline{\psi(t,y_l)}\partial_t(\psi(t,x_j)\overline{\psi(t,y_j)}) \\
&= -\tfrac{1}{2}\sum_{j=1}^{k} \prod_{\substack{l=1\\ l\neq j}}^{k} \psi(t,x_l)\overline{\psi(t,y_l)}(\Delta_{x_j} - \Delta_{y_j})\psi(t,x_j)\overline{\psi(t,y_j)} \\
&\quad + \sum_{j=1}^{k} \rho_k(t,X_k,Y_k) \int_{\mathbf{R}^d} (V(x_j - z) - V(y_j - z))|\psi(t,z)|^2 dz \\
&= -\tfrac{1}{2}\sum_{j=1}^{k}(\Delta_{x_j} - \Delta_{y_j}) \prod_{l=1}^{k} \psi(t,x_l)\overline{\psi(t,y_l)} \\
&\quad + \sum_{j=1}^{k} \int_{\mathbf{R}^d} (V(x_j - z) - V(y_j - z))\rho_k(t,X_k,Y_k)|\psi(t,z)|^2 dz \\
&= -\tfrac{1}{2}\sum_{j=1}^{k}(\Delta_{x_j} - \Delta_{y_j})\rho_k(t,X_k,Y_k) \\
&\quad + \sum_{j=1}^{k} \int_{\mathbf{R}^d} (V(x_j - z) - V(y_j - z))\rho_{k+1}(t,X_k,z,Y_k,z)dz.
\end{aligned}$$

Therefore, the sequence $(\rho_k(t, X_k, Y_k))_{k\geq 1}$ is a solution of the infinite quantum mean field hierarchy.

It remains to prove that all limit points $(D_k)_{k\geq 1}$ of the sequence $(D_{N:k})_{k\geq 1}$ as $N \to \infty$ fall in some uniqueness class for the infinite quantum mean field hierarchy. If this is the case, each limit point $(D_k)_{k\geq 1}$, being a solution of the infinite quantum mean field hierarchy with the same initial data as $(\rho_k)_{k\geq 1}$, must coincide with it for all $t \in \mathbf{R}$. By compactness and uniqueness of the limit point, one concludes that

$$D_{N:k}(t) \to \rho_k(t) := |\psi(t,\cdot)^{\otimes k}\rangle\langle\psi(t,\cdot)^{\otimes k}|, \qquad \text{for each } k \geq 1$$

in the weak-* topology of $L^\infty(\mathbf{R}; \mathscr{L}^1(L^2((\mathbf{R}^d)^k)))$ as $N \to \infty$.

Since the integral kernels of the operators ρ_k are of factorized form, i.e.

$$\rho_k(t, X_k, Y_k) = \prod_{j=1}^{k} \psi(t, x_j)\overline{\psi(t, y_j)}, \qquad k \geq 1,$$

this last statement is the quantum analogue of the propagation of chaos studied above in the context of the mean field limit in classical statistical mechanics.

In order to carry out this program, two things need to be checked:

(a) one needs an estimate for the interaction terms

$$D_{k+1} \to [V_{j,k+1}, D_{k+1}]_{:k}$$

for $j = 1, \ldots, k$, and

(b) one needs an estimate for the growth of

$$D_k \quad \text{as } k \to \infty.$$

In the next section, we present an abstract framework explaining the role of each estimates.

1.11.2 A Tool for Studying Infinite Hierarchies

We first review a seemingly unrelated result, namely the Nirenberg-Ovcyannikov abstract analogue of the Cauchy-Kovalevska theorem. More precisely, we recall below a variant of this theorem due to Nishida.

1.11.2.1 The Abstract Cauchy-Kovalevska Theorem

The Nirenberg-Ovcyannikov abstract Cauchy-Kovalevska theory bears on the Cauchy problem for differential equations of the form

$$\dot{u}(t) = F(t, u(t)),$$

in situations where one can apply neither the Cauchy-Lipschitz theorem nor the Peano existence theorem (see Chap. 1 in [60]). Such situations arise typically when the solution $u(t)$ belongs to function spaces that are not invariant under the mapping $v \mapsto F(t, v)$. We first introduce the functional setting and present the main result obtained independently by Nirenberg [82] and Ovcyannikov [84].

The setting involves a family (a scale) of Banach spaces $(B_r)_{r>0}$ indexed by the positive real numbers. The norm in B_r is denoted by $\|\cdot\|_r$. This scale of Banach spaces is nonincreasing, in the sense that, for each $r, r' > 0$, one has

$$(HB_r) \qquad r' \le r \Rightarrow B_r \subset B_{r'} \quad \text{with} \quad \|v\|_{r'} \le \|v\|_r \quad \text{for each } v \in B_r.$$

Next, we specify how the time-dependent "vector field" F behaves with respect to the scale of Banach spaces $(B_r)_{r>0}$.

Our first assumption bears on the joint dependence of F in both of its arguments:

(a) there exists $r_0 > 0$ s.t.

$$(HF_a) \qquad \begin{aligned} &0 < r' < r < r_0 \Rightarrow F \in C(\mathbf{R} \times B_r, B_{r'}) \\ &\qquad\qquad \text{and } B_r \ni v \mapsto F(t, v) \in B_{r'} \text{ is linear for all } t \in \mathbf{R}. \end{aligned}$$

A second assumption bears on the dependence of F in its second argument, uniformly in the time variable t:

(b) there exists $C > 0$ s.t.

$$(HF_b) \qquad 0 < r' < r < r_0 \Rightarrow \|F(t, v)\|_{r'} \le C\frac{\|v\|_r}{r - r'}.$$

We state only the uniqueness part of the Nirenberg-Ovcyannikov theorem, and refer the interested reader to [83] for a the complete statement—that is an existence theorem as well, and moreover bears on possibly nonlinear "vector fields" F.

Theorem 1.11.2 *For all $r \in (0, r_0)$ and all $\alpha > 0$, the only solution u of the Cauchy problem*

$$\begin{cases} \dot{u}(t) = F(t, u(t)), \\ u(0) = 0, \end{cases}$$

such that $u \in C^1((-\alpha, \alpha), B_r)$ is $u = 0$.

In order to fully appreciate the meaning of assumption (HF_b), we briefly explain how this theorem is related to the classical Cauchy-Kovalevska theorem.

Consider the PDE with unknown $u \equiv u(t, z) \in \mathbf{C}$,

$$\partial_t u(t, z) = P(t, z, \partial_z)u(t, z),$$

where $t \in \mathbf{R}$ and $z \in \mathbf{C}$. It is assumed that P is a 1st order linear differential operator of the form

$$P(t, z, \partial_z)u(t, z) := A(t, z)\partial_z u(t, z) + B(t, z)u(t, z),$$

with coefficients A and B bounded and analytic on $(-T, T) \times \Omega_{r_0}$, where $\Omega_r \subset \mathbf{C}$ is the open strip defined as follows:

$$\Omega_r := \{z \in \mathbf{C} \text{ s.t. } |\operatorname{Im}(z)| < r\}.$$

Denote by B_r the space of bounded holomorphic functions on Ω_r with

$$\|v\|_r := \sup_{z \in \Omega_r} |v(z)|.$$

We recall that, for each function f that is holomorphic and bounded in the strip Ω_r, one has

$$|f^{(k)}(z)| \leq \frac{\sup_{z \in \Omega_r} |f(z)|}{(r - r')^k} \qquad \text{for all } z \in \Omega_{r'} \text{ and } k \geq 1,$$

because of Cauchy's estimates for holomorphic functions (see for instance Chap. 4, Sect. 2.3 in [3]).

Therefore, one has

$$\begin{aligned}
\|P(t, \cdot, \partial_z) u(t, \cdot)\|_{r'} &\leq \sup_{z \in \Omega_{r_0}} |A(t, z)| \sup_{z \in \Omega_{r'}} |\partial_z u(t, z)| \\
&\quad + \sup_{z \in \Omega_{r_0}} |B(t, z)| \sup_{z \in \Omega_{r'}} |u(t, z)| \\
&\leq \sup_{z \in \Omega_{r_0}} |A(t, z)| \frac{\sup_{z \in \Omega_r} |u(t, z)|}{r - r'} \\
&\quad + \sup_{z \in \Omega_{r_0}} |B(t, z)| \sup_{z \in \Omega_r} |u(t, z)| \\
&\leq C \frac{\sup_{z \in \Omega_r} |u(t, z)|}{r - r'}
\end{aligned}$$

with

$$C := \sup_{z \in \Omega_{r_0}} |A(t, z)| + r_0 \sup_{z \in \Omega_{r_0}} |B(t, z)|.$$

In other words, the 1st order differential operator $P(t, z, \partial_z)$ satisfies assumption (HF_b) as a consequence of the Cauchy inequalities for bounded holomorphic functions in the strip Ω_r.

Observe that differential operators $Q(t, z, \partial_z)$ of order higher than 1 do not satisfy assumption (HF_b), but an inequality of the form

$$\|Q(t, \cdot, \partial_z) u(t, \cdot)\|_{r'} \leq C \frac{\sup_{z \in \Omega_r} |u(t, z)|}{(r - r')^k}, \qquad \text{with } k = \text{order of } Q.$$

This is a weaker inequality that the one in (HF_b), which follows again from Cauchy's estimates, this time on higher order derivatives of holomorphic functions.

In fact, this is not surprising: the Cauchy-Kovalevska theorem guarantees the local existence and uniqueness of a local analytic solution of the Cauchy problem for the PDE

$$\begin{cases} \partial_t u(t,z) = Q(t,z,\partial_z)u(t,z) \\ u\big|_{t=0} = u^{in} \end{cases}$$

defined for $|t| < T$ and $|\operatorname{Im}(z)| < r^*$ for some $T > 0$ and some $r^* > 0$. This is obviously impossible in general if Q is of order higher than 1. Consider indeed the simplest possible example where $Q(t,z,\partial_z) = \partial_z^2$. It is well known that the Cauchy problem for the backward heat equation is not well-posed in any one of the class B_r. Indeed, for each $t > 0$, the semigroup $e^{t\partial_z^2}$ maps the set of distributions with compact support on $\mathbf{R}$ into the set of functions that are entire holomorphic (i.e. holomorphic on $\mathbf{C}$).

1.11.2.2 Application of the Abstract Cauchy-Kovalevska Theorem to Infinite Hierarchies

In this section, we give an abstract theorem for proving the uniqueness of the solution of infinite hierarchies based on the Cauchy-Kovalevska theorem presented above.

Our setting involves a sequence of Banach spaces $(E_n)_{n\ge 1}$. For each $n \ge 1$, the norm in E_n is denoted by $|\cdot|_n$.

Along with the Banach spaces E_n, our setting also involves

(a) a strongly continuous group of isometries $U_n(t)$ defined on E_n for each $n \ge 1$, and

(b) a sequence of bounded linear operators $L_{n,n+1} : E_{n+1} \to E_n$ defined for each $n \ge 1$.

It will be assumed that the sequence of operators $L_{n,n+1}$ satisfies the following bound: there exists $C > 0$ such that

$$(HL) \qquad \|L_{n,n+1}\|_{\mathscr{L}(E_{n+1},E_n)} \le Cn, \qquad \text{for each } n \ge 1.$$

In the setting so defined, we consider the infinite hierarchy of differential equations

$$\dot{u}_n(t) = U_n(t)L_{n,n+1}U_{n+1}(-t)u_{n+1}(t), \qquad n \ge 1.$$

Theorem 1.11.3 *Let $t^* > 0$ and let $u_n \in C^1([0,t^*],E_n)$ for each $n \ge 1$ be a solution of*

$$\begin{cases} \dot{u}_n(t) = U_n(t)L_{n,n+1}U_{n+1}(-t)u_{n+1}(t), & n \ge 1, \\ u_n(0) = 0. \end{cases}$$

Assume there exists $R > 0$ such that

$$(Hu) \qquad \sup_{0\le t\le t^*} \|u_n(t)\|_n \le R^n \qquad \text{for all } n \ge 1.$$

Then

$$u_n(t) = 0 \quad \textit{for all } t \in [0, t^*] \textit{ and each } n \geq 1.$$

Proof We apply the abstract Cauchy-Kovalevska theorem in the following setting. First we choose the scale of Banach spaces: for each $r > 0$, we set

$$B_r := \left\{ v = (v_n)_{n\geq 1} \in \prod_{n\geq 1} E_n \text{ s.t. } \|v\|_r := \sum_{n\geq 1} r^n |v_n|_n < \infty \right\}.$$

Obviously, if $0 < r' < r$, one has

$$B_r \subset B_{r'},$$

and

$$\|v\|_{r'} = \sum_{n\geq 1} r'^n |v_n|_n \leq \sum_{n\geq 1} r^n |v_n|_n = \|v\|_r$$

for each $v \in B_r$.

The mapping F is defined on sequences $v = (v_n)_{n\geq 1}$ as follows:

$$F(t, v) := (U_n(t) L_{n,n+1} U_{n+1}(-t) v_{n+1})_{n\geq 1}.$$

With this definition, observing that

$$nr'^n \leq r'^n + r'^{n-1} r + \ldots + r' r^{n-1} + r^n = \frac{r^{n+1} - r'^{n+1}}{r - r'},$$

assumption (HL) implies that

$$\begin{aligned} \|F(t, v)\|_{r'} &\leq C \sum_{n\geq 1} nr'^n |v_n|_n \leq C \sum_{n\geq 1} \frac{r^{n+1} - r'^{n+1}}{r - r'} |v_n|_n \\ &\leq \frac{C}{r - r'} \sum_{n\geq 1} r^{n+1} |v|_{n+1} \leq \frac{C\|v\|_r}{r - r'} \end{aligned}$$

for each $r, r' > 0$ such that $r' < r$, where C is the constant that appears in the control of the norm of $L_{n,n+1}$ in condition (HL).

This implies that F satisfies assumption (HF$_b$). As for assumption (HF$_a$), it follows from assumption (HF$_b$) and the fact that $U_n(t)$ is a linear group of isometries on E_n for each $n \geq 1$.

Thus, if a sequence $u(t) = (u(t)_n)_{n\geq 1}$ with $u(t)_n \in E_n$ for each $n \geq 1$ and each $t \in [0, t^*]$ satisfies the growth condition (Hu) and the infinite hierarchy, then $u(t) \in B_r$ for all $t \in [0, t^*]$ and for each $r \in (0, 1/R)$.

We deduce from the differential equation that $u \in C^1([0, t^*], B_r)$ if $0 < r < 1/R$. By the abstract Cauchy-Kovalevska theorem, we conclude that $u(t) = 0$ for each $t \in [0, t^*]$, which means that

$$u_n(t) = 0 \qquad \text{for each } t \in [0, t^*] \text{ and each } n \geq 1.$$

This is precisely the expected uniqueness property.

Notice that assumption (Hu) is essential for the uniqueness property above. Here is an easy counterexample, showing that this assumption cannot be dispensed with.

Consider the infinite hierarchy of ODEs

$$\begin{cases} \dot{y}_k(t) = k y_{k+1}(t), & k \geq 1, \\ y_k(0) = y_k^{in}, \end{cases}$$

This hierarchy has obvious factorized solutions: if x is the solution of the Riccati equation

$$\begin{cases} \dot{x}(t) = x(t)^2, \\ x(0) = x^{in}, \end{cases}$$

i.e.

$$x(t) = \frac{x^{in}}{1 - t x^{in}},$$

then

$$y_k(t) = x(t)^k, \quad k \geq 1$$

is a solution of the infinite hierarchy with initial data

$$y_k^{in} := (x^{in})^k, \quad k \geq 1.$$

The infinite hierarchy with $y_k^{in} = 0$ for all $k \geq 0$ has a unique solution satisfying the growth condition (Hu), which is $y_k(t) = 0$ for all $t \in \mathbf{R}$ and $k \geq 1$.

Consider now the function E defined by

$$E(t) = \begin{cases} e^{-1/t} & \text{if } t > 0 \\ 0 & \text{if } t \leq 0 \end{cases}$$

that is of class C^∞ on the real line and satisfies $E^{(n)}(0) = 0$ for all $n \geq 0$. Then the formula

$$y_k(t) = \frac{1}{(k-1)!} E^{(k-1)}(t), \quad t \in \mathbf{R} \text{ and } k \geq 1,$$

also defines a solution of the infinite hierarchy.

That this solution does not grows exponentially as $k \to \infty$ corresponds with the fact that E is not an analytic function on the line, since it vanishes on the half-line without being identically 0, which contradicts the principle of isolated zeros.

References for this section are [6, 101].

1.11.3 *Application to the Hartree Limit in the Bounded Potential Case*

Finally we explain how to prove Theorem 1.11.1 with the formalism discussed above.

Proof (Proof of Theorem 1.11.1)
Step 1: First we write the infinite quantum mean field hierarchy

$$i\dot{D}_j(t) = -\tfrac{1}{2}\sum_{k=1}^{j}[\Delta_j, D_j(t)] + \sum_{k=1}^{j}[V_{k,j+1}, D_{j+1}]_{:j}, \quad j \geq 1$$

in the form

$$\dot{u}_j(t) = U_j(t)L_{j,j+1}U_{j+1}(-t)u_{j+1}(t), \quad j \geq 1,$$

where

$$L_{j,j+1}D_{j+1} := \sum_{k=1}^{j}[V_{k,j+1}, D_{j+1}]_{:j}$$

while

$$U_j(t)D_j := e^{-\frac{1}{2}it(\Delta_1+\ldots+\Delta_j)}D_j e^{\frac{1}{2}it(\Delta_1+\ldots+\Delta_j)}.$$

As before, Δ_k designates the Laplacian acting on the kth variable in $L^2((\mathbf{R}^d)^j)$, for each $k = 1, \ldots, j$, while

$$u_j(t) := U_j(t)D_j(t).$$

Step 2: Next we choose appropriate norms and function spaces so as to apply Theorem 1.11.3. Define $E_j := \mathcal{L}^1(L^2((\mathbf{R}^d)^j))$ with trace norm denoted by $\|\cdot\|_1$ (or $\|\cdot\|_{1,j}$ in cases where the previous notation would lead to ambiguities).

The operator $U_j(t)$ defined in Step 1 is an isometry of E_j for all $j \geq 1$ and all $t \in \mathbf{R}$, since it consists of conjugating elements of E_j with the unitary group associated to the j-particle free Schrödinger equation. Indeed, we recall the following elementary result.

Lemma 1.11.1 *Let $\mathfrak{H}$ be a separable Hilbert space. For each $A \in L^1(\mathfrak{H})$ and each unitary operator U on $\mathfrak{H}$, then $UAU^* \in \mathcal{L}^1(\mathfrak{H})$ and one has*

$$\|UAU^*\|_1 = \|A\|_1.$$

Exercise Prove the lemma above. (Hint: use the definition of the trace-norm.)

Now we need to control the interaction term.

First, since $V \in L^\infty$, one has

$$\begin{aligned}\|[V_{k,j+1}, D_{j+1}]\|_1 &\le \|V_{k,j+1}D_{j+1}\|_1 + \|D_{j+1}V_{k,j+1}\|_1 \\ &\le 2\|V_{k,j+1}\|\|D_{j+1}\|_1 \le 2\|V\|_{L^\infty}\|D_{j+1}\|_1\end{aligned}$$

for each $(D_j)_{j\ge1}$ with $D_j \in \mathcal{L}^1(L^2((\mathbf{R}^d)^j))$ for each $j \ge 1$.

Next, we use the following auxiliary result.

Lemma 1.11.2 *Set $\mathfrak{H} = L^2(\mathbf{R}^{m+n})$ and $\mathfrak{H}_1 = L^2(\mathbf{R}^m)$. Let $K \in \mathcal{L}^1(\mathfrak{H})$ be an integral operator of the form*

$$K\phi(x,z) = \int_{\mathbf{R}^{m+n}} k(x,z,y,w)\phi(y,w)dydw.$$

Let $K_{z,w}$ be the operator defined on $\mathfrak{H}_1$ by

$$K_{z,w}\psi(x) = \int_{\mathbf{R}^m} k(x,z,y,w)\psi(y)dy$$

for a.e. $z, w \in \mathbf{R}^n$. Then
(1) the map $h \mapsto [z \mapsto K_{z,z+h}]$ belongs to $C(\mathbf{R}^n_h, L^1(\mathbf{R}^n_z, \mathcal{L}^1(\mathfrak{H}_1)))$;
(2) for a.e. $z \in \mathbf{R}^n$ the operator $K_{z,z}$ is trace-class on $L^2(\mathbf{R}^m)$ and

$$\operatorname{trace}\left|\int_{\mathbf{R}^n} K_{z,z}dz\right| \le \int_{\mathbf{R}^n} \operatorname{trace}|K_{z,z}|dz \le \|K\|_1.$$

This result is an amplification of Lemma 1.9.1—see Lemma 2.1 in [7].

We apply the lemma above to

$$K = [V_{k,j+1}, D_{j+1}],$$

which is an integral operator with integral kernel

$$\begin{aligned}&k(X_j, x_{j+1}, Y_j, y_{j+1}) \\ &= (V(x_k - x_{j+1}) - V(y_k - y_{j+1}))D_{j+1}(X_j, x_{j+1}, Y_j, y_{j+1}).\end{aligned}$$

This shows that

$$\|[V_{k,j+1}, D_{j+1}]_{:j}\|_1 = \operatorname{trace}\left|\int_{\mathbf{R}^d} K_{z_{j+1},z_{j+1}} dz_{j+1}\right|$$
$$\leq \|K\|_1 \leq 2\|V\|_{L^\infty}\|D_{j+1}\|_1.$$

Therefore we conclude that

$$\|L_{j,j+1}\|_{\mathscr{L}(E_{j+1},E_j)} \leq 2j\|V\|_{L^\infty},$$

which means that the infinite quantum mean field hierarchy with bounded potential satisfies assumption (HL) in Theorem 1.11.3.

Step 3: Finally we control the growth of D_j as $j \to \infty$, where D_j is a limit point of $D_{N:j}$ in $L^\infty(\mathbf{R}_+; \mathscr{L}^1(L^2((\mathbf{R}^d)^j)))$ weak-* as $N \to \infty$. First we recall that, for each $N \geq 1$ and each $t \geq 0$, one has

$$D_N(t) = D_N(t)^* \geq 0, \quad \text{and } \|D_N(t)\|_1 = \operatorname{trace} D_N(t) = 1.$$

Applying the lemma above shows that

$$\|D_{N:j}(t)\|_1 \leq 1 \qquad \text{for all } t \geq 0,\ N, j \geq 1.$$

(We recall the convention $D_{N:j} = 0$ whenever $j > N$.) Since

$$D_{N:j} \rightharpoonup D_j(t) \text{ in } L^\infty(\mathbf{R}_+; \mathscr{L}^1(L^2((\mathbf{R}^d)^j))) \text{ weak-*},$$

we conclude that

$$\|D_j(t)\|_1 \leq 1 \qquad \text{for all } t \geq 0,\ j \geq 1.$$

Step 4: One can apply Theorem 1.11.3 since the infinite hierarchy under consideration satisfies the bound (HL) with $C = 2\|V\|_{L^\infty}$, while each limit point D_j of the sequence $D_{N:j}$ in $L^\infty(\mathbf{R}_+; \mathscr{L}^1(L^2((\mathbf{R}^d)^j)))$ as $N \to \infty$ satisfies the growth estimate (Hu) with $R = 1$.

Indeed, setting

$$v_j(t) := U_j(t)(D_j(t) - |\psi(t,\cdot)^{\otimes j}\rangle\langle\psi(t,\cdot)^{\otimes j}|)$$

we know that the sequence $(v_j)_{j\geq 1}$ is a solution of the infinite mean field hierarchy, which satisfies

$$\|v_j(t)\|_{1,j} \leq \|D_j(t)\|_{1,j} + \|\,|\psi(t,\cdot)^{\otimes j}\rangle\langle\psi(t,\cdot)^{\otimes j}|\,\|_{1,j} \leq 1 + \|\psi(t,\cdot)\|_{L^2}^{2j} = 2.$$

Since $v_j(0) = 0$ for each $j \geq 1$, we conclude from Theorem 1.11.3 that $v_j(t) = 0$ for each $j \geq 1$ and each $t \geq 0$. In other words,

$$D_j(t) = |\psi(t,\cdot)^{\otimes j}\rangle\langle\psi(t,\cdot)^{\otimes j}|$$

for each $j \geq 1$, which is precisely the desired result.

Notice that the verification of the growth condition (Hu) on the sequence $(D_j)_{j\geq 1}$ is trivial in this problem. This is not always the case, and we shall see some examples below, where checking this crucial condition can be a serious difficulty.

References for this section are [7, 96].

1.12 Other Mean Field Limits in Quantum Mechanics

1.12.1 Derivation of the Schrödinger-Poisson Equation

The quantum analogue of the Vlasov-Poisson system is the following variant of the Hartree equation, where the interaction potential is the repulsive Coulomb potential (between particles of equal charges):

$$V(z) = \frac{1}{4\pi|z|}$$

in space dimension $d = 3$. It takes the form

$$\begin{cases} i\partial_t\psi(t,x) = -\frac{1}{2}\Delta_x\psi(t,x) + U(t,x)\psi(t,x), & x \in \mathbf{R}^3, \\ U(t,x) = \frac{1}{4\pi}\displaystyle\int_{\mathbf{R}^3} \frac{|\psi(t,y)|^2}{|x-y|}dy, \end{cases}$$

or equivalently

$$\begin{cases} i\partial_t\psi(t,x) = -\frac{1}{2}\Delta_x\psi(t,x) + U(t,x)\psi(t,x), & x \in \mathbf{R}^3, \\ -\Delta_x U(t,x) = |\psi(t,x)|^2. \end{cases}$$

This last form of the mean field equation explains why, as mentioned above, the Hartree equation in this case is also known as the "Schrödinger-Poisson" equation.

Although the Coulomb potential is unbounded near the origin, the Schrödinger-Poisson equation can be derived following the strategy described above: see [32] for this very interesting result. In that case, the right choice of spaces E_n is as follows. Define

$$S_j := \sqrt{I - \Delta_{x_j}}, \qquad j = 1,\ldots,N.$$

With the same notation as above, the sequence of Banach spaces in Theorem 1.11.3 is defined as follows:

$$E_n := \{D \in \mathscr{L}^1(L^2((\mathbf{R}^3)^n)) \mid S_1 \dots S_n D S_1 \dots S_n \in \mathscr{L}^1(L^2((\mathbf{R}^3)^n))\},$$

for all $n \geq 1$, with norm

$$\|D\|_{E_n} = \operatorname{trace} |S_1 \dots S_n D S_1 \dots S_n|.$$

With this choice of spaces, the inequality in assumption (HL) in Theorem 1.11.3 follows from Hardy's inequality

$$\int_{\mathbf{R}^3} \frac{f(x)^2}{|x|^2} dx \leq C \int_{\mathbf{R}^3} |\nabla f(x)|^2 dx$$

(see Lemma 7.1 in [32]).

The estimate (Hu) follows from the conservation laws

$$\frac{d}{dt} \operatorname{trace}(\mathcal{H}_N^m D_N(t)) = 0 \quad \text{for all } m \geq 0,$$

where

$$\mathcal{H}_N := -\tfrac{1}{2} \sum_{k=1}^{N} \Delta_{x_k} + \frac{1}{N} \sum_{1 \leq k < l \leq N} V_{kl}$$

(with V_{kl} denoting the multiplication by $V(x_k - x_l)$) and

$$D_N(t) = |e^{it\mathcal{H}_N} (\psi^{in})^{\otimes N}\rangle\langle e^{it\mathcal{H}_N} (\psi^{in})^{\otimes N}|.$$

A crucial step in the proof of the growth estimate (Hu) is to compare powers of the N-particle Hamiltonian $\mathcal{H}_N$ with powers of the N-particle free Schrödinger operator

$$L_N := -\tfrac{1}{2} \sum_{k=1}^{N} \Delta_{x_k}$$

—see Propositions 4.1 and 5.1 in [32]. Verifying the growth condition (Hu) is perhaps the most technical parts of [32].

A more direct proof, by a somewhat different argument, was proposed later in [87]—see Sect. 1.12.4.

1.12.2 Derivation of the Nonlinear Schrödinger Equation

One can consider the mean field limit with even more singular interactions, viz. $V(z) = \delta_0(z)$. Equivalently, one can prove the mean field limit with a potential whose range shrinks to zero as the number of particles tends to infinity. Specifically, one can consider the N-particle Schrödinger operator in space dimension 1:

$$\mathcal{H}_N := -\tfrac{1}{2}\sum_{k=1}^{N} \partial_{x_k}^2 + \frac{1}{N}\sum_{k,l=1}^{N} N^{\gamma} U(N^{\gamma}(x_k - x_l))$$

where $U \geq 0$ belongs to the Schwartz class $\mathscr{S}(\mathbf{R})$, and $\gamma \in (0, 1)$.

The corresponding mean field equation is the cubic nonlinear Schrödinger equation

$$i\partial_t \psi(t,x) = -\tfrac{1}{2}\partial_x^2 \psi(t,x) + a|\psi(t,x)|^2\psi(t,x), \quad x \in \mathbf{R},$$

where

$$a = \int_{\mathbf{R}} U(z)dz.$$

In this case again, the strategy outlined in Theorem 1.11.3 applies with some appropriate choice of norms. The right choice of Banach spaces in this case is as follows:

$$E_n := \{D \in \mathscr{L}^2(L^2(\mathbf{R}^n)) \mid S_1 \dots S_n D S_1 \dots S_n \in \mathscr{L}^2(L^2(\mathbf{R}^n))\},$$

with norm

$$\|D\|_{E_n} = (\operatorname{trace}(|S_1 \dots S_n D S_1 \dots S_n|^2))^{1/2},$$

where, as in the previous example,

$$S_j := \sqrt{I - \partial_{x_j}^2}, \quad j = 1, \dots N.$$

Controling the interaction operator in the nth equation of the infinite mean field hierarchy associated to the Schrödinger operator $\mathcal{H}_N$ is essentially equivalent to controling n integral operators on $L^2(\mathbf{R}^n)$ with integral kernels of the form

$$D_{n+1}(x_1, \dots, x_n, x_1, y_1, \dots, y_n, x_1)$$

in terms of $\|D_{n+1}\|_n$.

If D_n is the density matrix associated to a factorized wave function, i.e. if

$$D_n(x_1, \dots, x_n, y_1, \dots, y_n) = \psi(x_1) \dots \psi(x_n)\overline{\psi(y_1)} \dots \overline{\psi(y_n)},$$

then $D_n \in E_n$ if and only if $\psi \in H^1(\mathbf{R})$ and $\|D_n\|_n = \|\psi\|_{H^1}^{2n}$. In this case,

$$\begin{aligned}D_{n+1}&(x_1, \dots, x_n, x_1, y_1, \dots, y_n, x_1)\\&= \psi(x_1)|\psi(x_1)|^2\psi(x_2)\dots\psi(x_n)\overline{\psi(y_1)}\dots\overline{\psi(y_n)}\end{aligned}$$

and the estimate in assumption (HL) reduces to the fact that the Sobolev space $H^1(\mathbf{R})$ is an algebra. (Indeed, we recall that

$$u, v \in H^1(\mathbf{R}) \Rightarrow uv \in H^1(\mathbf{R}),$$

with

$$\|uv\|_{H^1} \leq C\|u\|_{H^1}\|v\|_{H^1} \quad \text{for all } u, v \in H^1(\mathbf{R})$$

for some positive constant C. This elementary fact is most easily checked in terms of the Fourier transforms of u and v.)

Proving the bound in assumption (HL) for general density matrices (and not only in the factorized case) is a rather straightforward generalization of the (classical) proof that $H^1(\mathbf{R})$ is an algebra. Proving the bound in assumption (Hu) is somewhat more technical and will not be discussed here.

We refer the interested reader to [2]—see also [1]—for a detailed statement of the result and a complete proof thereof.

The analogous result in space dimension higher than 1 is considerably more involved—notice that $H^1(\mathbf{R}^d)$ is not an algebra whenever $d \geq 2$, so that the argument above for the bound in assumption (HL) is no longer valid. The proof of the mean field limit for interaction potentials shrinking to a Dirac measure can be found in [30, 33, 34, 88]. An interesting approach to the uniqueness problem in the corresponding infinite hierarchy can be found in [66].

A general reference for this and the previous section is [38]. However Gérard's survey in [38] predates Pickl's contributions to the subject, which are described in Sect. 1.12.4.

1.12.3 The Time-Dependent Hartree-Fock Equations

In the quantum mean field theories considered so far, the asymptotic N-particle wave function always approached a factorized state of the form

$$\Psi_N(t, x_1, \dots, x_N) \sim \prod_{k=1}^{N} \psi(t, x_k)$$

in the limit as $N \to \infty$. In fact, we have assumed that the initial state is already of this form, i.e.

$$\Psi_N(0, x_1, \ldots, x_N) = \prod_{k=1}^{N} \psi^{in}(x_k)$$

in all the situations considered so far. In particular, Ψ_N is symmetric in its space variables, i.e.

$$\Psi_N(t, x_{\sigma(1)}, \ldots, x_{\sigma(N)}) = \Psi_N(t, x_1, \ldots, x_N)$$

for all $\sigma \in \mathfrak{S}_N$ and all $x_1, \ldots, x_N \in \mathbf{R}^d$. In other words, all the mean field theories considered above apply to the case of bosons; the case of fermions requires a separate study.

In the case of fermions, one should replace the initial factorized state with

$$\Psi_N(0, x_1, \ldots, x_N) = \frac{1}{\sqrt{N!}} \det(\psi_k^{in}(x_l))_{1\le k,l\le N}$$

where $(\psi_1, \ldots, \psi_N)$ is an orthonormal system in $L^2(\mathbf{R}^d)$, i.e.

$$\int_{\mathbf{R}^d} \overline{\psi_k^{in}(x)} \psi_l^{in}(x) dx = \delta_{kl}, \quad k, l = 1, \ldots, N.$$

This special form of N-particle wave function is referred to as a *Slater determinant*.

Elementary computations show that Slater determinants satisfy

$$\int_{(\mathbf{R}^d)^N} |\det(\psi_k^{in}(x_l))_{1\le k,l\le N}|^2 dx_1 \ldots dx_N = N!$$

so that $\Psi_N(0, \ldots)$ is normalized in $L^2((\mathbf{R}^d)^N)$.

Moreover, one easily checks that the density matrix associated to a Slater determinant has its first marginal given by

$$D_{N:1}(0) := |\Psi_N(0, \ldots)\rangle\langle\Psi_N(0, \ldots)|_{:1} = \frac{1}{N} \sum_{k=1}^{N} |\psi_k^{in}\rangle\langle\psi_k^{in}|.$$

In other words, $ND_{N:1}(0)$ is the orthogonal projector on $\text{span}(\{\psi_1, \ldots, \psi_N\})$, so that

$$D_{N:1}(0)^* = D_{N:1}(0) = ND_{N:1}(0)^2 \ge 0, \quad \text{and } \operatorname{trace}(D_{N:1}(0)) = 1.$$

Higher order marginals of the density matrix associated with a Slater determinant are given in terms of the first marginal by the following expressions:

$$D_{N:n}(0) = \frac{N^n(N-n)!}{N!} D_{N:1}(0)^{\otimes n} \sum_{\sigma \in \mathfrak{S}_n} (-1)^{\text{sign}(\sigma)} U_\sigma$$

for all $n = 1, \dots, N$, where we recall that, for each permutation $\sigma \in \mathfrak{S}_n$, the operator U_σ is defined by

$$U_\sigma \Phi_n(x_1, \dots, x_n) = \Phi_n(x_{\sigma^{-1}(1)}, \dots, x_{\sigma^{-1}(n)})$$

for all $\Phi_n \in L^2((\mathbf{R}^d)^n)$.

Consider the N-particle Schrödinger operator in the mean field scaling, i.e.

$$\mathcal{H}_N := -\tfrac{1}{2}\sum_{k=1}^{N} \Delta_k + \frac{1}{N-1}\sum_{1\le k<l\le N}^{N} V_{kl}$$

for all $N > 1$, where we recall that V_{kl} designates the multiplication by $V(x_k - x_l)$ acting on $L^2((\mathbf{R}^d)^N)$.

The expected mean field equation in the case of fermions is the time-dependent Hartree-Fock (TDHF) equation, written below in the language of operators:

$$i\dot{F}(t) = [-\tfrac{1}{2}\Delta, F(t)] + [V_{12}, F(t)^{\otimes 2}(I - U_{12})]_{:1}.$$

Assuming that $F(t)$ is an integral operator on $L^2(\mathbf{R}^d)$ with integral kernel $F(t, x, y)$, the equation above can be recast in terms of integral kernels as follows:

$$\begin{aligned} &i\partial_t F(t, x, y) = -\tfrac{1}{2}(\Delta_x - \Delta_y)F(t, x, y) \\ &+ \int_{\mathbf{R}^d} (V(x-z) - V(y-z))(F(t, x, y)F(t, z, z) - F(t, z, y)F(t, x, z))dz, \end{aligned}$$

or equivalently

$$\begin{aligned} i\partial_t F(t, x, y) = &- \tfrac{1}{2}(\Delta_x - \Delta_y)F(t, x, y) \\ &+ F(t, x, y)\int_{\mathbf{R}^d} (V(x-z) - V(y-z))F(t, z, z)dz \\ &- \int_{\mathbf{R}^d} (V(x-z) - V(y-z))F(t, x, z)F(t, z, y)dz. \end{aligned}$$

The existence and uniqueness of the solution of the Cauchy problem for the TDHF equation is extensively studied in [18, 19, 27].

The mean field limit in the case of N fermions with 2-body bounded interaction is summarized in the following statement.

Theorem 1.12.1 *Assume that V is an even function belonging to $L^\infty(\mathbf{R}^d)$, and let $(\psi_1, \dots, \psi_N)$ be an orthonormal system in $L^2(\mathbf{R}^d)$. Let*

$$\Psi_N(t, \cdot) = e^{-it\mathcal{H}_N}\Psi_N^{in},$$

where

$$\mathcal{H}_N := -\tfrac{1}{2}\sum_{k=1}^{N}\Delta_{x_k} + \frac{1}{N}\sum_{1\le k<l\le N} V(x_k - x_l),$$

and

$$\Psi_N^{in}(x_1,\dots,x_N) = \frac{1}{\sqrt{N!}}\det(\psi_k(x_l))_{1\le k,l\le N}.$$

is the Slater determinant built on the orthonormal system $(\psi_1,\dots,\psi_N)$.

Then the density operator $D_N(t) := |\Psi_N(t)\rangle\langle\Psi_N(t)|$ *satisfies*

$$D_{N:n}(t) - \frac{N^n(N-n)!}{N!}D_{N:1}(t)^{\otimes n}\sum_{\sigma\in\mathfrak{S}_n}(-1)^{\mathrm{sign}(\sigma)}U_\sigma \to 0$$

in $\mathscr{L}^1(L^2((\mathbf{R}^d)^n))$ *as* $N\to\infty$ *for each* $n\ge 1$, *and*

$$D_{N:1}(t) - F_N(t) \to 0$$

in $\mathscr{L}^1(L^2(\mathbf{R}^d))$ *as* $N\to\infty$, *where* $F_N(t)$ *is the solution of the Cauchy problem*

$$\begin{cases} i\dot{F}_N(t) = [-\tfrac{1}{2}\Delta, F_N(t)] + [V_{12}, F_N(t)^{\otimes 2}(I - U_{12})|_{:1},\\ F_N(0) = \dfrac{1}{N}\displaystyle\sum_{k=1}^{N}|\psi_k^{in}\rangle\langle\psi_k^{in}|. \end{cases}$$

See [5, 8, 9] for proofs of this result and of more general theorems in the same direction.

There is another, more familiar formulation of the TDHF equation. Seek the operator $F(t)$ in the form

$$F(t) = \frac{1}{N}\sum_{k=1}^{N}|\psi_k(t,\cdot)\rangle\langle\psi_k(t,\cdot)|,$$

or equivalently its integral kernel in the form

$$F(t,x,y) = \frac{1}{N}\sum_{k=1}^{N}\psi_k(t,x)\overline{\psi_k(t,y)},$$

where $(\psi_1(t,\cdot),\dots,\psi_N(t,\cdot))$ is an orthonormal system in $L^2(\mathbf{R}^d)$. Then the functions $\psi_k(t,\cdot)$ should satisfy the system of PDEs

$$i\partial_t\psi_k(t,x) = -\tfrac{1}{2}\Delta_x\psi(t,x) + \psi_k(t,x)\int_{\mathbf{R}^d} V(x-z)\frac{1}{N}\sum_{l=1}^{N}|\psi_l(t,z)|^2 dz$$

$$-\frac{1}{N}\sum_{k=1}^{N}\psi_l(x)\int_{\mathbf{R}^d} V(x-z)\overline{\psi_l(t,z)}\psi_k(t,z)dz, \quad 1 \le k \le N.$$

In this formulation, the functions ψ_k indexed by $k = 1, \ldots, N$ are referred to as "molecular orbitals" in the context of quantum chemistry.

These equations are to be compared with the Hartree equation discussed above, i.e.

$$i\partial_t\psi(t,x) = -\tfrac{1}{2}\Delta_x\psi(t,x) + \psi(t,x)\int_{\mathbf{R}^d} V(x-z)|\psi(t,z)|^2 dz.$$

Obviously the term

$$\psi_k(t,x)\int_{\mathbf{R}^d} V(x-z)\frac{1}{N}\sum_{l=1}^{N}|\psi_l(t,z)|^2 dz$$

is the analogue in the Hartree-Fock case of the term

$$\psi(t,x)\int_{\mathbf{R}^d} V(x-z)|\psi(t,z)|^2 dz$$

in Hartree's equation. The additional term in the Hartree-Fock equation, i.e.

$$-\frac{1}{N}\sum_{k=1}^{N}\psi_l(x)\int_{\mathbf{R}^d} V(x-z)\overline{\psi_l(t,z)}\psi_k(t,z)dz$$

is called the "exchange interaction integral" and is special to the case of fermions: see for instance [69] (Problem 1 on p. 233).

Approaching the N-particle wave function of a system of fermions by a Slater determinant is obviously the most natural idea. However, other choices are also possible—and often made in practice, especially for the purpose of numerical computations in the context of quantum chemistry. One such method is the theory of multi-configuration time-dependent Hartree-Fock equations [11], where the N-particle wave function is approximated by a linear combination of Slater determinants. For the first detailed analysis of the multi-configuration ansatz, which has a very rich mathematical structure, we refer the interested reader to [4].

Hartree-Fock equations are used primarily in the context of quantum chemistry. Therefore, one has to take into account the following interactions

(a) electron-electron,
(b) electron-nuclei,
(c) nuclei-nuclei.

In many situations, the nuclei are considered either as fixed (as in [71]), or as macroscopic objects governed by a system of ODEs (as in [24]), while the quantum description of the electron system involves the Hartree-Fock equations. In that case, the only part of the Hamiltonian that is approximated by a nonlinear, self-consistent mean field interaction is the repulsive force (a) between electrons.

All these interactions involve the Coulomb potential; the derivation of the time dependent Hartree-Fock theory from the N-particle Schrödinger equation with Coulomb interactions remains an open problem at the time of this writing.

1.12.4 Pickl's Approach to Quantum Mean Field Limits

There are other approaches to the mean field limit in quantum mechanics—see for instance the work of Rodnianski and Schlein [92], which proposes an error estimate for the mean field limit, that uses the formalism of Fock spaces in second quantization, as well as the more recent references [35, 36].

In the present section, we give a brief presentation of yet another approach of the mean field limit that avoids the technicalities of BBGKY hierarchies as well as the formalism of Fock spaces. The discussion below is based on [87].

Pickl's idea is to consider some appropriate quantity that measures the distance between the first marginal $D_{N:1}(t)$ of the N-particle density matrix $D_N(t)$ and the single particle density matrix $|\psi(t,\cdot)\rangle\langle\psi(t,\cdot)|$ built on the solution ψ of the Hartree equation.

Let Ψ_N be the solution of the N-particle Schrödinger equation

$$\left\{\begin{aligned} &i\partial_t\Psi_N(t,x_1,\ldots,x_N)=\mathcal{H}_N\Psi_N(t,x_1,\ldots,x_N),\quad x_1,\ldots,x_N\in\mathbf{R}^d,\\ &\Psi_N\big|_{t=0}=(\psi^{in})^{\otimes N},\end{aligned}\right.$$

with

$$\mathcal{H}_N:=-\tfrac{1}{2}\sum_{k=1}^{N}\Delta_{x_k}+\frac{1}{N}\sum_{1\le k<l\le N}V(x_k-x_l).$$

Let ψ be the solution of Hartree's equation

$$\left\{\begin{aligned} &i\partial_t\psi(t,x)=-\tfrac{1}{2}\Delta_x\psi(t,x)+(V\star_x|\psi|^2)\psi(t,x),\quad x\in\mathbf{R}^d,\\ &\psi\big|_{t=0}=\psi^{in}.\end{aligned}\right.$$

Consider the quantity

$$\mathscr{E}_N(t) := \text{trace}(D_{N:1}(t)(I - |\psi(t,\cdot)\rangle\langle\psi(t,\cdot)|)).$$

Pickl's derivation of the mean field limit is based on proving that

$$\mathscr{E}_N(t) \to 0 \quad \text{as } N \to \infty.$$

This estimate implies that $D_{N:1}(t)$ converges to $|\psi(t,\cdot)\rangle\langle\psi(t,\cdot)|$ in operator norm, in view of the lemma below.

Henceforth we denote by $\mathfrak{H}_N$ the N-particle Hilbert space $\mathfrak{H}_N := L^2((\mathbf{R}^d)^N)$ for each $N \geq 1$, and we set

$$\rho(t) := |\psi(t,\cdot)\rangle\langle\psi(t,\cdot)|.$$

Lemma 1.12.1 *Let $N \geq 1$ and let Ψ_N be a symmetric element of $\mathfrak{H}_N$ satisfying the condition $\|\Psi_N\|_{\mathfrak{H}_N} = 1$. Let $D_N := |\Psi_N\rangle\langle\Psi_N|$ and let $\psi \in \mathfrak{H}_1$ satisfy $\|\psi\|_{\mathfrak{H}_1} = 1$. Then*

$$\text{trace}(D_{N:1}(I - |\psi\rangle\langle\psi|)) \to 0 \text{ as } N \to \infty$$

if and only if

$$D_{N:1} \to |\psi\rangle\langle\psi| \text{ in operator norm as } N \to \infty.$$

This elementary result is statement (a) in Lemma 2.3 of [87]. In some sense, it can be considered as a variant of the exercise at the end of Sect. 1.9. Notice that $\Psi_N(t,\cdot)$ is a symmetric function of $x_1,\ldots,x_N$ for all $t \in \mathbf{R}$ by Lemma 1.8.1, and because of the choice of the initial data $\Psi_N\big|_{t=0} = (\psi^{in})^{\otimes N}$, which is itself a symmetric function of $x_1,\ldots,x_N$.

Next we seek to control the evolution of $\mathscr{E}_N(t)$. We recall that, for $k,l = 1,\ldots,N$, the notation V_{kl} designates the operator on $\mathfrak{H}_N$ defined as

$$(V_{kl}\Phi_N)(x_1,\ldots,x_N) := V(x_k - x_l)\Phi_N(x_1,\ldots,x_N).$$

Since $\text{trace}(D_{N:1}(t)) = 1$, one has

$$\begin{aligned}
\dot{\mathscr{E}}_N(t) &= -\frac{d}{dt}\text{trace}(D_{N:1}(t)\rho(t)) \\
&= -\text{trace}(\dot{D}_{N:1}(t)\rho(t)) - \text{trace}(D_{N:1}(t)\dot{\rho}(t)) \\
&= -\text{trace}(i[\tfrac{1}{2}\Delta, D_{N:1}(t)]\rho(t)) - \text{trace}(iD_{N:1}(t)[\tfrac{1}{2}\Delta, \rho(t)]) \\
&+ \frac{N-1}{N}\text{trace}(i[V_{12}, D_{N:2}(t)]_{:1}\rho(t)) + \text{trace}(iD_{N:1}(t)[V_{12}, \rho(t)^{\otimes 2}]_{:1}).
\end{aligned}$$

First

$$[\tfrac{1}{2}\Delta, D_{N:1}(t)]\rho(t) + D_{N:1}(t)[\tfrac{1}{2}\Delta, \rho(t)] = [\tfrac{1}{2}\Delta, D_{N:1}(t)\rho(t)],$$

so that

$$\operatorname{trace}([\tfrac{1}{2}\Delta, D_{N:1}(t)]\rho(t)) + \operatorname{trace}(D_{N:1}(t)[\tfrac{1}{2}\Delta, \rho(t)])$$
$$= \operatorname{trace}([\tfrac{1}{2}\Delta, D_{N:1}(t)\rho(t)]) = 0.$$

Next

$$\operatorname{trace}([V_{12}, D_{N:2}(t)]_{:1}\rho(t)) = \operatorname{trace}([V_{12}, D_{N:2}(t)](\rho(t)\otimes I)),$$
$$\operatorname{trace}(D_{N:1}(t)[V_{12}, \rho(t)^{\otimes 2}]_{:1}) = \operatorname{trace}(D_{N:2}(t)([V_{12}, \rho(t)^{\otimes 2}]_{:1}\otimes I)).$$

Therefore

$$\dot{\mathscr{E}}_N(t) = \tfrac{N-1}{N}\operatorname{trace}(i[V_{12}, D_{N:2}(t)]\rho(t)\otimes I) + \operatorname{trace}(iD_{N:2}(t)[V_{12}, \rho(t)^{\otimes 2}]_{:1})$$
$$= -i\operatorname{trace}(D_{N:2}(t)[\tfrac{N-1}{N}V_{12} - (V\star|\psi(t,\cdot)|^2)\otimes I, \rho(t)\otimes I]).$$

The result of this computation is summarized in the following lemma.

Lemma 1.12.2 *With the same notation as above, one has*

$$\dot{\mathscr{E}}_N(t) = \mathscr{F}_N(t)$$

where

$$\mathscr{F}_N(t) := 2\operatorname{Im}\operatorname{trace}\,(D_{N:2}(t)(\rho(t)\otimes I)A((I-\rho(t))\otimes I))$$

and

$$A := \tfrac{N-1}{N}V_{12} - (V\star|\psi(t,\cdot)|^2)\otimes I.$$

The core of Pickl's argument is the following estimate for $\mathscr{F}_N(t)$ in terms of $\mathscr{E}_N(t)$ and N.

Proposition 1.12.1 *Assuming that $r \geq 1$ and denoting by $r' = \frac{r}{r-1}$ the dual Hölder exponent of r, one has*

$$|\mathscr{F}_N(t)| \leq 10\|V\|_{L^{2r}}\|\psi(t,\cdot)\|_{L^{2r'}}(\mathscr{E}_N(t) + \tfrac{1}{N}).$$

The proof of Proposition 1.12.1 is sketched below. The interested reader is referred to the original paper [87] for more details.

For the sake of notational simplicity, we set

$$p(t) := I - \rho(t).$$

Elementary computations show that

$$\begin{aligned}\mathscr{F}_N(t) &= 2\,\mathrm{Im}\,\mathrm{trace}(D_{N:2}(t)\rho(t)^{\otimes 2}A(p(t)\otimes\rho(t)))\\ &+ 2\,\mathrm{Im}\,\mathrm{trace}(D_{N:2}(t)\rho(t)^{\otimes 2}Ap(t)^{\otimes 2})\\ &+ 2\,\mathrm{Im}\,\mathrm{trace}(D_{N:2}(t)(\rho(t)\otimes p(t))Ap(t)^{\otimes 2}).\end{aligned}$$

The term

$$\mathrm{Im}\,\mathrm{trace}(D_{N:2}(t)(\rho(t)\otimes p(t))A(p(t)\otimes\rho(t)))$$

vanishes identically by symmetry of $\Psi_N(t,\cdot)$.

The first term on the right hand side is mastered as follows: observe that

$$(I\otimes\rho(t))A(I\otimes\rho(t)) = -\tfrac{1}{N}(V\star|\psi(t,\cdot)|^2)\otimes\rho(t)$$

—where $V\star|\psi(t,\cdot)|^2$ designates the operator on $\mathfrak{H}_1 = L^2(\mathbf{R}^d)$ defined by

$$\phi\mapsto (V\star|\psi(t,\cdot)|^2)\phi.$$

Therefore

$$\begin{aligned}&|\,\mathrm{trace}(D_{N:2}(t)\rho(t)^{\otimes 2}A(p(t)\otimes\rho(t)))|\\ &= |\,\mathrm{trace}(D_{N:2}(t)(\rho(t)\otimes I)(I\otimes\rho(t))A(I\otimes\rho(t))(p(t)\otimes I))|\\ &\le \|(\rho(t)\otimes I)(I\otimes\rho(t))A(I\otimes\rho(t))(p(t)\otimes I)\|\,\|D_{N:2}\|_1\\ &\le \|(\rho(t)\otimes I)(I\otimes\rho(t))A(I\otimes\rho(t))\|.\end{aligned}$$

Then, observing that A is self-adjoint since V is a real-valued, even function, one has

$$\begin{aligned}&\|(\rho(t)\otimes I)(I\otimes\rho(t))A(I\otimes\rho(t))\|^2\\ &= \|(\rho(t)\otimes I)((I\otimes\rho(t))A(I\otimes\rho(t)))^2(\rho(t)\otimes I)\|\\ &= \frac{1}{N^2}\|\langle\psi(t,\cdot)|(V\star|\psi(t,\cdot)|^2)^2\psi(t,\cdot)\rangle\rho(t)\otimes\rho(t)\|\\ &\le \frac{1}{N^2}\int_{\mathbf{R}^d}(V\star|\psi(t,\cdot)|^2)^2(x)|\psi(t,x)|^2dx\\ &\le \frac{1}{N^2}\|(V\star|\psi(t,\cdot)|^2)^2\|_{L^r}\|\psi(t,\cdot)^2\|_{L^{r'}}\\ &= \frac{1}{N^2}\|V\star|\psi(t,\cdot)|^2\|^2_{L^{2r}}\|\psi(t,\cdot)\|^2_{L^{2r'}}\\ &\le \frac{1}{N^2}\|V\|^2_{L^{2r}}\|\psi(t,\cdot)\|_{L^{2r'}}.\end{aligned}$$

Indeed,

$$\|V\star|\psi(t,\cdot)|^2\|_{L^{2r}}\le\|V\|_{L^{2r}}\||\psi(t,\cdot)|^2\|_{L^1}=\|V\|_{L^{2r}}$$

since the solution of Hartree's equation satisfies $\|\psi(t,\cdot)\|_{L^2}=1$ for all $t\in\mathbf{R}$.

The third term satisfies

$$\begin{aligned}&|\operatorname{trace}(D_{N:2}(t)(\rho(t)\otimes p(t))Ap(t)^{\otimes 2})|\\ &\le\|(\rho(t)\otimes I)A(p(t)\otimes I)\|\|(I\otimes p(t))D_{N:2}(t)(I\otimes p(t))\|_1\\ &=\|(\rho(t)\otimes I)A(p(t)\otimes I)\|\operatorname{trace}((I\otimes p(t))D_{N:2}(t)(I\otimes p(t)))\\ &\le\|(\rho(t)\otimes I)A\|\operatorname{trace}(D_{N:1}p(t))\\ &=\|(\rho(t)\otimes I)A\|\mathscr{E}_N(t).\end{aligned}$$

Now

$$\begin{aligned}\|(\rho(t)\otimes I)A\|^2&=\|(\rho(t)\otimes I)A^2(\rho(t)\otimes I)\|\\ &\le\left(2\|V^2\|_{L^r}+2\|(V\star|\psi(t,\cdot)|^2)^2\|_{L^r}\right)\||\psi(t,\cdot)|^2\|_{L^{r'}}\\ &=\left(2\|V\|_{L^{2r}}^2+2\|V\star|\psi(t,\cdot)|^2\|_{L^{2r}}^2\right)\|\psi(t,\cdot)\|_{L^{2r'}}^2\\ &\le 4\|V\|_{L^{2r}}^2\|\psi(t,\cdot)\|_{L^{2r'}}^2,\end{aligned}$$

again because the solution ψ of Hartree's equation satisfies $\|\psi(t,\cdot)\|_{L^2}=1$ for all $t\in\mathbf{R}$.

Controlling the second term in $\mathscr{F}_N(t)$ is slightly more complicated.

We shall need the following elements of notation. First we denote by $p(t)$ the projection $p(t)=I_1-\rho(t)=I_1-|\psi(t,\cdot)\rangle\langle\psi(t,\cdot)|$.

For each $j=1,\dots,N$, we set $I_j:=I_{\mathfrak{H}_j}$, and for each $T\in\mathscr{L}(\mathfrak{H}_1)$, we define $T_{j,N}\in\mathscr{L}(\mathfrak{H}_N)$ for $j=1,\dots,N$ as

$$T_{j,N}:=I_{j-1}\otimes T\otimes I_{N-j}.$$

For each $a\in\{0,1\}^{\{1,\dots,N\}}$ and each $\pi\in\mathscr{L}(\mathfrak{H}_1)$ such that $\pi^*=\pi=\pi^2$, we define

$$P_N[a,\pi]:=\prod_{j=1}^N\pi_{j,N}^{1-a(j)}(I-\pi_{j,N})^{a(j)}.$$

Lemma 1.12.3 *Let*

$$M_N(t)=\frac{1}{N}\sum_{j=1}^N p_{j,N}(t).$$

Then

$$M_N(t)=\sum_{k=1}^N\frac{k}{N}\Pi_{k,N}[\rho(t)],$$

where

$$\Pi_{k,N}[\rho(t)] := \sum_{\substack{a\in\{0,1\}^{\{1,\ldots,N\}} \\ a(1)+\ldots+a(N)=k}} P_N[a,\rho(t)].$$

Besides

$$\Pi_{j,N}[\rho(t)]^* = \Pi_{j,N}[\rho(t)], \quad \textit{and } \Pi_{j,N}[\rho(t)]\Pi_{k,N}[\rho(t)] = \delta_{jk}\Pi_{j,N}[\rho(t)]$$

for each $j, k = 1, \ldots, N$, *and*

$$\sum_{k=0}^{N} \Pi_{k,N}[\rho(t)] = I_{\mathfrak{H}_N}.$$

In other words, the relation

$$M_N(t) = \sum_{k=0}^{N} \frac{k}{N} \Pi_{k,N}[\rho(t)]$$

is the spectral decomposition of the self-adjoint element $M_N(t)$ of $\mathscr{L}(\mathfrak{H}_N)$. These elementary computations are summarized in formula (6) of [87].

Define, for each $\alpha > 0$,

$$M_N^{\alpha}(t) = \sum_{k=1}^{N} \left(\frac{k}{N}\right)^{\alpha} \Pi_{k,N}[\rho(t)]$$

and

$$M_N^{-\alpha}(t) := \sum_{k=1}^{N} \left(\frac{k}{N}\right)^{-\alpha} \Pi_{k,N}[\rho(t)],$$

that is the pseudo-inverse of $M_N^{\alpha}(t)$ extended by 0 on $\mathrm{Ker}(M_N(t))$. In other words

$$M_N^{\alpha}(t)M_N^{-\alpha}(t) = I_N - \Pi_{0,N}[\rho(t)]$$

In particular

$$M_N^{\alpha}(t)M_N^{-\alpha}(t)p_{j,N}(t) = p_{j,N}(t)$$

for each $j = 1, \ldots, N$.

After this observation, the term that remains to be estimated, i.e.

$$\mathrm{trace}(D_{N:2}(t)\rho(t)^{\otimes 2}A(I-\rho(t))^{\otimes 2})$$

is recast as

$$\operatorname{trace}(D_N(t)\rho_{1,N}(t)\rho_{2,N}(t)AM_N^{1/2}(t)M_N^{-1/2}(t)p_{1,N}(t)p_{2,N}(t)).$$

Then

$$\begin{aligned}|\operatorname{trace}(D_N(t)\rho_{1,N}(t)\rho_{2,N}(t)AM_N^{1/2}(t)M_N^{-1/2}(t)p_{1,N}(t)p_{2,N}(t))|\\ \le \|M_N^{-1/2}(t)p_{1,N}(t)p_{2,N}(t)\Psi_N(t,\cdot))\|_{\mathfrak{H}_N}\\ \times\|M_N^{1/2}(t)A\rho_{1,N}(t)\rho_{2,N}(t)\Psi_N(t,\cdot))\|_{\mathfrak{H}_N}.\end{aligned}$$

The second factor on the right hand side is mastered by observing that

$$\begin{aligned}&\|M_N^{1/2}(t)A\rho_{1,N}(t)\rho_{2,N}(t)\Psi_N(t,\cdot))\|^2_{\mathfrak{H}_N}\\ &=\langle\Psi_N(t,\cdot)|\rho_{1,N}(t)\rho_{2,N}(t)AM_N(t)A\rho_{1,N}(t)\rho_{2,N}(t)\Psi_N(t,\cdot))\rangle\\ &=\tfrac{N-2}{N}\langle\Psi_N(t,\cdot)|\rho_{1,N}(t)\rho_{2,N}(t)Ap_{3,N}(t)A\rho_{1,N}(t)\rho_{2,N}(t)\Psi_N(t,\cdot))\rangle\\ &\quad+\tfrac{2}{N}\langle\Psi_N(t,\cdot)|\rho_{1,N}(t)\rho_{2,N}(t)Ap_{1,N}(t)A\rho_{1,N}(t)\rho_{2,N}(t)\Psi_N(t,\cdot))\rangle\end{aligned}$$

since $\Psi_N(t,\cdot)$ is symmetric. The first summand is bounded by

$$\begin{aligned}\tfrac{N-2}{N}\|\rho_{1,N}(t)A^2\rho_{1,N}(t)\|_{\mathscr{L}(\mathfrak{H}_N)}\|p_{3,N}(t)\Psi_N(t)\|^2_{\mathfrak{H}_N}\\ \le\|\rho_{1,N}(t)A^2\rho_{1,N}(t)\|_{\mathscr{L}(\mathfrak{H}_N)}\mathscr{E}_N(t)\\ \le 4\|V\|^2_{L^{2r}}\|\psi(t,\cdot)\|^2_{L^{2r'}}\mathscr{E}_N(t),\end{aligned}$$

while the second is bounded by

$$\tfrac{2}{N}\|\rho_{1,N}(t)A^2\rho_{1,N}(t)\|_{\mathscr{L}(\mathfrak{H}_N)}\le\tfrac{8}{N}\|V\|^2_{L^{2r}}\|\psi(t,\cdot)\|^2_{L^{2r'}}.$$

Finally, we use again the fact that $\Psi_N(t,\cdot)$ is symmetric to conclude that

$$\begin{aligned}N(N-1)\|M_N^{-1/2}(t)p_{1,N}(t)p_{2,N}(t)\Psi_N(t,\cdot))\|^2_{\mathfrak{H}_N}\\ =N(N-1)\langle\Psi_N(t,\cdot),M_N^{-1}(t)p_{1,N}(t)p_{2,N}(t)\Psi_N(t,\cdot)\rangle\\ =2\sum_{1\le j<k\le N}\langle\Psi_N(t,\cdot),M_N^{-1}(t)p_{j,N}(t)p_{k,N}(t)\Psi_N(t,\cdot)\rangle\\ \le N^2\langle\Psi_N(t,\cdot),M_N^{-1}(t)M_N^2(t)\Psi_N(t,\cdot)\rangle\\ =N^2\langle\Psi_N(t,\cdot),M_N(t)\Psi_N(t,\cdot)\rangle\\ =N^2\mathscr{E}_N(t).\end{aligned}$$

Putting all these estimates together establishes the bound in Proposition 1.12.1.

Applying Gronwall's estimate, we deduce from Proposition 1.12.1 the following result, stated in space dimension 3 for the sake of simplicity. The assumptions on the potential V have been adapted to match those in Kato's Theorem 1.8.1.

Theorem 1.12.2 *Let V be a real-valued even function such that*

$$V \in L^{2r}(\mathbf{R}^3) \quad \text{and } V\big|_{\mathbf{R}^3\setminus B(0,R)} \in L^\infty(\mathbf{R}^3 \setminus B(0,R))$$

for some $r \geq 1$ and some $R > 0$.

Let $\psi^{in} \in L^2(\mathbf{R}^3)$ be such that $\|\psi^{in}\|_{L^2} = 1$, and assume that the Cauchy problem for Hartree's equation with initial data ψ^{in} has a unique solution $\psi \in C(\mathbf{R}; L^{2r'}(\mathbf{R}^3))$, where $r' = \frac{r}{r-1}$ is the dual Hölder exponent of r. Let Ψ_N be the solution of the Cauchy problem for the N-particle Schrödinger equation with initial data $(\psi^{in})^{\otimes N}$, and $D_N(t) := |\Psi_N(t,\cdot)\rangle\langle\Psi_N(t,\cdot)|$.

Then, for each $t \in \mathbf{R}$, one has

$$D_{N:1}(t) \to |\psi(t,\cdot)\rangle\langle\psi(t,\cdot)| \text{ in operator norm as } N \to +\infty,$$

with the estimate

$$\begin{aligned}&\operatorname{trace}(D_{N:1}(t)(I - |\psi(t,\cdot)\rangle\langle\psi(t,\cdot)|))\\ &\quad \leq \frac{1}{N}\left(\exp\left|\int_0^t 10\|V\|_{L^{2r}}\|\psi(s,\cdot)\|_{L^{2r'}}ds\right| - 1\right).\end{aligned}$$

Some remarks are in order after this rather remarkable result.

The first obvious advantage of Pickl's argument is its simplicity. The proof of Proposition 1.12.1 and the resulting Theorem 1.12.2 provide a much simpler derivation of the mean field limit for potentials with a Coulomb singularity at the origin than the Erdös-Yau proof [32] based on the BBGKY hierarchy. According to Theorem 3.1 in [39], if $\psi^{in} \in H^k(\mathbf{R}^3)$, then the unique mild solution ψ of Hartree's equation belongs to $C(\mathbf{R}_+; H^k(\mathbf{R}^3))$ under the same assumption on V as in Theorem 1.12.2. Therefore $\psi \in C(\mathbf{R}; L^{2r'}(\mathbf{R}^3))$ by Sobolev's embedding theorem provided that k is chosen large enough. In fact, even stronger singularities at the origin than Coulomb in the interaction potential can be handled with this method.

A second advantage of Pickl's method is its versatility. For instance, it can also be applied to the derivation of the Gross-Pitaevski equation: see [88]. In this latter reference, Pickl uses appropriate nonlinear functions of the operator $M_N(t)$, which are defined in terms of the spectral decomposition provided by Lemma 1.12.3. The same type of method also applies to study the convergence rate in the mean field limit [67] in a more general and systematic way than in Theorem 1.12.2.

On the other hand, unlike methods based on the BBGKY hierarchy, Pickl's argument [87] seems to be limited to pure states and to bosonic particles so far. Therefore, in spite of its appealing simplicity, Pickl's method does not really supersedes all previous derivations of the mean field limits in quantum mechanics.

Besides, even if BBGKY hierarchies are a rather cumbersome mathematical object, they are ubiquitous in the field of nonequilibrium statistical physics, and appear in the treatment of quantum as well as classical mechanical models. This accounts for the importance of these hierarchies in these lectures.

1.13 Afterword

These notes are an updated and (much) expanded version of my earlier (2003) lectures [43]. I express my gratitude to my colleagues A. Muntean, J. Rademacher and A. Zagaris for their kind invitation to the NDNS+ 2012 Summer School at Universiteit Twente in Enschede. The first part of these notes (Sections 1–6) were also presented in a joint course with L. Desvillettes at the 37th Summer School on Mathematical Physics in Ravello (2012). I am also very grateful to Profs. T. Ruggeri and S. Rionero for the most enjoyable hospitality of the Ravello school. I also wish to thank the participants, especially L. Oparnica, M. Pavic, M. Renger and V. Ricci for their positive feedback.

As in [43], the core of the material presented here is the mean field limit of N-particle systems governed by the equations of classical mechanics in the case where the N-particle empirical measure is an exact weak solution of the mean field PDE, and the mean field limit of the N-particle Schrödinger equation.

Unlike in [43], I have chosen to introduce both the approach based on empirical measures and the method based on the BBGKY hierarchy on the mean field limit of the N-particle systems in classical mechanics with Lipschitz continuous interaction kernel K. The discussion of the propagation of chaos in Sect. 1.6 is much more detailed here than in [43]. The connection between the empirical measure and the hierarchy approach is hopefully clarified by the discussion in Sects. 1.6.2, 1.7.1 and 1.7.3.

The most recent developments—in particular results obtained since the publication of [43]—are surveyed in Sects. 1.7 and 1.12. However, I have chosen to give less many details on the derivation of the time-dependent Hartree-Fock equations from the N-particle Schrödinger equation. The discussion of the time-dependent Hartree-Fock equations in the present notes is limited to a presentation of the equations and to the statement of the main result obtained on this problem. My earlier notes [43] contain a sketch of the proof of the mean field limit leading to the time-dependent Hartree-Fock equations; the interested reader is referred to [43] for more details in this direction.

I have limited myself to a discussion of evolution problems. Yet there are many important results bearing on steady problems: see for instance the work of Caglioti-Lions-Marchioro-Pulvirenti [23] and Kießling [65] on the Onsager statistical theory of vortices in incompressible fluids. I have also neglected to discuss the similarities and the differences between the mean field limit of large particles systems and the convergence of particle methods for mean field equations. The interested reader will find information on these issues in Chap. 5 of [76] in the context of the vortex formulation of the Euler equations for two dimensional incompressible fluids. See also, in the context of Vlasov type equations, [13, 28, 105] and the references therein.

I have also deliberately chosen to avoid discussing results that would involve both the mean field limit (as the particle number N tends to infinity) and the classical limit of quantum mechanics (assuming that the typical action of an individual particle is large compared to Planck's constant $\hbar$). One of the reasons for this choice is that this part of the theory is technically more involved and perhaps not as complete as what has been presented above. There are many important contributions of in this directions, and the interested reader should have a look at [55, 80], to quote only a few references exploring that topic. Nevertheless, there remain some outstanding open questions in this direction. For instance, even for smooth potentials, it is yet unknown at the time of this writing[21] whether the mean field limit of the N-body Schrödinger equation leading to Hartree's equation is "uniform in the Planck constant $\hbar$" (viewing $\hbar$ as a vanishingly small parameter, as is customary in all mathematical discussions of the classical limit of quantum mechanics[22]).

Another conspicuous omission in the second part of these notes is the formalism of Fock spaces and quantum field theory. This mathematical setting allows considering data where the particle number can be infinite—more precisely, the particle number N is an unbounded operator in this formalism. The mean field limit can be formulated in this setting: see for instance [9, 35, 36], together with the papers [40, 55] already mentioned above, involving the classical limit in addition to the case of infinitely many particles. Interestingly, there is a striking analogy between the classical limit of quantum mechanics formulated in terms of Egorov's theorem (Theorem 25.3.5 in [59]) and the point of view introduced in [35, 36] for the quantum mean field limit: see [35] on p. 1024.

I am indebted in various ways to several colleagues for the material presented in these notes. First I wish to thank my collaborators R. Adami, C. Bardos, B. Ducomet, L. Erdös, A. Gottlieb, N. Mauser, V. Ricci, A. Teta and H.-T. Yau, with whom I have worked on mean field limits on various occasions. Besides, the material presented here is in (many) places far from original, and the present notes owe much to earlier presentations of mean field limits—such as [100] or unpublished notes by C. Villani, as well as lectures by P.-L. Lions at the Collège de France. I also learned a lot from a series of lectures by C. Mouhot presenting his joint work with S. Mischler and B. Wennberg [78, 79] on quantitative estimates for the propagation of chaos. Most of what I know on the derivation of Vlasov type equations from classical particle dynamics comes from discussions with J. Batt, H. Neunzert, J. Wick on various occasions in the late 80s. I learned of Dobrushin's beautiful estimate [29] from a lecture given by M. Pulvirenti at Ecole Normale Supérieure in 1997.

Another key idea in these lectures is the use of the Nirenberg-Ovcyannikov abstract version of the Cauchy-Kovalevska theorem to handle BBGKY type hierarchies. I got acquainted with this idea in a seminar given by S. Ukai at the seminar of Prof. Y.

[21] This problem was mentioned to me by M. Pulvirenti in October 2013.

[22] After completion of these lecture notes, there has been some progress on this problem: see [45, 47]. Earlier results on this problem can be also found in [50] and in [85].

Sone in 1998 in Kyoto. Ukai used this technique in a slightly different context (the derivation of the Boltzmann equation from the dynamics of a large number of hard spheres in the Boltzmann-Grad limit), and for a slightly different purpose (he wanted to simplify the uniform stability estimates in Lanford's argument). With his usual modesty, Ukai presented his result as a failed attempt at simplifying Lanford's notoriously intricate proof, and did not immediately published it—I came across his paper [101] only very recently. In the case of the infinite mean field hierarchies for quantum N-particle dynamics, I find the argument based on the abstract variant the Cauchy-Kovalevska theorem much more elegant and illuminating than the conventional argument based on the Duhamel series as in Lanford's proof, and thus I have chosen to present the former in these notes, as in [43]. The argument based on the Duhamel series can be found in many places—see for instance the proof of Theorem 5.7 on p. 610 in [96].

S. Ukai was at the origin of several fundamental results on the mathematical analysis of kinetic models—such as the existence and uniqueness of global solutions of the Boltzmann equation for all initial data near Maxwellian equilibrium, the existence and uniqueness of global solutions of the Vlasov-Poisson system [102] in space dimension 2, the hydrodynamic limit of the Boltzmann equation leading to the incompressible Navier-Stokes equations in the regime of "small" initial data. In view of his influence on some of the topics discussed above, these notes are dedicated to his memory.

References

1. Adami, R., Bardos, C., Golse, F., Teta, A.: Towards a rigorous derivation of the cubic NLSE in dimension one. Asymptot. Anal. **40**, 93–108 (2004)
2. Adami, R., Golse, F., Teta, A.: Rigorous derivation of the cubic NLS in dimension one. J. Stat. Phys. **127**, 1193–1220 (2007)
3. Ahlfors, L.V.: Complex analysis. An introduction to the theory of analytic functions of onecomplex variable, 3rd edn. International Series in Pure and Applied Mathematics. McGraw-Hill Book Co., New York (1978)
4. Bardos, C., Catto, I., Mauser, N., Trabelsi, S.: Setting and analysis of the multi-configuration time-dependent Hartree-Fock equations. Arch. Ration. Mech. Anal. **198**, 273–330 (2010)
5. Bardos, C., Ducomet, B., Golse, F., Gottlieb, A., Mauser, N.: The TDHF approximation for Hamiltonians with m-particle interaction potentials. Commun. Math. Sci. suppl. **1**, 1–9 (2007)
6. Bardos, C., Erdös, L., Golse, F., Mauser, N., Yau, H.-T.: Derivation of the Schrödinger-Poisson equation from the quantum N-body problem. C. R. Math. Acad. Sci. Paris **334**, 515–520 (2002)
7. Bardos, C., Golse, F., Mauser, N.: Weak coupling limit of the N particles Schrödinger equation. Methods Appl. Anal. **7**, 275–293 (2000)
8. Bardos, C., Golse, F., Gottlieb, A., Mauser, N.: Mean field dynamics of fermions and the time-dependent Hartree-Fock equation. J. de Math. Pures et Appl. **82**, 665–683 (2003)
9. Bardos, C., Golse, F., Gottlieb, A., Mauser, N.: Accuracy of the time-dependent Hartree-Fock approximation for uncorrelated initial states. J. Stat. Phys. **115**, 1037–1055 (2004)
10. Bardos, C., Golse, F., Gottlieb, A., Mauser, N.: On the derivation of nonlinear Schrödinger and Vlasov equations. In: Dispersive Transport Equations and Multiscale Models. Minneapolis, MN (2000), pp. 1–23. IMA Vol. Math. Appl. 136, Springer, New York, NY (2004)

11. Bardos, C., Mauser, N.: One particle equations for many particle quantum systems: the MCTHDF method. Quart. Appl. Math. **68**, 43–59 (2010)
12. Basdevant, J.-L., Dalibard, J.: Quant. Mech. Springer, Berlin (2005)
13. Batt, J.: N-particle approximation to the nonlinear vlasov-poisson system. Nonlin. Anal. **47**, 1445–1456 (2000)
14. Billingsley, J.: Convergence of Probability Measures, 2nd edn. Wiley, New York (1999)
15. Bouchut, F., Golse, F., Pallard, C.: On classical solutions of the 3D Vlasov-Maxwell system: a simplified proof of the Glassey-Strauss theorem. Arch. Rational Mech. Anal. **170**, 1–15 (2003)
16. Bouchut, F., Golse, F., Pallard, C.: Nonresonant smoothing for coupled wave + transport equations; applications to the Vlasov-Maxwell system. Rev. Mat. Iberoamericana **20**, 865–892 (2004)
17. Bouchut, F., Golse, F., Pulvirenti, M.: Kinetic equations and asymptotic theory. Edited and with a foreword by L. Desvillettes et B. Perthame. Gauthier-Villars, Editions Scientifiques et Médicales Elsevier, Paris (2000)
18. Bove, A., DaPrato, G., Fano, G.: An existence proof for the Hartree-Fock time-dependent problem with bounded two-body interaction. Commun. Math. Phys. **37**, 183–191 (1974)
19. Bove, A., DaPrato, G., Fano, G.: On the Hartree-Fock time-dependent problem. Commun. Math. Phys. **49**, 25–33 (1976)
20. Braun, W., Hepp, K.: The Vlasov dynamics and its fluctuations in thc $1/N$ limit of interacting classical particles. Commun. Math. Phys. **56**, 101–113 (1977)
21. Breiman, L.: Probability. Addison Wesley, Reading (1968)
22. Brezis, H.: Analyse fonctionnelle. Théorie et Applications. Masson, Paris (1987)
23. Caglioti, E., Lions, P.-L., Marchioro, C., Pulvirenti, M.: A special class of flows for two-dimensional Euler equations: a statistical mechanics description. Commun. Math. Phys. **143**, 501–525 (1992)
24. Cancès, E., Le Bris, C.: On the time-dependent Hartree-Fock equations coupled with a classical nuclear dynamics. Math. Models Methods Appl. Sci. **9**, 963–990 (1999)
25. Cercignani, C.: On the Boltzmann equation for rigid spheres. Transport Theory Statist. Phys. **2**, 211–225 (1972)
26. Cercignani, C., Illner, R., Pulvirenti, M.: The mathematical theory of dilute gases. Appl. Math. Sci. **106** Springer-Verlag, New York, NY (1994)
27. Chadam, J.M., Glassey, R.T.: Global existence of solutions to the Cauchy problem for time-dependent Hartree equations. J. Math. Phys. **16**, 1122–1130 (1975)
28. Cottet, G.-H., Raviart, P.-A.: On particlc-in-cells methods for the Vlasov-Poisson equations. Transp. Theory Stat. Phys. **15**, 1–31 (1986)
29. Dobrushin, R.L.: Vlasov equations. Funct. Anal. Appl. **13**, 115–123 (1979)
30. Elgart, A., Erdös, L., Schlein, B., Yau, H.-T.: Gross-Pitaevskii equation as the mean field limit of weakly coupled bosons. Arch. Ration. Mech. Anal. **179**, 265–283 (2006)
31. Elskens, Y., Kießling, M.K.-H., Ricci, V.: The Vlasov limit for a system of particles which interact with a wave field. Commun. Math. Phys. **285**, 673–712 (2009)
32. Erdös, L., Yau, H.-T.: Derivation of the nonlinear Schrödinger equation from a many body Coulomb system. Adv. Theor. Math. Phys. **5**, 1169–1205 (2001)
33. Erdös, L., Schlein, B., Yau, H.-T.: Derivation of the cubic non-linear Schrödinger equation from quantum dynamics of many-body systems. Invent. Math. **167**, 515–614 (2007)
34. Erdös, L., Schlein, B., Yau, H.-T.: Derivation of the Gross-Pitaevskii equation for the dynamics of Bose-Einstein condensate. Ann. Math. **2**(172), 291–370 (2010)
35. Fröhlich, J., Knowles, A., Schwarz, S.: On the mean-field limit of bosons with Coulomb two-body interaction. Commun. Math. Phys. **288**, 1023–1059 (2009)
36. Fröhlich, J., Knowles, A.: A microscopic derivation of the time-dependent Hartree-Fock equation with Coulomb two-body interaction. J. Stat. Phys. **145**, 23–50 (2011)
37. Gallagher, I., Saint-Raymond, L., Texier, B.: From Newton to Boltzmann: hard sphere and short range potentials. Eur. Math. Soc., Zürich, Zürich Lectures in Advanced Mathematics (2013)

38. Gérard, P.: Equations de champ moyen pour la dynamique quantique d'un grand nombre de particules (d'après Bardos, Erdös, Golse, Gottlieb, Mauser, Yau). Séminaire Bourbaki. Vol. 2003/2004. Astérisque 299, Exp. no. 930, 147–164 (2005)
39. Ginibre, J., Velo, G.: On a class of nonlinear Schrödinger equations with nonlocal interactions. Math. Z. **170**, 109–145 (1980)
40. Ginibre, J., Velo, G.: The classical field limits of scattering field theory for non-relativistic many-boson systems 1 & 2. Commun. Math. Phys. **66**, 37–76 and 68, 45–68 (1979)
41. Glassey, R.T.: The Cauchy problem in kinetic theory. Society for Industrial and Applied Mathematics (SIAM), Philadelphia (1996)
42. Göll, M., Verbitskiy, E.: Homoclinic points of principal algebraic actions. This contribution
43. Golse, F.: The mean-field limit for the dynamics of large particle systems. Journées Equations aux Dérivées Partielles (Forges-les-Eaux), Exp. no. 9, Univ. de Nantes, Nantes (2003)
44. Golse, F.: The mean-field limit for a regularized Vlasov-Maxwell dynamics. Commun. Math. Phys. **310**, 789–816 (2012)
45. Golse, F., Mouhot, C., Paul, T.: On the mean field and classical limits of quantum mechanics. Commun. Math. Phys. to appear. arXiv:1502.06143
46. Golse, F., Mouhot, C., Ricci, V.: Empirical measures and mean field hierarchies. Kinet. Relat. Models **6**, 919–943 (2013)
47. Golse, F., Paul, T.: The Schrödinger equation in the mean-field and semiclassical regime. preprint arXiv:1510.06681
48. Goodman, J., Hou, T., Lowengrub, J.: Convergence of the point vortex method for the 2-D Euler equations. Comm. Pure Appl. Math. **43**, 415–430 (1990)
49. Goodman, J., Hou, T.: New stability estimates for the 2-D vortex method. Commun. Pure Appl. Math. **44**, 1015–1031 (1991)
50. Graffi, S., Martinez, A., Pulvirenti, M.: Mean-field approximation of quantum systems and classical limit. Math. Models Methods Appl. Sci. **13**, 59–73 (2003)
51. Grünbaum, F.A.: Propagation of chaos for the Boltzmann equation. Arch. Rat. Mech. Anal. **42**, 323–345 (1971)
52. Hauray, M.: Wasserstein distances for vortices approximation of Euler-type equations. Math. Models Methods Appl. Sci. **19**, 1357–1384 (2009)
53. Hauray, M., Jabin, P.-E.: N-particle approximation of the Vlasov equations with singular potential. Arch. Rational Mech. Anal. **183**, 489–524 (2007)
54. Hauray, M., Jabin, P.-E.: Propagation of chaos for particle approximations of Vlasov equations with singular forces. Ann. Sci. Ecole Normale Sup. **48**, 891–940 (2015)
55. Hepp, K.: The classical limit for quantum mechanical correlation functions. Commun. Math. Phys. **35**, 265–277 (1974)
56. Hewitt, E., Savage, L.J.: Symmetric measures on cartesian products. Trans. Amer. Math. Soc. **80**, 470–501 (1955)
57. Hörmander, L.: The analysis of linear partial differential operators. I. Distribution theory and Fourier analysis. Classics in Mathematics. Springer, Berlin (2003)
58. Hörmander, L.: The analysis of linear partial differential operators. III. Pseudo-differential operators. Classics in Mathematics. Springer, Berlin (2007)
59. Hörmander, L.: The analysis of linear partial differential operators. IV. Fourier integral operators. Classics in Mathematics. Springer, Berlin (2009)
60. Hörmander, L.: Lectures on nonlinear hyperbolic differential equations. Mathématiques and Applications, vol. 26. Springer, Berlin (1997)
61. Horowitz, J., Karandikar, R.: Mean rates of convergence of empirical measures in the Wasserstein metric. J. Comput. Appl. Math. **55**, 261–273 (1994)
62. Kac, M.: Foundations of kinetic theory. In: Proceedings of the Third Berkeley Symposium on Mathematical Statistics and Probability, 1954–1955, vol. III, pp. 171–197. University of California Press, Berkeley and Los Angeles, CA (1956)
63. Kato, T.: Fundamental properties of Hamiltonian operators of Schrödinger type. Trans. Amer. Math. Soc. **70**, 195–211 (1951)

64. Kato, T.: Perturbation theory for linear operators. Classics in Mathematics. Springer, Berlin (1995)
65. Kießling, M.K.-H.: Statistical mechanics of classical particles with logarithmic interactions. Commun. Pure Appl. Math. **46**, 27–56 (1993)
66. Klainerman, S., Machedon, M.: On the uniqueness of solutions to the Gross-Pitaevskii Hierarchy. Commun. Math. Phys. **279**, 169–185 (2008)
67. Knowles, P., Pickl, P.: Mean-field dynamics: singular potentials and rate of convergence. Commun. Math. Phys. **298**, 101–138 (2010)
68. Krylov, N.S.: Works on the Foundations of Statistical Physics. Princeton University Press, Princeton (1979)
69. Landau, L.D., Lifshitz, E.M.: Quantum mechanics: non-relativistic theory. Course of Theoretical Physics, vol. 3. Translated from the Russian by J.B. Sykes, J.S. Bell. Addison-Wesley Series in Advanced Physics. Addison-Wesley Publishing Co., Inc, Reading, MA (1958)
70. Lanford, O.E.: Time evolution of large classical systems. In: Dynamical Systems, Theory and Applications (Rencontres, Battelle Res. Inst., Seattle, Wash., 1974). Lecture Notes in Physics, vol. 38, pp. 1–111. Springer, Berlin (1975)
71. Lieb, E.H., Simon, B.: The Hartree-Fock theory for Coulomb systems. Commun. Math. Phys. **53**, 185–194 (1977)
72. Lions, P.-L.: Mean field games. Course at the Collège de France, 2007–2008. Summary of the course and video of the lectures of November 9 and 16, 2007. http://www.college-de-france.fr/site/pierre-louis-lions/
73. Lions, P.-L., Perthame, B.: Propagation of moments and regularity for the 3-dimensional Vlasov-Poisson system. Invent. Math. **105**, 415–430 (1991)
74. Loeper, G.: Uniqueness of the solution to the Vlasov-Poisson system with bounded density. J. Math. Pures Appl. **9**(86), 68–79 (2006)
75. Malliavin, P., Airault, H., Kay, L., Letac, G.: Integration and probability. Graduate Texts in Mathematics. Springer, New York (1995)
76. Marchioro, C., Pulvirenti, M.: Mathematical Theory of Incompressible Nonviscous Fluids. Springer, New York (1994)
77. Mielke, A.: On evolutionary Γ-convergence for gradient systems. This contribution
78. Mischler, S., Mouhot, C.: Kacs program in kinetic theory. Invent. Math. **193**, 1–147 (2013)
79. Mischler, S., Mouhot, C., Wennberg, B.: A new approach to quantitative propagation of chaos for drift, diffusion and jump processes. Probab. Theory Related Fields **161**, 1–59 (2015)
80. Narnhoffer, H., Sewell, G.: Vlasov hydrodynamics of a quantum mechanical model. Commun. Math. Phys. **79**, 9–24 (1981)
81. Neunzert, H., Wick, J.: Die Approximation der Lösung von Integro-Differentialgleichungen durch endliche Punktmengen. Lecture Notes in Mathematics, vol. 395, pp. 275–290. Springer, Berlin (1974)
82. Nirenberg, L.: An abstract form of the nonlinear Cauchy-Kowalewski theorem. J. Diff. Geometry **6**, 561–576 (1972)
83. Nishida, T.: A note on a theorem of Nirenberg. J. Diff. Geometry **12**, 629–633 (1977)
84. Ovcyannikov, L.V.: A nonlinear Cauchy problem in a scale of Banach spaces. (Russian). Dokl. Akad. Nauk SSSR **200**, 789–792 (1971)
85. Pezzoti, F., Pulvirenti, M.: Mean-field limit and semiclassical expansion of a quantum particle system. Ann. Henri Poincaré **10**, 145–187 (2009)
86. Pfaffelmoser, K.: Global classical soutions of the Vlasov-Poisson system in three dimensions for general initial data. J. Diff. Eq. **95**, 281–303 (1992)
87. Pickl, P.: A simple derivation of mean field limits for quantum systems. Lett. Math. Phys. **97**, 151–164 (2011)
88. Pickl, P.: Derivation of the time dependent Gross-Pitaevskii equation without positivity condition on the interaction. J. Stat. Phys. **140**, 76–89 (2010)
89. Rachev, S., Rüschendorf, L.: Mass transportation problems, vol. I. Theory. Probability and its Applications. Springer, New York (1998)

90. Rein, G.: Global weak solutions of the relativistic Vlasov-Maxwell system revisited. Commun. Math. Sci. **2**, 145–158 (2004)
91. Rein, G.: Collisionless kinetic equations from astrophysicsthe Vlasov-Poisson system. In: Handbook of differential equations: evolutionary equations, vol. III, pp. 383–476. Handb. Differ. Eqn. Elsevier/North-Holland, Amsterdam (2007)
92. Rodnianski, I., Schlein, B.: Quantum fluctuations and rate of convergence towards mean field dynamics. Commun. Math. Phys. **291**, 31–61 (2009)
93. Scardia, L.: Continuum limits of discrete models by Γ-convergence. This contribution
94. Schochet, S.: The point-vortex method for periodic weak solutions of the 2-D Euler equations. Commun. Pure Appl. Math. **49**, 911–965 (1996)
95. Shatah, J., Struwe, M.: Geometric wave equations. Courant Lecture Notes in Mathematics, 2. American Mathematical Society, Providence, RI (1998)
96. Spohn, H.: Kinetic equations from hamiltonian dynamics. Rev. Mod. Phys. **52**, 600–640 (1980)
97. Spohn, H.: On the Vlasov hierarchy. Math. Methods Appl. Sci. **3**, 445–455 (1981)
98. Spohn, H.: Large scale dynamics of interacting particles. Springer, London (2012)
99. Stroock, D.W., Varadhan, S.R.S.: Multidimensional Diffusion Processes. Springer, Berlin (2006)
100. Sznitman, A.-S.: Topics in propagation of chaos. In: Ecole d'été de Probabilités de Saint-Flour XIX–1989, Lecture Notes in Mathematics, vol. 1464, pp. 165–251. Springer, Berlin (1991)
101. Ukai, S.: The Boltzmann-Grad limit and Cauchy-Kovalevskaya theorem. In: Recent topics in mathematics moving toward science and engineering. Japan J. Indust. Appl. Math. **18**, 383–392 (2001)
102. Ukai, S., Okabe, T.: On classical solutions in the large in time of two dimensional Vlasov's equation. Osaka J. Math. **15**, 245–261 (1978)
103. Villani, C.: Topics in Optimal Transportation. American Math. Soc, Providence RI (2003)
104. Villani, C.: Optimal Transport: Old and New. Springer, Berlin (2009)
105. Wollman, S.: On the approximation of the Vlasov-Poisson system by particle methods. SIAM J. Numer. Anal. **37**, 1369–1398 (2000)

Chapter 2
Continuum Limits of Discrete Models via Γ-Convergence

Lucia Scardia

In memory of my beloved mother Maria Rosaria Indino.

Abstract These lecture notes are a step-by-step guide to Γ-convergence and its applications to the upscaling of discrete systems. In many cases of interest—atoms, defects in metals, crowds—one has to face the challenging problem of deriving a macroscopic model to describe the *collective* behaviour of the interacting particles. Γ-convergence is a mathematically rigorous approach to the upscaling, and has been successfully applied to several problems in material science. The focus of this contribution is on one-dimensional chains of particles, and the starting point is the interaction energy of the system. The nature of the interaction depends on the application and we will consider the case of convex and non convex potentials and both short and long-range interactions.

Keywords Γ-Convergence · Discrete-to-continuum upscaling · Lennard-Jones · Dislocations

2.1 Introduction

Systems of interacting particles are ubiquitous: crowds of people, birds flocking, fish swarming, atoms, dislocations in metals, vortices in superconductors are examples of discrete collections of particles—different in nature and in scale—which aggregate and have an interesting collective behaviour (Fig. 2.1).

L. Scardia (✉)
Department of Mathematical Sciences, University of Bath, Bath, UK
e-mail: l.scardia@bath.ac.uk

A. Muntean et al. (eds.), *Macroscopic and Large Scale Phenomena: Coarse Graining, Mean Field Limits and Ergodicity*, Lecture Notes in Applied Mathematics and Mechanics 3, DOI 10.1007/978-3-319-26883-5_2

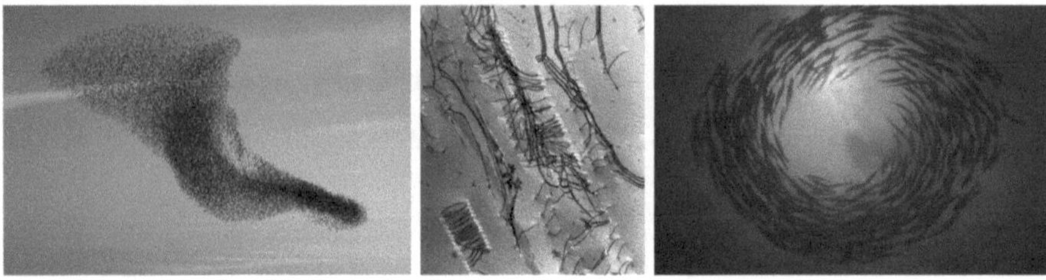

Fig. 2.1 Examples of interacting discrete systems: birds flocking, dislocations and fish swarming

Dislocations, in particular, are defects at the microscopic scale which collectively, at the macroscopic scale, determine how metals deform permanently. To understand this collective behaviour one has to bridge the scales between a microscopic, individual-based description of the system, and a macroscopic, group-led one. In other words, one has to perform a discrete-to-continuum *upscaling*, also referred to as *coarse-graining*, or *averaging* in different disciplines.

Upscaling is a *selection* mechanism in the zoo of available macroscopic theories, often derived phenomenologically and poorly justified. Since the interaction forces between particles at the discrete level are generally well (or better) understood, deriving a macroscopic model from discrete interactions is the best way to ensure that the macroscopic models are physically sound. Over the years, and in different communities, several approaches have been used, e.g. formal expansions, matched asymptotics [12, 17, 19], and statistical mechanics approaches similar to the derivation of the BBGKY hierarchy explained in Chap. 1 [15, 16]. So, what *method* should be used for the upscaling?

A mathematically rigorous way to bridge the scales is via Γ-convergence, a converge concept for the energies of the system. The Γ-convergence approach has many advantages. In particular, it ensures that minimisers of the discrete problem are *close* to minimisers of the limit problem; hence the limit model is a good approximation of the original one, as far as minimisers are concerned. This property makes this approach very well-suited for tackling problems in materials science. Fracture mechanics [6, 9, 10, 18, 23], spin systems [1, 2], magnetomechanics [24] and dislocations [14, 22, 25] are some of the many examples where this approach has been successful. Moreover, although Γ-convergence is a convergence of energies and equilibrium configurations—hence of *static* problems—it can be combined with gradient flow or rate-independent techniques, as described above, in order to treat dynamics (see e.g. [5, 20, 21]).

In these lecture notes we are going to focus on the static case, and explain the basic concepts and difficulties of the rigorous upscaling by Γ-convergence for a one-dimensional chain of interacting particles, and for both short-range and long-range interactions.

In the systems we are going to consider the particles are ordered and interact via a pairwise interaction potential. This means that we are assuming that there is an underlying *reference lattice* that enforces the ordering of the particles, and in particular implies that a particle has always the same neighbours. Although this assumption has some limitations, it is reasonable for the systems we study. The nature of the potential depends on the applications. For inter-atomic interactions it is typically the Lennard-Jones potential, which blows up when the atoms collide and decays rapidly by increasing the inter-atomic distance. For Coulomb interactions, the potential can model attraction or repulsion, depending on the sign of the charges. Moreover, it can be convex (like in a linear spring model, where it is quadratic) or non-convex (like the Lennard-Jones potential).

The aim of the lecture notes is to walk through the difficulties inherent to the study of discrete problems by facing one difficulty per time. We will mainly limit ourselves to nearest-neighbour interactions. In this simplified setting we will first consider superlinear energies, for which the energy functional is naturally defined on a reflexive space, and we will analyse both the case of convex and non-convex potentials. Then we will move away from the superlinearity assumption and focus on the Lennard-Jones interaction potential. The final part of the lecture notes will be devoted to the case of more interactions (next-to-nearest neighbours and more) and to the case of dislocation interactions.

2.2 Preliminaries

In this section we collect a number of preliminaries that will be needed in the rest of the lecture notes. More precisely, we are going to define Γ-convergence, and the more technical concepts of coerciveness and relaxation.

2.2.1 Γ-Convergence

This first subsection is devoted to the definition and some properties of *Γ-convergence*. For a more comprehensive introduction to this topic the interested reader can have a look at the excellent book by Dal Maso [13].

Definition 2.2.1 Let (X, d) be a metric space, let $E_n : X \to \mathbb{R}$ be a sequence of functionals for $n \in \mathbb{N}$, and let $E : X \to \mathbb{R}$. We say that E_n Γ-converges to E as $n \to +\infty$ if the following conditions are satisfied:

(i) For all $x \in X$ and for every $(x_n) \subset X$, $x_n \to x$:

$$\liminf_{n\to+\infty} E_n(x_n) \geq E(x);$$

(ii) For all $x \in X$ there exists a *recovery* sequence $(\bar{x}_n) \subset X$, $\bar{x}_n \to x$:

$$\limsup_{n\to+\infty} E_n(\bar{x}_n) \leq E(x).$$

Remark 2.2.1 Condition (i) is usually referred to as *liminf inequality* or lower bound; (ii) is referred to as *limsup inequality* or upper bound. If we think of E_n as the energy associated to a discrete system of n particles and of E as the continuum limit model, then (i) means that for a given macroscopic configuration x, the discrete energy corresponding to *any* configuration x_n close to x is larger than the limit energy computed at x. Property (ii), on the other hand, ensures that the lower bound is *sharp*, namely that it can be attained on a special discrete configuration $\bar{x}_n$.

Remark 2.2.2 Notice that the Γ-limit will depend on the convergence used in the space X. In particular, if we endow X with the distances d and d' and assume that d is stronger than d' (i.e., that convergence in d implies convergence in d') then the $\Gamma(d)$-limit will be larger than the $\Gamma(d')$-limit.

Γ-convergence enjoys very nice properties concerning the behaviour of minimisers and minimum values which other, more natural definitions of convergence (e.g. the pointwise convergence) do not enjoy.

Theorem 2.2.1 *Let $E_n, E : X \to \mathbb{R}$, and assume the following:*

(a) $E_n \xrightarrow{\Gamma} E$ *as* $n \to +\infty$*;*
(b) x_n *is a minimiser of* E_n *for every* $n \in \mathbb{N}$*;*
(c) $x_n \to x$.

Then x is a minimiser of E.

Proof We claim that $E(x) \le E(y)$ for every $y \in X$.

Let $y \in X$; by assumption (a) there exists a recovery sequence $\bar{y}_n \to y$ satisfying property (ii) in Definition 2.2.1. Therefore

$$E(y) \ge \limsup_{n\to+\infty} E_n(\bar{y}_n) \overset{(b)}{\ge} \limsup_{n\to+\infty} E_n(x_n) \ge \liminf_{n\to+\infty} E_n(x_n) \ge E(x),$$

where the last inequality follows from (c) and from property (i) in Definition 2.2.1.

Example 2.2.1 Consider the sequence $f_n : \mathbb{R} \to \mathbb{R}$ defined as

$$f_n(x) = \begin{cases} -nx & 0 \le x \le \frac{1}{n} \\ -2 + nx & \frac{1}{n} \le x \le \frac{2}{n} \\ 0 & x \notin \left[0, \frac{2}{n}\right]. \end{cases}$$

Clearly the minimum value of f_n is $\min f_n = -1$, reached at the point $\frac{1}{n}$.

Note that the pointwise limit of the sequence f_n (as $n \to +\infty$) is the function $f_p \equiv 0$ for every $x \in \mathbb{R}$. In particular for the pointwise limit f_p the conclusion of Theorem 2.2.1 does not hold true, since $\min f_p = 0 \neq -1$. Hence, the pointwise limit is not the right object if we are interested in the behaviour of minimisers and minimum values.

It is easy to check that the Γ-limit of f_n as $n \to +\infty$ is the function f_∞ defined as

$$f_\infty(x) = \begin{cases} -1 & x = 0 \\ 0 & x \neq 0, \end{cases}$$

which instead has the same minimum value of f_n.

In conclusion, Γ-convergence is the natural convergence concept if one is interested in the asymptotic behaviour of minimisers. Note however that Theorem 2.2.1 states that if we want to infer, from Γ-convergence, that minimisers of the energies E_n converge to a minimiser of the limit energy E, we need to prove property (c). More precisely, we need to prove that minimisers of E_n *converge* to some limit configurations. Next section focuses on this point.

2.2.2 *Coerciveness*

The concept of coerciveness for an energy is intimately related to its minimisation. More precisely, it is related to the concept of minimising sequences, i.e., sequences along which the functional reaches its minimum value. If such a sequence has a limit, then the limit is the natural candidate for being the minimum point of the limit energy. The fact that such a sequence of energy minimisers converges depends on the bounds that the energy provides; and this is the essence of coerciveness.

Given the relation between Γ-convergence and convergence of minimisers, it is clear that coerciveness is also related to Γ-convergence. More precisely, the choice of the convergence in the space X in Definition 2.2.1 is determined by the coerciveness of the functionals E_n.

The definition of *coerciveness* will be given for functionals defined in general topological spaces (which are not necessarily metric spaces). The reason for that is that we are going to work often with energy bounds ensuring only weak convergence of the sequences.

Definition 2.2.2 Let X be a topological space and let $E : X \to \mathbb{R}$. The functional E is said to be (sequentially) coercive if for every t there exists a (sequentially) compact set K_t such that

$$S_t := \{x \in X : E(x) \leq t\} \subset K_t.$$

Recall that a set K is said to be *sequentially compact* if every sequence in K_t has a converging subsequence to a point in K_t.

Remark 2.2.3 Since the definition of the coerciveness in Definition 2.2.2 is very abstract, there is a characterisation of coerciveness that is used in practice when X is a reflexive Banach space:

F is coercive in the weak topology of X if and only if $F(x)$ tends to infinity as $\|x\| \to +\infty$.

This gives a simple criterion for the weak coerciveness. In fact, if the functional satisfies the lower bound $F(x) \geq C\|x\|$, with a constant $C > 0$, then it is weakly coercive. This means that if we can prove some *a priori* bounds for the energy with respect to some norm, then we directly have compactness, in a weak or in a strong sense, as we will see.

The concrete meaning of the definition and of the remark above will be made clear in the following, more familiar examples.

Example 2.2.2 Recall that $L^p(0, 1)$ and $W^{1,p}(0, 1)$, for $p > 1$, are reflexive Banach spaces; recall also that a bounded sequence in a reflexive space has a weakly convergent subsequence (see, e.g. [11]).

Let us consider the following examples, where $p > 1$.

(a) The functional $F : L^p(0, 1) \to \mathbb{R}$ defined as

$$F(u) = \int_0^1 |u|^p ds$$

is weakly sequentially coercive.

(b) The functional $F : W^p(0, 1) \to \mathbb{R}$ defined as

$$F(u) = \int_0^1 |u'|^p ds + \int_0^1 |u|^p ds$$

is weakly sequentially coercive in $W^{1,p}$ and strongly sequentially coercive in L^p.

2.2.2.1 Weak Compactness in L^1

This section is very technical and might be skipped on a first reading. The results stated here, however, will be heavily used in the applications studied in Sect. 2.3 and therefore are needed to make these lecture notes self-contained.

Here we are going to consider the problem of compactness in L^1 and in $W^{1,1}$, which is mathematically more subtle since the unit ball in these spaces is not weakly compact (as the spaces are not reflexive). The next two theorems offer a good way to check the coerciveness of a functional in L^1 by means of some useful (and typically easier to check) characterizations.

We will state the theorems in the general setting of functionals defined in $L^1(\Omega)$, for $\Omega \subset \mathbb{R}^N$ and $N \geq 1$.

Theorem 2.2.2 (Dunford-Pettis) Let $\mathscr{F} \subset L^1(\Omega)$, with Ω bounded. The following statements are equivalent:

(1) $\mathscr{F}$ is relatively compact in the weak topology of L^1 (i.e., every sequence has a weakly convergent subsequence);
(2) the family $\mathscr{F}$ is uniformly integrable, i.e., for every $\varepsilon > 0$ there exists $c > 0$ such that:

$$\int_{|f|>c} |f| ds < \varepsilon \quad \text{for every } f \in \mathscr{F}.$$

Remark 2.2.4 There are a number of characterisations of the *uniform integrability* that might be easier to check in practice. One of them is that the family $\mathscr{F}$ is bounded and uniformly (in $\mathscr{F}$) absolutely continuous.

We recall that we say that the *absolute continuity* of the integrand holds uniformly in $\mathscr{F}$ if for every $\varepsilon > 0$ there exists $\delta > 0$ such that for every measurable set A with $|A| < \delta$:

$$\int_A |f| ds < \varepsilon \quad \text{for every } f \in \mathscr{F}.$$

In the formula above, $|A|$ denotes the (Lebesgue) measure of the set A.

Another useful criterion is the following:

Theorem 2.2.3 *Let $\mathscr{F} \subset L^1(\Omega)$, with Ω bounded. Assume that there exists $\psi : \mathbb{R} \to [0, +\infty)$ continuous such that*

$$\lim_{|z| \to +\infty} \frac{\psi(z)}{|z|} = +\infty,$$

and with the property that there exists $M > 0$ such that

$$\int_\Omega \psi(|f|) ds < M \quad \textit{for every } f \in \mathscr{F}.$$

Then the family $\mathscr{F}$ is uniformly integrable.

Proof By the superlinearity assumption on ψ it follows that for every a there exists a radius R such that, for every $|z| > R$:

$$\frac{\psi(z)}{|z|} > a.$$

So, if $f \in \mathscr{F}$ and if $s \in \{\xi \in \Omega : |f(\xi)| > R\}$, then

$$\frac{\psi(f(s))}{|f(s)|} > a.$$

In particular, since ψ is nonnegative:

$$\int_{\Omega} \psi(|f(\xi)|)d\xi \geq \int_{|f|>R} \psi(|f(s)|)ds \geq a \int_{|f|>R} |f(s)|ds.$$

Now, let $\varepsilon > 0$ be fixed and let a be such that $\frac{M}{a} < \varepsilon$. Then, by assumption and by the previous inequalities:

$$M \geq \int_{\Omega} \psi(|f(\xi)|)d\xi \geq a \int_{|f|>R} |f(s)|ds.$$

Hence

$$\frac{M}{a} \geq \int_{|f|>R} |f(s)|ds,$$

implying the claim.

Remark 2.2.5 The converse implication follows if ψ is convex.

Corollary 2.2.1 *Under the assumptions of Theorem 2.2.3 it follows that the family $\mathscr{F}$ is in fact relatively compact with respect to the weak convergence in L^1.*

2.2.3 Lower Semicontinuity and Relaxation

Lower semicontinuity is, together with coerciveness, the key assumption ensuring the existence of minimisers.

Also in this section we will assume that X is a general topological space.

Definition 2.2.3 A functional $F : X \to \mathbb{R}$ is said to be lower semicontinuous if for every t the sublevels

$$S_t := \{x \in X : F(x) \leq t\}$$

are closed.

The following sequential definition is more used in practice (and coincides with Definition 2.2.3 for metric spaces).

Definition 2.2.4 A functional $F : X \to \mathbb{R}$ is said to be sequentially lower semicontinuous if for every $x \in X$ and for every sequence (x_j) converging to x as $j \to +\infty$:

$$F(x) \leq \liminf_{j \to +\infty} F(x_j).$$

Example 2.2.3 Consider the following cases:

- The functional in Example 2.2.2 (a) is strongly continuous and weakly lower semicontinuous in L^p;
- The functional in Example 2.2.2 (b) is strongly continuous and weakly lower semicontinuous in $W^{1,p}$.

Note that the integrand in both cases is a convex function. This is not a coincidence: As we will see, for integral functionals, lower semicontinuity is strictly related to the convexity of the integrand.

The connection between lower semicontinuity and coerciveness on the one hand, and existence of minimisers on the other hand is made clear by the following theorem.

Theorem 2.2.4 *Let $F : X \to \mathbb{R}$ be a coercive and lower semicontinuous functional. Then there exists a minimiser of F.*

Proof Let (x_j) be a minimising sequence for F, namely $\lim_{j\to\infty} F(x_j) = \inf_X F$. By coerciveness we deduce that x_j converges to some $x \in X$. The fact that x is a minimiser of F follows by lower semicontinuity, since

$$\inf_X F = \lim_{j\to\infty} F(x_j) \geq F(x).$$

Remark 2.2.6 (Lower semicontinuity and Γ-convergence) The Γ-limit defined by properties (i) and (ii) of Definition 2.2.1 is lower semicontinuous.

Remark 2.2.7 Given a sequence of functionals $F_n : X \to \mathbb{R}$ it is possible to define the following functionals:

$$F'(x) := \inf\{\liminf_{n\to+\infty} F_n(y_n) : y_n \to x\}$$
$$F''(x) := \inf\{\limsup_{n\to+\infty} F_n(y_n) : y_n \to x\}.$$

The functionals F' and F'' are lower semicontinuous and the sequence F_n Γ-converges if and only if $F' = F''$ (and the common value is the Γ-limit of F_n).

If a functional F is not lower semicontinuous one can construct a lower semicontinuous functional related to F, as in Definition 2.2.5.

Definition 2.2.5 Given a functional $F : X \to \mathbb{R}$, its *lower semicontinuous envelope* (or *relaxed* functional) is defined as

$$\mathrm{sc}^- F(x) := \sup\{G(x) | G : X \to \mathbb{R}, G \text{ lower semicontinuous}, G \leq F\}.$$

Remark 2.2.8 Given a functional $F : X \to \mathbb{R}$, we can define the sequence $F_n := F$ for every $n \in \mathbb{N}$. Then the Γ-limit of the sequence F_n coincides with the relaxation of F. The functional $\text{sc}^- F$ is therefore characterized by the same properties of the Γ-limit.

From the point of view of Γ-convergence, the 'right' space would be the space of lower semicontinuous functionals.

It is possible to prove that F and $\text{sc}^- F$ have the same 'ground states' in some sense. Therefore, from the point of view of the minimisation, the energies F and $\text{sc}^- F$ are the same.

2.2.3.1 Lower Semicontinuity in L^1

As for the coerciveness, also for the lower semicontinuity in the case of non-reflexive spaces requires some care and will be dealt with separately.

The proof of the next theorem is quite technical and can be skipped by a reader who is not familiar with convergence of measures. The lower semicontinuity result will be heavily used in the applications in the next sections.

Theorem 2.2.5 *Let* $\psi : \mathbb{R} \to \mathbb{R}$ *be a convex function such that* $\psi(z) = +\infty$ *for* $z < 0$, $\lim_{z\to 0^+} \psi(z) = +\infty$ *and* $\lim_{z\to+\infty} \psi(z) = \beta$, *for some* $\beta \in \mathbb{R}$.

For $u \in W^{1,1}(0, 1)$ *satisfying the Dirichlet boundary conditions* $u(0) = 0$ *and* $u(1) = \ell > 0$ *we define the functional*

$$u \mapsto F(u) = \int_0^1 \psi(u'(s))ds;$$

the functional F *is lower semicontinuous with respect to the strong convergence in* L^1.

Proof Let $(u_j) \subset W^{1,1}(0, 1)$ and let $u \in W^{1,1}(0, 1)$ be such that $u_j(0) = 0 = u(0)$, $u_j(1) = \ell = u(1)$ and $u_j \to u$ strongly in $L^1(0, 1)$. We claim that

$$\liminf_{j\to+\infty} F(u_j) \geq F(u).$$

We assume further that $F(u_j) \leq c$ (otherwise the claim is trivial).

From the assumptions on F it follows that u_j are increasing; moreover, they are bounded uniformly by ℓ by the monotonicity. For the derivative we have the following uniform bound (where we use that functions in $W^{1,1}(0, 1)$ are absolutely continuous, see e.g. [11]):

$$\int_0^1 |u_j'(s)|ds = \int_0^1 u_j'(s)ds = u_j(1) - u_j(0) = \ell.$$

Therefore the sequence (u_j) is bounded in $W^{1,1}$ and consequently the sequence $(u_j)'$ regarded as a sequence of measures converges weakly* to u'. By [3, Proposition 5.1–Theorem 5.2] we have that the functional F on absolutely continuous measures (with respect to the Lebesgue measure) is lower semicontinuous with respect to the weak convergence of measures, which implies the thesis. (To apply the theorem just observe that $\psi_\infty = 0$, where ψ_∞ denotes the *recession* function of the convex function ψ.)

2.3 Discrete Systems: Short-Range Interactions

As a first step we establish the setup of one-dimensional discrete models. This general setup will be used also in Sect. 2.4.

2.3.1 *One-Dimensional Discrete Setting*

Consider $n+1$ points, for $n \in \mathbb{N}$, representing $n+1$ particles in a one-dimensional chain. Set $\lambda_n := \frac{1}{n}$. We define a reference lattice, which, roughly speaking, provides an ordering of the particles. In the reference lattice the $n+1$ particles occupy $n+1$ uniformly spaced points in the interval [0,1] (with no loss of generality). More precisely, the positions of the particles in the reference lattice are $\lambda_n i = \frac{i}{n}$, for $i \in \{0, \dots, n\}$. We denote with I_n the reference lattice, i.e.,

$$I_n := [0, 1] \cap \lambda_n \mathbb{Z}.$$

The real, *current* positions of the particles will be described by means of a function defined on I_n, denoted by $u : I_n \to \mathbb{R}$. In practice, u is nothing but a collection of $n+1$ ordered values, which are the evaluations of u on I_n. In Sects. 2.3.2.2 and 2.3.3 the values $\{u_i : i = 0, \dots, n\}$ represent the *displacement* that each of the particles undergoes. Therefore, the current position of the i-th particle is $\lambda_n i + u_i$. In Sect. 2.3.4 the function u is instead a *deformation*, hence the values $\{u_i\}$ are the positions of the particles in the current, deformed configuration.

The function u (displacement or deformation) can be equivalently regarded as a continuous piecewise affine function (affine interpolation of the $n+1$ values). In formulas, given $\{u_i : i = 0, \dots, n\}$, we define $\tilde{u} : [0, 1] \to \mathbb{R}$ as

$$\tilde{u}(s) := u_i + \frac{u_{i+1} - u_i}{\lambda_n}(s - \lambda_n i), \quad s \in (\lambda_n i, \lambda_n (i+1)).$$

Notice that

$$\tilde{u}'(s) := \frac{u_{i+1} - u_i}{\lambda_n}, \quad s \in (\lambda_n i, \lambda_n (i+1)).$$

Fig. 2.2 Sketch of nearest neighbour interactions in the reference configuration

We denote with $\mathscr{A}_n(0, 1)$ the set of discrete values of u or, equivalently, the set of piecewise affine functions $\tilde{u}$ defined as above. Notice that $\mathscr{A}_n(0, 1) \subset W^{1,1}(0, 1)$ and $\mathscr{A}_n(0, 1) \subset W^{1,\infty}(0, 1)$.

Remark 2.3.1 It is possible to provide an alternative description of the current positions of the $n + 1$ particles $\{u_i : i = 0, \dots, n\}$ in terms of *empirical measure*. The empirical measure associated to the current configuration is μ_n defined as

$$\mu_n := \frac{1}{n} \sum_{i=0}^{n} \delta_{u_i}.$$

2.3.2 Nearest-Neighbour Interactions

In this section we focus on *nearest-neighbour* interactions, as sketched in Fig. 2.2 for the reference lattice.

The interaction energy for particles interacting with their nearest neighbours via an interaction potential φ is a functional $E_n : \mathscr{A}_n(0, 1) \to \mathbb{R}$ defined as

$$E_n(\{u_i\}) := \lambda_n \sum_{i=0}^{n-1} \varphi\left(\frac{u_{i+1} - u_i}{\lambda_n}\right).$$

The pairwise interaction described by the functional E_n depends only on the ratio between the difference of the displacements of two consecutive particles (or the deformed distance between consecutive particles) and their distance in the reference lattice. In what follows we will consider different expressions for the potential φ, corresponding to different types of interaction.

As we noticed, $\mathscr{A}_n(0, 1) \subset W^{1,1}(0, 1)$. Nevertheless, as we will see, the most natural setting for this type of functionals is $L^1(0, 1)$. Therefore we will use the extension of the functional to $L^1(0, 1)$, (still denoted by E_n) defined as

$$E_n(\{u_i\}) := \begin{cases} \lambda_n \displaystyle\sum_{i=0}^{n-1} \varphi\left(\frac{u_{i+1} - u_i}{\lambda_n}\right) & u \in \mathscr{A}_n(0, 1), \\ +\infty & u \in L^1(0, 1) \backslash \mathscr{A}_n(0, 1). \end{cases} \tag{2.3.1}$$

Remark 2.3.2 The extension in (2.3.1) has a great advantage. In fact, while the original functionals E_n are defined on the domains $\mathscr{A}_n$ that depend on n, the extended functionals E_n are defined on a common domain $L^1(0, 1)$ independent of n.

2.3.2.1 Rewriting the Energy as an Integral

Using the identification of the discrete function $\{u_i\}$ with its piecewise affine interpolation $\tilde{u}$ we can rewrite the sum in E_n as an integral. In fact

$$\varphi\left(\frac{u_{i+1}-u_i}{\lambda_n}\right)=\frac{1}{\lambda_n}\int_{\lambda_n i}^{\lambda_n(i+1)}\varphi(\tilde{u}'(s))ds,$$

and therefore

$$\lambda_n\sum_{i=0}^{n-1}\varphi\left(\frac{u_{i+1}-u_i}{\lambda_n}\right)=\int_0^1\varphi(\tilde{u}'(s))ds. \tag{2.3.2}$$

2.3.2.2 Convex and Superlinear Interaction

In this section we consider discrete energies E_n with a convex and superlinear interaction potential φ. The model case is $\varphi(z)=z^2$, corresponding to spring-type interactions. Notice that such a convex interaction favours configurations where the relative displacement is zero. In particular, in the case of clamped boundary conditions $u_0=u_n=0$, the minimum energy configuration is the reference configuration, i.e., $u_i=0$ for every i.

We want to study the Γ-convergence of the energies E_n in the 'many particles' limit $n\to+\infty$. The goal is to obtain a continuum, upscaled energy.

Finding the Γ-limit is a procedure that consists of two main steps, as we already pointed out. The first step is to determine the *natural* convergence in the domain of the functional (the space X in Definition 2.2.1). The second step is to prove the lower and upper bounds (i) and (ii) in Definition 2.2.1.

We start by focusing on the second step and we skip the first step for now. In Theorem 2.3.1 we state and prove Γ-convergence with respect to the strong convergence in $W^{1,1}(0,1)$. In what follows we will see that this is not the natural convergence to be used, and we will determine the natural convergence by studying the coerciveness of the functionals E_n in (2.3.1).

Theorem 2.3.1 *Let E_n be defined as in* (2.3.1) *and assume that φ is continuous, convex and nonnegative. The Γ-limit of E_n with respect to the strong convergence in $W^{1,1}$ is the functional E defined for $u\in W^{1,1}(0,1)$ as*

$$E(u)=\int_0^1\varphi(u'(s))ds.$$

Proof *(Liminf inequality)* Let $u\in W^{1,1}(0,1)$ and $\{u_i^n\}\subset\mathscr{A}_n(0,1)$ such that $\{u_i^n\}\to u$ in $W^{1,1}$. (Notice that the convergence is meant in terms of the convergence of the piecewise affine interpolations $\tilde{u}^n$.) We claim that

$$\liminf_{n\to+\infty} E_n(\{u_i^n\}) \geq E(u).$$

Using the identification (2.3.2), this is equivalent to proving that

$$\liminf_{n\to+\infty} \int_0^1 \varphi((\tilde{u}^n)'(s))ds \geq \int_0^1 \varphi(u'(s))ds.$$

First of all, we reduce to a subsequence $\tilde{u}^j := \tilde{u}^{n_j}$ of $\tilde{u}^n$ such that

$$\liminf_{n\to+\infty} \int_0^1 \varphi((\tilde{u}^n)'(s))ds = \lim_{j\to+\infty} \int_0^1 \varphi((\tilde{u}^j)'(s))ds.$$

Clearly, the subsequence $\tilde{u}^j$ converges to u in $W^{1,1}$ as $j \to +\infty$, hence in particular $(\tilde{u}^j)'$ converges to u' in L^1. Consequently, there exists a further subsequence $\tilde{u}^{j_k}$ such that $(\tilde{u}^{j_k})'$ converges to u' a.e. on $(0, 1)$ as $k \to +\infty$. Being φ continuous, this implies that

$$\lim_{k\to+\infty} \varphi\left((\tilde{u}^{j_k})'\right) = \varphi(u') \quad \text{a.e.}$$

It follows that

$$\int_0^1 \lim_{k\to+\infty} \varphi\left((\tilde{u}^{j_k})'(s)\right) ds = \int_0^1 \varphi(u'(s))ds. \tag{2.3.3}$$

By Fatou's Lemma (φ measurable and nonnegative) it follows that

$$\int_0^1 \lim_{k\to+\infty} \varphi\left((\tilde{u}^{j_k})'(s)\right) ds \leq \liminf_{k\to+\infty} \int_0^1 \varphi\left((\tilde{u}^{j_k})'(s)\right) ds;$$

by (2.3.3) this implies that

$$\int_0^1 \varphi(u'(s))ds \leq \liminf_{k\to+\infty} \int_0^1 \varphi\left((\tilde{u}^{j_k})'(s)\right) ds,$$

and hence the claim. Notice that in the proof of the lower bound we did not make use of the assumption that φ is convex.

Limsup inequality. Let $u \in W^{1,1}(0,1)$; we define the recovery sequence as the evaluations of u on I_n (well defined since u is continuous), i.e.,

$$u_i^n := u(\lambda_n i).$$

Also, since u is absolutely continuous (being in $W^{1,1}(0,1)$), the Fundamental Theorem of Calculus can be applied to u and gives

$$\frac{u_{i+1}^n - u_i^n}{\lambda_n} = \frac{1}{\lambda_n} \int_{\lambda_n i}^{\lambda_n(i+1)} u'(s)ds.$$

Then, for the functionals E_n we have

$$E_n(\{u_i^n\}) = \lambda_n \sum_{i=0}^{n} \varphi\left(\frac{u_{i+1}^n - u_i^n}{\lambda_n}\right) = \lambda_n \sum_{i=0}^{n} \varphi\left(\frac{1}{\lambda_n} \int_{\lambda_n i}^{\lambda_n(i+1)} u'(s)ds\right)$$
$$\overset{(*)}{\leq} \lambda_n \sum_{i=0}^{n} \frac{1}{\lambda_n} \int_{\lambda_n i}^{\lambda_n(i+1)} \varphi(u'(s))ds = \int_0^1 \varphi(u'(s))ds,$$

where the $(*)$ inequality follows by Jensen's inequality.

Remark 2.3.3 Notice that by the convexity assumption the functionals E_n also Γ-converge to E with respect to the weak convergence in $W^{1,1}$.

The recovery sequence is clearly the same, since a strongly convergent sequence converges weakly to the same limit. The only difference in the proof is in the liminf-inequality, where one has to prove the lower semicontinuity of the functionals with respect to the weak convergence in $W^{1,1}$. Thanks to the convexity of φ, this stronger lower-semicontinuity property follows directly from Theorem 2.3.2:

Theorem 2.3.2 *Let X be a Banach space, let $F : X \to \mathbb{R}$ be convex. Then F is strongly lower semicontinuous if and only if is weakly lower semicontinuous.*

Now we focus on the step we skipped so far, namely on the identification of the *natural* convergence for the energies E_n. This will be the result of a coerciveness result for the functionals E_n. In Theorem 2.3.3 we consider the general case of a nonnegative and continuous energy density φ with superlinear growth, and analyse the coerciveness of the corresponding integral functional.

Theorem 2.3.3 *Let E_n be defined as in* (2.3.1) *and assume that φ is continuous and nonnegative and satisfies*

$$\lim_{|z|\to+\infty} \frac{\varphi(z)}{|z|} = +\infty.$$

Let $(u^n) \subset W^{1,1}(0,1)$ *be such that* $\|u^n\|_{L^1(0,1)} \leq c$ *and such that*

$$E_n(u^n) := \int_0^1 \varphi((u^n)'(s))ds \leq c.$$

Then there exists $u \in W^{1,1}(0,1)$ *such that* $u^n \rightharpoonup u$ *weakly in* $W^{1,1}$.

Proof The assumptions on φ imply that the sequence (u^n) is bounded in $W^{1,1}$; hence, by the compact embedding of $W^{1,1}(0,1)$ into $L^1(0,1)$, there exists a subsequence of u^n (not re-labelled) and $u \in L^1(0,1)$ such that

$$u^n \to u \quad \text{strongly in } L^1(0,1).$$

The superlinearity of φ also implies that the sequence $(u^n)'$ is uniformly integrable, by Theorem 2.2.3. This means, by Theorem 2.2.2, that there exists $g \in L^1(0,1)$ such that

$$(u^n)' \rightharpoonup g \quad \text{weakly in } L^1(0,1).$$

Hence $u \in W^{1,1}(0,1)$ and $u' = g$, as we obtain by passing to the limit $n \to +\infty$ in the following equality, valid for test functions $\phi \in C_c^1(0,1)$:

$$\int_0^1 u^n(s)\phi'(s)ds = -\int_0^1 (u^n)'(s)\phi(s)ds.$$

Remark 2.3.4 Notice that in the proof of the coerciveness no convexity is assumed. The main role of the convexity is to guarantee the lower semicontinuity of the integral functional with respect to weak convergence (see Theorem 2.3.2).

Remark 2.3.5 The conclusion of Theorem 2.3.3 follows also if the a-priori L^1-bound for the sequence u^n is replaced by some Dirichlet boundary conditions (which imply the L^1-bound by the Poincaré Inequality).

Theorem 2.3.3 suggests that the natural convergence for the discrete functionals E_n in (2.3.1) is the weak convergence in $W^{1,1}$ (instead of the strong convergence used in Theorem 2.3.1). Nevertheless, since it is always desirable to deal with a metric space, we prefer to consider the Γ-convergence with respect to the strong convergence in L^1 (which is implied by the weak convergence in $W^{1,1}$, thanks to the compactness embedding of $W^{1,1}(0,1)$ in $L^1(0,1)$).

Therefore, the most natural Γ-convergence result is the following:

Theorem 2.3.4 *Let* E_n *be defined as in* (2.3.1) *and assume that* φ *is continuous, convex, nonnegative and superlinear at infinity, i.e.,*

$$\lim_{|z|\to+\infty} \frac{\varphi(z)}{|z|} = +\infty.$$

Then the Γ-limit of E_n with respect to the strong convergence in L^1 is the functional E defined for $u \in L^1(0,1)$ as

$$E(u) = \begin{cases} \int_0^1 \varphi(u'(s))ds & u \in W^{1,1}(0,1), \\ +\infty & u \in L^1(0,1) \backslash W^{1,1}(0,1). \end{cases} \tag{2.3.4}$$

Proof (*Compactness of sequences with bounded energy*) Let $\{u_i^n\} \subset \mathscr{A}_n(0,1)$ be such that $E_n(\{u_i^n\}) \leq c$ and assume, additionally, that $||\tilde{u}^n||_{L^1} \leq c$, where $\tilde{u}^n$ denotes the piecewise affine interpolation of $\{u_i^n\}$.[1] By Theorem 2.3.3 it follows that there exists $u \in W^{1,1}(0,1)$ such that

$$\tilde{u}^n \rightharpoonup u \quad \text{weakly in } W^{1,1}.$$

This shows that the natural convergence is the weak convergence in $W^{1,1}$, and hence that the domain of the Γ-limit is $W^{1,1}(0,1)$.

Liminf inequality. Let $u \in W^{1,1}(0,1)$ and $\{u_i^n\} \subset \mathscr{A}_n(0,1)$ such that $\tilde{u}^n \to u$ in L^1. We claim that

$$\liminf_{n\to+\infty} E_n(\tilde{u}^n) \geq E(u).$$

We can assume that $E_n(\tilde{u}^n) \leq c$, otherwise the claim is trivial. By the previous step the energy bound implies that $\tilde{u}^n \rightharpoonup u$ weakly in $W^{1,1}$. Then the claim follows by the lower semicontinuity of the integral functional

$$v \mapsto \int_0^1 \varphi(v'(s))ds$$

with respect the weak convergence in $W^{1,1}$, which in its turn follows by the convexity of φ.

Limsup inequality. The recovery sequence is the same as in Theorem 2.3.1.

[1]This bound can be guaranteed by some additional term in the discrete energy, or by Dirichlet boundary conditions for the admissible displacements, e.g. $u_0^n = 0 = u_n^n$ for every n.

2.3.3 Non-convex and Superlinear Interaction

In this section we drop the convexity assumption for the interaction potential φ. The Γ-limit will involve a convexification process at a small scale, which is smaller than the macroscopic one, but larger than the microscopic scale λ_n of the lattice.

Definition 2.3.1 (*Convex envelope*) For a function φ the lower semicontinuous and *convex envelope* of φ, denoted by φ^{**}, is the largest convex and lower semicontinuous function below φ. If φ is lower semicontinuous then the definition of φ^{**} simplifies to:

$$\varphi^{**}(z) = \min\left\{t\varphi(z_1) + (1-t)\varphi(z_2) : t \in [0,1], z = tz_1 + (1-t)z_2\right\}. \quad (2.3.5)$$

Remark 2.3.6 (*Geometric interpretation of the convex envelope*) Consider a continuous function φ. If φ is convex, then for every z

$$\varphi^{**}(z) = \varphi(z),$$

since the optimal choice in (2.3.5) is $z_1 = z_2 = z$.

If instead φ is not convex, then the equality $\varphi^{**} = \varphi$ is not anymore true for every z. Consider for instance the case of the double-well potential $\varphi(z) = (z^2-1)^2$. Then

$$\varphi^{**}(z) = \begin{cases} \varphi(z) & z \le -1, z \ge 1, \\ 0 & -1 \le z \le 1. \end{cases}$$

This follows easily by noticing that for every $-1 \le z \le 1$ the optimal choice in (2.3.5) is $z_1 = -1$ and $z_2 = 1$. See Fig. 2.3.

Next theorem is a Γ-convergence result for E_n, without the convexity assumption on φ.

Theorem 2.3.5 *Let E_n be defined as in* (2.3.1) *and assume that φ is continuous, nonnegative and superlinear at infinity i.e.,*

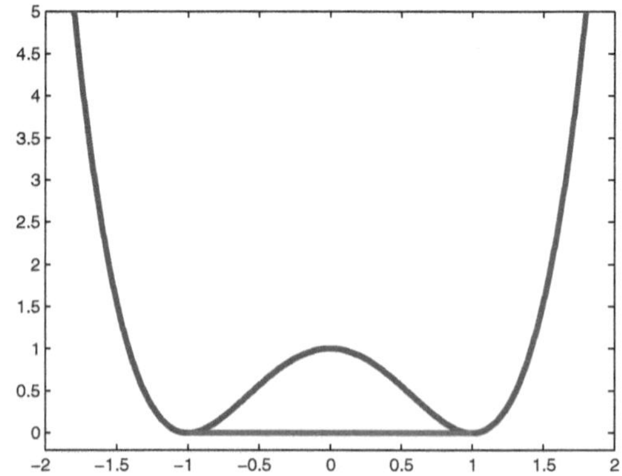

Fig. 2.3 Plot of the function φ (in *blue*). The convexification of φ differs from φ only in (0, 1) and its plot in (0, 1) is shown in *red* (color online)

$$\lim_{|z|\to+\infty} \frac{\varphi(z)}{|z|} = +\infty.$$

Then the Γ-limit of E_n with respect to the strong convergence in L^1 is the functional E defined for $u \in L^1(0,1)$ as

$$E(u) = \begin{cases} \int_0^1 \varphi^{**}(u'(s))\,ds & u \in W^{1,1}(0,1), \\ +\infty & u \in L^1(0,1)\setminus W^{1,1}(0,1), \end{cases}$$

where φ^{**} denotes the convex and lower semicontinuous envelope of φ.

Proof (*Compactness of sequences with bounded energy*) Since the proof of the coerciveness of E_n in Theorem 2.3.4 did not make use of the convexity of φ, the result in this case is the same, i.e., the natural convergence is the weak convergence in $W^{1,1}$, and hence the domain of the Γ-limit is $W^{1,1}(0,1)$.

Liminf inequality. Let $u \in W^{1,1}(0,1)$ and $\{u_i^n\} \subset \mathscr{A}_n(0,1)$ be such that $\tilde{u}^n \to u$ in L^1. Since $\varphi^{**} \leq \varphi$, we have the bound

$$E_n(\{u_i^n\}) \geq \int_0^1 \varphi^{**}((\tilde{u}^n)'(s))ds. \tag{2.3.6}$$

The lower bound for the energies E_n follows by the lower semicontinuity (with respect the weak convergence in $W^{1,1}$) of the functional in the right-hand side of (2.3.6).

Limsup inequality. For the construction of the recovery sequence we will use a density argument. We first recall that piecewise affine functions are dense in $W^{1,1}(0,1)$ with respect to the strong convergence in $W^{1,1}$. Moreover, the Γ-limit functional E is continuous with respect to the strong convergence in $W^{1,1}$. The continuity of E follows by the general result:

Let X be a normed space and let $F : X \to \mathbb{R}$ be a bounded convex function on $B = \{x \in X : \|x\| \leq 1\}$. Then F is continuous at $x = 0$.

The continuity of E and the density of piecewise affine functions in $W^{1,1}(0,1)$ allow us to restrict our attention to the case of piecewise affine functions u. Moreover, the case of piecewise affine functions can be obtained from the simpler case of linear functions by superposition. In conclusion, it is sufficient to consider only the case of linear functions, i.e., $u(s) = zs$.

Let then $u(s) = zs$, and let t, z_1, z_2 be given by (2.3.5) so that

$$\varphi^{**}(z) = t\varphi(z_1) + (1-t)\varphi(z_2).$$

The idea is to construct for every n a *zig-zag function* with slopes z_1 and z_2, oscillating at a small scale δ_n, with $\delta_n \to 0$ as $n \to +\infty$ (which we need in order to have a recovery sequence), but with $\delta_n \gg \lambda_n$.

Let therefore $\ell_n \in \mathbb{N}$ be such that $\ell_n \to +\infty$ and $\lambda_n \ell_n \to 0$ as $n \to +\infty$ and define $\delta_n := \lambda_n \ell_n$. Let $K_n \in \{0, \dots, \ell_n\}$ be so that $\frac{K_n}{\ell_n} \to t$, and define the recovery sequence $\{u_i^n\} \subset \mathscr{A}_n(0,1)$ as $\tilde{u}^n(s) := zs + \tilde{v}^n(s)$, where $\{v_i^n\} \subset \mathscr{A}_n(0,1)$ are $\lambda_n \ell_n$-periodic, with derivative (almost everywhere) given by

$$(\tilde{v}^n)'(s) = \begin{cases} z_1 - z \; 0 \le s < \lambda_n K_n, \\ z_2 - z \; \lambda_n K_n \le s < \lambda_n \ell_n. \end{cases}$$

Notice that, by definition,

$$(\tilde{u}^n)'(s) = \begin{cases} z_1 \; 0 \le s < \lambda_n K_n, \\ z_2 \; \lambda_n K_n \le s < \lambda_n \ell_n. \end{cases}$$

Moreover, $\|\tilde{v}^n\|_{L^\infty(0,1)} = \lambda_n K_n |z_1 - z| \le \lambda_n \ell_n |z_1 - z| \to 0$ as $n \to +\infty$, and therefore $\tilde{u}^n \to u$ uniformly. (Notice that $\tilde{u}^n$ converges to u also strongly in $W^{1,1}$, so it is admissible for any convergence we might want to consider.)

It remains to check the convergence of the energies E_n along the sequence $\tilde{u}^n$. We have

$$\begin{aligned} E_n(\{u_i^n\}) &= \int_0^1 \varphi((\tilde{u}^n)'(s))ds \\ &= \lambda_n K_n \left(\frac{1}{\lambda_n \ell_n}\right) \varphi(z_1) + \lambda_n(\ell_n - K_n) \left(\frac{1}{\lambda_n \ell_n}\right) \varphi(z_2). \end{aligned}$$

Moreover, by the definition of ℓ_n and K_n it follows that

$$\lim_{n \to +\infty} E_n(\{u_i^n\}) = t\varphi(z_1) + (1-t)\varphi(z_2) = \varphi^{**}(z) = \int_0^1 \varphi^{**}(u'(s))ds,$$

which proves that $\{u_i^n\}$ is a recovery sequence for E_n.

2.3.4 Lennard-Jones Interactions

In the previous section we showed that dropping the convexity assumption for the potential φ, but keeping the superlinearity at infinity, had the solely effect of giving rise to a convexification of the limit energy (necessary to have lower semicontinuity, which the Γ-limit has to satisfy).

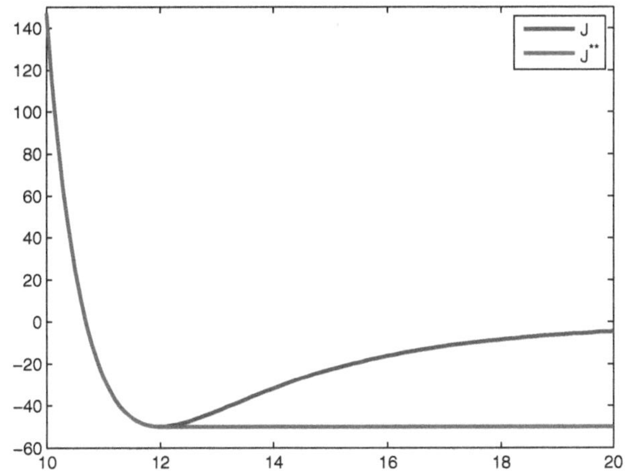

Fig. 2.4 Lennard-Jones potential defined as in (2.3.7) with $\alpha = 50$ and $z_0 = 12$

A different situation arises when the growth condition is weakened—or dropped. In this case the compactness becomes more subtle and limit deformations might develop discontinuities (notice that so far the limit deformations were in $W^{1,1}(0, 1)$, therefore continuous). This is clearly the right type of interaction to be considered for problems in fracture mechanics. A fracture, indeed, corresponds to a discontinuity of the deformation, and hence it is natural to consider weaker growth conditions for the energy to allow jumps to occur with a finite cost, in terms of elastic energy.

The typical energy potential used in this case is the *Lennard-Jones interaction* potential (see Fig. 2.4)

$$J(z) = \alpha\left(\frac{z_0}{z^{12}}\right) - 2\alpha\left(\frac{z_0}{z^6}\right), \quad z > 0, \tag{2.3.7}$$

where z_0 is the minimum point (location of the well) and $-\alpha$ is the minimum value (depth of the well). Note that J can be negative (unlike in the previous cases) and moreover

$$\lim_{z\to+\infty} \frac{J(z)}{z} = 0.$$

More in general, we will consider Lennard-Jones-type energies. The assumptions we require for $J : \mathbb{R} \to \mathbb{R}$ are listed below.

(LJ-1) *Non-interpenetration condition:*

$$J(z) = +\infty \quad \text{for every } z < 0;$$

(LJ-2) *Strong repulsion:*

$$\lim_{z\to 0^+} J(z) = +\infty;$$

(LJ-3) *Regularity and behaviour at infinity.* The potential J is smooth in $(0, \infty)$; moreover the strength of the interaction decays at infinity, i.e.,

$$\lim_{z \to +\infty} J(z) = 0.$$

(LJ-4) *Structure of the potential.* There exists a unique minimiser $\gamma > 0$ and there exists $\delta > 0$ such that J is strictly convex in $(0, \gamma + \delta)$ and strictly concave in $(\gamma + \delta, +\infty)$. In particular, by assumption (LJ-3) we have that $J(\gamma) < 0$.

The discrete energy is defined for functions $\{u_i^n\} \subset \mathscr{A}_n(0, 1)$ as

$$E_n(\{u_i^n\}) = \lambda_n \sum_{i=0}^{n} J\left(\frac{u_{i+1}^n - u_i^n}{\lambda_n}\right). \tag{2.3.8}$$

We will see that unlike in Sects. 2.3.2.2 and 2.3.3, where the superlinearity of the potential ensured the weak compactness in $W^{1,1}$ of sequences with bounded energy, in this case one can only get a bound in $W^{1,1}$, which is not enough for weak compactness (since $W^{1,1}$ is not reflexive). Moreover, the $W^{1,1}$ bound for sequences with bounded energy does not follow from growth conditions of the potential J, but only on the non-interpenetration condition (LJ-1). That assumption, in particular, forces admissible deformations to be increasing. And sequences of increasing functions enjoy nice compactness properties, as we are going to show.

Note that in Sects. 2.3.2.2 and 2.3.3, since u plays the role of a displacement and not of a deformation, the difference quotient in the argument of φ had no sign restriction. Also, for the convex potential $\varphi(z) = z^2$ the minimum energy configuration (under appropriate Dirichlet boundary conditions) is the reference configuration. In the case of Lennard-Jones interactions the energy is minimised by the $\lambda_n\gamma$-equispaced configuration (under appropriate boundary conditions).

To simplify what follows, we restrict the space of admissible sequences $\{u_i^n\}$ to the following subset of $\mathscr{A}_n(0, 1)$:

$$\mathscr{A}_n^{bc}(0, 1) := \{u_i \in \mathscr{A}_n(0, 1) : u_0 = 0, u_n = \ell\}. \tag{2.3.9}$$

Theorem 2.3.6 *Let E_n be defined as in* (2.3.8), *and let $(u^n) \subset \mathscr{A}_n^{bc}(0, 1)$ be such that $E_n(u^n) \leq c$. Then the sequence (u^n) is bounded in $W^{1,1}$.*

Proof Notice that, since the sequence $\{u_i^n\}$ has bounded energy and by assumption (LJ-1) $J(z) = +\infty$ for $z < 0$, the difference quotients $(u_{i+1}^n - u_i^n)/\lambda_n$ have to be nonnegative. Moreover, since $\mathscr{A}_n(0, 1) \subset W^{1,1}(0, 1)$, the sequence $(\tilde{u}^n)$ (namely the affine interpolation of $\{u_i^n\}$) is continuous. Therefore for every n the function $\tilde{u}^n$ is continuous and increasing.

The monotonicity and the common boundary conditions $\tilde{u}^n(0) = 0$ and $\tilde{u}^n(1) = \ell$ give the uniform bound $\|\tilde{u}^n\|_{L^\infty(0,1)} \leq \ell$.

For the first derivatives notice that, by the Fundamental Theorem of Calculus, which can be applied since functions in $W^{1,1}(0, 1)$ are absolutely continuous,

$$\int_0^1 (\tilde{u}^n)'(s)ds = \tilde{u}^n(1) - \tilde{u}^n(0) = \ell.$$

Moreover, by the monotonicity of $\tilde{u}^n$, $(\tilde{u}^n)' \geq 0$ (a.e.); hence

$$\int_0^1 |(\tilde{u}^n)'(s)|ds = \int_0^1 (\tilde{u}^n)'(s)ds = \tilde{u}^n(1) - \tilde{u}^n(0) = \ell. \tag{2.3.10}$$

The bound (2.3.10), together with the uniform bound on $(\tilde{u}^n)$ guarantees the desired $W^{1,1}$-bound for the sequence $(\tilde{u}^n)$.

Corollary 2.3.1 *Under the same assumptions of Theorem* 2.3.6 we can conclude that, in particular, $u^n \to u$ strongly in L^1, and that the limit function u is increasing (being limit almost everywhere of increasing functions).

Moreover, the derivatives $(u^n)'$, regarded as measures, converge *weakly** to a limit measure μ with bounded variation. Clearly, $\mu = u'$ in distributional sense.

We remark once more that the limit deformation u can have discontinuities, since it is only an L^1-function. We restrict our attention to a subclass of the limit deformations, namely on functions that are *piecewise* $W^{1,1}(0,1)$ $W^{1,1}(0,1)$. We follow Braides' notation (see e.g. [8]) and we denote such a space by $P\text{-}W^{1,1}(0,1)$. More precisely:

Definition 2.3.2 For every $p > 1$ we define

$$P\text{-}W^{1,p}(0,1) := W^{1,p}(0,1) + PC(0,1), \tag{2.3.11}$$

where $PC(0,1)$ denotes the space of piecewise constant functions.

Notice that, since the only functions that are both in $W^{1,p}(0,1)$ and $PC(0,1)$ are the constants, the decomposition in (2.3.11) is unique modulo a constant.

In practice, a given $u \in P\text{-}W^{1,p}(0,1)$ is the sum of a function $v \in W^{1,p}(0,1)$ (in particular continuous) and a piecewise constant function $w \in PC(0,1)$. This implies in particular that

$$u' = v' \text{ (a.e.)}, \quad S(u) = S(w), \quad u^{\pm}(x) = v(x) + w^{\pm}(x) \quad \forall x \in S(u).$$

In the formula above, $S(u)$ denotes the discontinuity set of u, and $u^{\pm}$ denote the right and left limits of u at a discontinuity point x.

Now we can identify the restriction of the Γ-limit of E_n to the space $P\text{-}W^{1,1}(0,1)$.

Theorem 2.3.7 *Let E_n be defined as in* (2.3.8), *and consider its restriction to the space $\mathscr{A}_n^{bc}(0,1)$. Then the Γ-limit of E_n with respect to the strong convergence in L^1 is the functional E defined for $u \in P\text{-}W^{1,1}(0,1)$ as*

$$E(u) = \begin{cases} \int_0^1 J^{**}(u'(s))ds & u(0) = 0, u(1) = \ell, u \text{ increasing}, \\ +\infty & \text{otherwise}. \end{cases}$$

Proof (*Liminf inequality*) Let $u \in P\text{-}W^{1,1}(0,1)$ be an increasing function satisfying the boundary conditions, and let $(\tilde{u}^n) \subset \mathscr{A}_n^{bc}(0,1)$ be such that $\tilde{u}^n \to u$ strongly in L^1 and $E_n(\tilde{u}^n) \le c$. Since $u \in P\text{-}W^{1,1}(0,1)$, there exist $N+1$ points $0 = y_0 < y_1 < \cdots < y_N = 1$ such that

$$(0,1)\backslash S(u) = \bigcup_{k=1}^{N} (y_{k-1}, y_k), \quad u_{|(y_{k-1},y_k)} \in W^{1,1}(y_{k-1}, y_k).$$

Moreover, by Theorem 2.2.5 the functional

$$u \in W^{1,1}(y_{k-1}, y_k) \mapsto \int_{y_{k-1}}^{y_k} J^{**}(u'(s))ds$$

is lower semicontinuous with respect to the strong convergence in L^1. This means that, for every $k = 1, \ldots, N$:

$$\liminf_{n\to\infty} \int_{y_{k-1}}^{y_k} J^{**}((\tilde{u}^n)'(s))ds \ge \int_{y_{k-1}}^{y_k} J^{**}(u'(s))ds.$$

Adding up the previous inequality for every k leads to

$$E(u) = \sum_{k=1}^{N} \int_{y_{k-1}}^{y_k} J^{**}(u'(s))ds \le \sum_{k=1}^{N} \liminf_{n\to\infty} \int_{y_{k-1}}^{y_k} J^{**}((\tilde{u}^n)'(s))ds$$

$$\le \liminf_{n\to\infty} \sum_{k=1}^{N} \int_{y_{k-1}}^{y_k} J^{**}((\tilde{u}^n)'(s))ds \le \liminf_{n\to\infty} \int_0^1 J^{**}((\tilde{u}^n)'(s))ds.$$

In the previous estimate we have used that, for sequences $\{a_n\}$ and $\{b_n\}$,

$$\liminf_{n\to\infty}(a_n + b_n) \ge \liminf_{n\to\infty} a_n + \liminf_{n\to\infty} b_n.$$

Finally, the claim follows from the bound $J^{**} \le J$.

Limsup inequality. Let $u \in P\text{-}W^{1,1}(0,1)$ be an increasing function satisfying the Dirichlet boundary conditions. Let $0 = y_0 < y_1 < \cdots < y_N = 1$ be $N+1$ points such that

$$(0,1)\backslash S(u) = \bigcup_{k=1}^{N}(y_{k-1}, y_k), \quad u_k := u_{|(y_{k-1},y_k)} \in W^{1,1}(y_{k-1}, y_k).$$

Let us fix $k \in \{1, \ldots, N\}$; we are going to construct a recovery sequence $\tilde{u}_k^n$ in the interval (y_{k-1}, y_k). The recovery sequence $\tilde{u}^n$ in [0,1] will be simply defined as $\tilde{u}^n_{|(y_{k-1},y_k)} := \tilde{u}_k^n$.

Since $u_k \in W^{1,1}(y_{k-1}, y_k)$, we can use density arguments to reduce to the case of a linear function with slope z_k, i.e., $u_k = z_k s$. If we take $\tilde{u}_k^n$ to be simply the affine interpolation of the values of u_k on the lattice $I_n \cap (y_{k-1}, y_k)$, then we will miss the convexification of J, as the limit will be $J(z_k)$ instead of $J^{**}(z_k)$.

Therefore we have to take a zig-zag sequence that approximates u_k in the strong convergence in $W^{1,1}(y_{k-1}, y_k)$ and with oscillations at scale $\delta_n \gg \lambda_n$ so that

$$\lim_{n\to\infty} \int_{y_{k-1}}^{y_k} J((\tilde{u}_k^n)'(s))ds = \int_{y_{k-1}}^{y_k} J^{**}(u'(s))ds.$$

The details of the construction are as in Theorem 2.3.5.

2.4 Discrete Systems: Long-Range Interactions

In the previous section we considered the case of particles interacting only with their nearest neighbours. Clearly in reality particles interact with all their neighbours, not only with the closest ones, and truncating the energy to the first neighbours can potentially lead to a wrong limit energy. Hence it should be avoided, even in the case of a rapidly decaying interaction potential, like the Lennard Jones potential for instance.

In this section we consider long-range interactions, starting with the case of *next-to-nearest neighbour* interactions illustrated in Fig. 2.5.

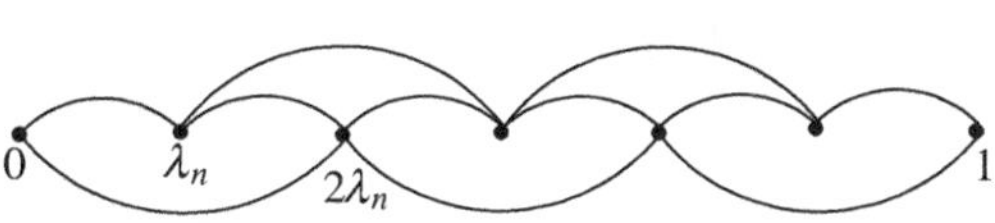

Fig. 2.5 Sketch of next-to-nearest neighbour interactions in the reference configuration

2.4.1 Next-to-Nearest-Neighbour Interactions

For given $\varphi^1, \varphi^2 : \mathbb{R} \to [0, \infty)$ we define

$$E_n(\{u_i\}) := \lambda_n \sum_{i=0}^{n-1} \varphi^1\left(\frac{u_{i+1} - u_i}{\lambda_n}\right) + \lambda_n \sum_{i=0}^{n-2} \varphi^2\left(\frac{u_{i+2} - u_i}{2\lambda_n}\right). \tag{2.4.1}$$

Like in the previous sections, we identify $\{u_i\}$ with its piecewise interpolation $\tilde{u}$. Also, as in (2.3.2), we rewrite the discrete energy (2.4.1) in integral form; for the nearest-neighbour interaction term we have

$$\lambda_n \sum_{i=0}^{n-1} \varphi^1\left(\frac{u_{i+1} - u_i}{\lambda_n}\right) = \int_0^1 \varphi^1(\tilde{u}'(s))ds.$$

For the next-to-nearest neighbour interaction term we first notice that, for $i = 0, \dots, n-2$,

$$\begin{aligned}\frac{u_{i+2} - u_i}{2\lambda_n} &= \frac{1}{2}\frac{u_{i+2} - u_{i+1}}{\lambda_n} + \frac{1}{2}\frac{u_{i+1} - u_i}{\lambda_n} \\ &= \frac{1}{2}\tilde{u}'(s + \lambda_n) + \frac{1}{2}\tilde{u}'(s), \quad s \in (\lambda_n i, \lambda_n(i+1)).\end{aligned} \tag{2.4.2}$$

Therefore, the second term in the energy (2.4.1) can be rewritten as

$$\lambda_n \sum_{i=0}^{n-2} \varphi^2\left(\frac{u_{i+2} - u_i}{2\lambda_n}\right) = \int_0^{1-\lambda_n} \varphi^2\left(\frac{\tilde{u}'(s + \lambda_n) + \tilde{u}'(s)}{2}\right) ds.$$

In conclusion, the discrete energy E_n can be rewritten as

$$E_n(u) = \int_0^1 \varphi^1(\tilde{u}'(s))ds + \int_0^{1-\lambda_n} \varphi^2\left(\frac{\tilde{u}'(s + \lambda_n) + \tilde{u}'(s)}{2}\right) ds. \tag{2.4.3}$$

The following Theorems 2.4.1 and 2.4.2 have been proved by Braides and collaborators (see e.g. [8]) and establish the Γ-convergence of E_n for convex and non convex interaction potentials.

Theorem 2.4.1 *Let E_n be defined as in* (2.4.1) *(and* (2.4.3)*) and assume that φ^1 and φ^2 are continuous, convex and nonnegative. Assume, moreover, that φ^1 is superlinear at infinity. Then the Γ-limit of E_n with respect to the strong convergence in L^1 is the functional E defined for $u \in L^1(0, 1)$ as*

$$E(u) = \begin{cases} \int_0^1 \varphi(u'(s))ds & u \in W^{1,1}(0,1) \\ +\infty & u \in L^1(0,1)\backslash W^{1,1}(0,1), \end{cases}$$

where $\varphi(t) = \varphi^1(t) + \varphi^2(t)$.

Proof (*Compactness of sequences with bounded energy*) Since the potential φ^2 is nonnegative, if a sequence $\{u_i^n\}$ has bounded energy E_n, then in particular the nearest-neighbour interaction term of the energy is bounded along the sequence. Therefore the convergence of the sequence follows directly from Theorem 2.3.4, under the additional assumption that the L^1-norm of the sequence is bounded. (Remember that this additional bound is guaranteed by Dirichlet boundary conditions for the sequence.) Recall that the sequence converges also weakly in $W^{1,1}$.

Liminf inequality. Let $u \in W^{1,1}(0,1)$ and let $\{u_i^n\} \subset \mathscr{A}_n(0,1)$ be such that $\tilde{u}^n \to u$ strongly in L^1 and $E_n(\tilde{u}^n) \le c$. From the compactness result we have that $\tilde{u}^n \rightharpoonup u$ weakly in $W^{1,1}$. Since the first term of the energy E_n is of the same type as the energy considered in Theorem 2.3.4, it follows directly that

$$\liminf_{n\to\infty} \lambda_n \sum_{i=0}^{n-1} \varphi^1\left(\frac{u_{i+1}^n - u_i^n}{\lambda_n}\right) \ge \int_0^1 \varphi^1(u'(s))ds.$$

For the second term we observe that $\bar{u}^n(s) := \frac{\tilde{u}^n(s)+\tilde{u}^n(s+\lambda_n)}{2}$ also converges weakly in $W^{1,1}$ to u. Therefore, for $\eta > 0$, by (2.4.3) and by the definition of $\bar{u}^n$ we have

$$\begin{aligned} \liminf_{n\to\infty} \lambda_n \sum_{i=0}^{n-2} \varphi^2\left(\frac{u_{i+2}^n - u_i^n}{2\lambda_n}\right) &\ge \liminf_{n\to\infty} \int_0^{1-\eta} \varphi^2((\bar{u}^n)'(s))ds \\ &\ge \int_0^{1-\eta} \varphi^2(u'(s))ds. \end{aligned}$$

The last inequality is a consequence of the lower semicontinuity of the functional

$$v \to \int_0^{1-\eta} \varphi^2(v'(s))ds$$

with respect to the weak convergence in $W^{1,1}$. Then the liminf inequality follows by the arbitrariness of η.

Limsup inequality. Let $u \in W^{1,1}(0,1)$. We define $u_i^n := u(\lambda_n i)$ for every $i = 0, \ldots, n$; by the Fundamental Theorem of Calculus and by Jensen's inequality

$$E_n(u^n) = \lambda_n \sum_{i=0}^{n-1} \varphi^1 \left(\fint_{\lambda_n i}^{\lambda_n(i+1)} u'(s)ds \right) + \lambda_n \sum_{i=0}^{n-2} \varphi^2 \left(\fint_{\lambda_n i}^{\lambda_n(i+2)} u'(s)ds \right)$$

$$\leq \sum_{i=0}^{n-1} \int_{\lambda_n i}^{\lambda_n(i+1)} \varphi^1(u'(s))ds + \frac{1}{2} \sum_{i=0}^{n-2} \int_{\lambda_n i}^{\lambda_n(i+2)} \varphi^2(u'(s))ds$$

$$\leq \int_0^1 \varphi^1(u'(s))ds + \int_0^1 \varphi^2(u'(s))ds.$$

We conclude this section by proving the analogue of Theorem 2.4.1 for non convex potentials.

Theorem 2.4.2 *Let E_n be defined as in* (2.4.1) (*and* (2.4.3)) *and assume that φ^1 and φ^2 are continuous and nonnegative. Assume, moreover, that φ^1 is superlinear at infinity. Then the Γ-limit of E_n with respect to the strong convergence in L^1 is the functional E defined for $u \in L^1(0,1)$ as*

$$E(u) = \begin{cases} \int_0^1 \varphi(u'(s))ds & u \in W^{1,1}(0,1) \\ +\infty & u \in L^1(0,1) \setminus W^{1,1}(0,1), \end{cases}$$

where $\varphi(z) = (\varphi^1 \# \varphi^2)^{**}(z)$, and $\varphi^1 \# \varphi^2$ denotes the inf-convolution *of the two potentials, defined as*

$$(\varphi^1 \# \varphi^2)(z) = \varphi^2(z) + \frac{1}{2} \inf\{\varphi^1(z_1) + \varphi^1(z_2) : z_1 + z_2 = 2z\}.$$

Remark 2.4.1 Notice that if φ^1 is convex, then $(\varphi^1 \# \varphi^2)(z) = \varphi^1(z) + \varphi^2(z)$.

Proof (*Compactness of sequences with bounded energy*) As in Theorem 2.4.1 the convergence follows directly from Theorem 2.3.4, since φ^2 is nonnegative.

Liminf inequality. Let $u \in W^{1,1}(0,1)$ and let $\tilde{u}^n \to u$ strongly in L^1, with $E_n(\tilde{u}^n) \leq c$. From the compactness result we have that $\tilde{u}^n \rightharpoonup u$ weakly in $W^{1,1}$.

In this case it is very convenient to rewrite the energy in an alternative way. We can look at the lattice I_n as obtained by the superposition of two lattices with double spacing $2\lambda_n$, with a relative λ_n-shift. For convenience we can write

$$I_n = I_n^{\text{odd}} \cup I_n^{\text{even}}, \quad I_n^{\text{odd}} \cap I_n^{\text{even}} = \emptyset,$$

where $I_n^{\text{odd}} = \{\lambda i : i \text{ odd}\}$, and $I_n^{\text{even}} = \{\lambda i : i \text{ even}\}$.

Therefore, splitting the sums in E_n into even and odd lattice,

$$
\begin{aligned}
&E_n(\tilde{u}^n)\\
&= \sum_{\substack{i=0\\ \text{odd}}}^{n-2} \lambda_n \left\{ \varphi^2\left(\frac{u^n_{i+2}-u^n_i}{2\lambda_n}\right) + \frac{1}{2}\varphi^1\left(\frac{u^n_{i+1}-u^n_i}{\lambda_n}\right) + \frac{1}{2}\varphi^1\left(\frac{u^n_{i+2}-u^n_{i+1}}{\lambda_n}\right)\right\}\\
&+ \sum_{\substack{i=0\\ \text{even}}}^{n-2} \lambda_n \left\{ \varphi^2\left(\frac{u^n_{i+2}-u^n_i}{2\lambda_n}\right) + \frac{1}{2}\varphi^1\left(\frac{u^n_{i+1}-u^n_i}{\lambda_n}\right) + \frac{1}{2}\varphi^1\left(\frac{u^n_{i+2}-u^n_{i+1}}{\lambda_n}\right)\right\}\\
&+ \frac{1}{2}\lambda_n\varphi^1\left(\frac{u^n_n-u^n_{n-1}}{\lambda_n}\right) + \frac{1}{2}\lambda_n\varphi^1\left(\frac{u^n_1-u^n_0}{\lambda_n}\right).
\end{aligned}
$$

Using the definition of the inf-convolution and the fact that $\varphi^1 \geq 0$ we have

$$
\begin{aligned}
E_n(\tilde{u}^n) &\geq \sum_{\substack{i=0\\ \text{odd}}}^{n-2} \lambda_n(\varphi^1\#\varphi^2)\left(\frac{u^n_{i+2}-u^n_i}{2\lambda_n}\right) + \sum_{\substack{i=0\\ \text{even}}}^{n-2} \lambda_n(\varphi^1\#\varphi^2)\left(\frac{u^n_{i+2}-u^n_i}{2\lambda_n}\right)\\
&\geq \sum_{\substack{i=0\\ \text{odd}}}^{n-2} \lambda_n(\varphi^1\#\varphi^2)^{**}\left(\frac{u^n_{i+2}-u^n_i}{2\lambda_n}\right) + \sum_{\substack{i=0\\ \text{even}}}^{n-2} \lambda_n(\varphi^1\#\varphi^2)^{**}\left(\frac{u^n_{i+2}-u^n_i}{2\lambda_n}\right).
\end{aligned}
$$

Therefore it only remains to write the sums as integrals, as usual. To this aim, we define *odd and even interpolants* on the lattices I_n^{odd} and I_n^{even}, say $\tilde{u}^n_{\text{odd}}$ and $\tilde{u}^n_{\text{even}}$. In formula,

$$
(u^n_{\text{odd}})'(s) = \frac{u^n_{i+2}-u^n_i}{2\lambda_n}, \quad s \in (\lambda_n i, \lambda_n(i+2)), \quad i \text{ odd},
$$

and analogously for u^n_{even}. The previous estimate implies that, for every $\eta > 0$,

$$
E_n(\tilde{u}^n) \geq \frac{1}{2}\int_\eta^{1-\eta} (\varphi^1\#\varphi^2)^{**}((\tilde{u}^n_{\text{odd}})'(s))ds + \frac{1}{2}\int_\eta^{1-\eta} (\varphi^1\#\varphi^2)^{**}((\tilde{u}^n_{\text{even}})'(s))ds.
$$

Notice that also such interpolants converge weakly in $W^{1,1}$ to u. The claim then follows by lower semicontinuity, since the integrands are convex.

Limsup inequality. As usual, via density arguments we can restrict to the case of a linear function $u(s) = zs$. Moreover, to focus on the inf-convolution rather than on the convexification we assume that $(\varphi^1\#\varphi^2)^{**}(z) = (\varphi^1\#\varphi^2)(z)$.

Let now $\eta > 0$, and z_1 and z_2 be such that $z_1 + z_2 = 2z$ and

$$
\varphi^2(z_2) + \frac{1}{2}\varphi^1(z_1) + \frac{1}{2}\varphi^1(z_2) < (\varphi^1\#\varphi^2)^{**}(z) + \eta. \tag{2.4.4}
$$

The idea for the recovery sequence in this case is that it should have constant slope z on the larger $2\lambda_n$-lattices I_n^{odd} and I_n^{even}, and slope alternatively z_1 and z_2 on the smaller λ_n-lattice I_n. In this way the energy term coming from the next-to-nearest neighbour interactions will be $\varphi^2(z)$, while half of the nearest neighbour interactions, say, the even ones, will contribute $\varphi^1(z_1)$, and the odd ones $\varphi^1(z_2)$. More in detail, the recovery sequence is

$$u_i^n = \begin{cases} \lambda_n i z & i \text{ odd}, \\ \lambda_n (i-1) z + \lambda_n z_1 & i \text{ even}. \end{cases}$$

It is easy to check that $u_{i+2}^n - u_i^n = 2\lambda_n z$ for both i odd and even. Instead

$$u_{i+1}^n - u_i^n = \begin{cases} \lambda_n z_2, & i \text{ odd}, \\ \lambda_n z_1, & i \text{ even}. \end{cases}$$

Therefore,

$$E_n(\{u_i^n\}) = \varphi^2(z) + \lambda_n \sum_{\substack{i=0\\ \text{odd}}}^{n-1} \varphi^1(z_2) + \lambda_n \sum_{\substack{i=0\\ \text{even}}}^{n-1} \varphi^1(z_1),$$

which implies the claim by (2.4.4).

2.4.2 *Long-Range Interactions: The Case of Dislocations*

In this section we consider long-range interactions. The particles in this case are *dislocations*, i.e., defects in the arrangement of the atoms in a crystal lattice of a metal, steel for instance. Dislocations are of great importance for their role as main carriers of the permanent or plastic deformation in metals. The main mechanism of plastic deformation is the relative slip of planes of atoms, and the presence and the motion of large numbers of dislocations make the slip energetically possible. Typically, the number of dislocations in a portion of metal is very large, and therefore a discrete model that keeps track of each dislocation is not useful in practice. Hence the need for *upscaled* models, in terms of a continuum, averaged quantity: The dislocation density.

Similarly as in the cases considered before, the starting point of our analysis is a one-dimensional discrete energy describing the pairwise interaction among dislocations. In this case, however, we will consider all interactions, not only nearest and next-to-nearest neighbour interactions. Actually, the interactions we consider are not among *single* dislocations, but among vertical *walls* of dislocations as in Fig. 2.6.

Each wall is h-periodic, where $h > 0$ is the spacing between consecutive dislocations within the wall. Therefore the discrete function $\{u_i\} \subset \mathscr{A}_n(0,1)$ represents the n deformed positions of the dislocations walls with respect to the reference lattice I_n.

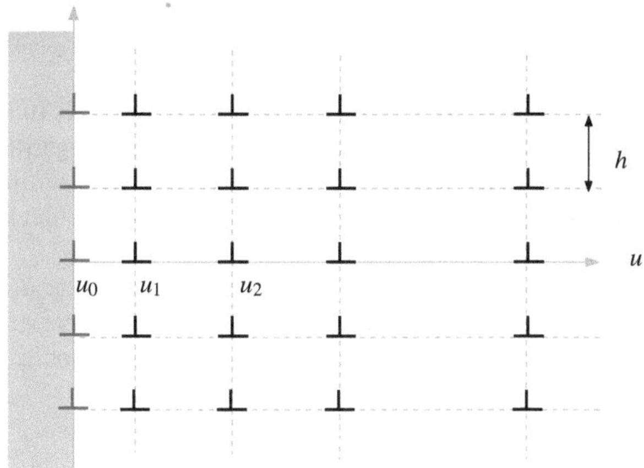

Fig. 2.6 The dislocation configuration. Infinite, *vertical walls* of equispaced dislocations are free to move in the *horizontal direction*. A wall of fixed dislocations is pinned at $u_0 = 0$ and acts as a repellent

The discrete dislocation energy is defined as

$$\mathscr{E}_n(\{u_i\}) := 2K \sum_{i=0}^{n} \sum_{\substack{j=0 \\ j \neq i}}^{n} V\left(\frac{u_i - u_j}{h}\right) + \sigma \sum_{i=0}^{n} u_i, \tag{2.4.5}$$

where $K > 0$ is a mechanical constant, $\sigma > 0$ is an applied load, and the interaction energy V is shown in Fig. 2.7 and is defined as

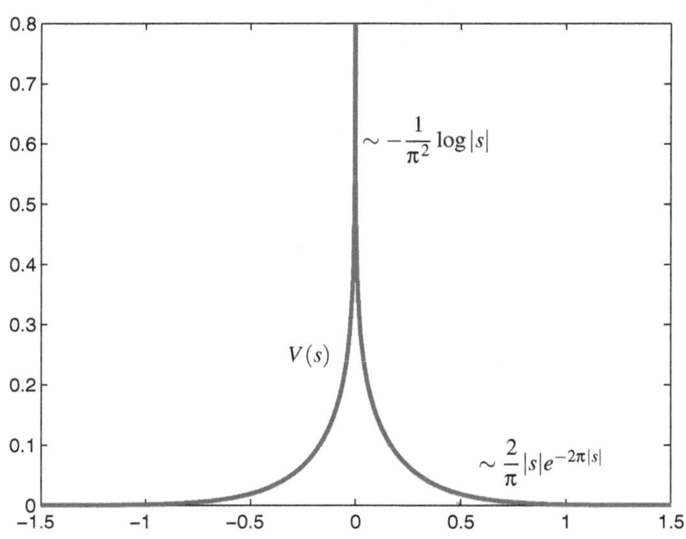

Fig. 2.7 The interaction energy V

$$V(s) := \frac{1}{\pi} s \coth \pi s - \frac{1}{\pi^2} \log(2 \sinh \pi s) = \frac{2}{\pi} \frac{|s|}{(e^{2\pi |s|} - 1)} - \frac{1}{\pi^2} \log(1 - e^{-2\pi |s|}).$$

Therefore the first term in the discrete energy $\mathscr{E}_n$ is fully repulsive: each pair of walls repels each other, with a potential that diverges logarithmically as the walls approach each other. The second term of the energy, accounting for the applied load, drives the walls to the left. Moreover, we impose a Dirichlet boundary condition for the left most wall, $u_0 = 0$, which models pinning.

In the special case where the vertical distance h between consecutive dislocations within a wall and the *average* distance between consecutive dislocations in the horizontal direction is of the same order, non-dimensionalising the energy leads to

$$E_n(\{u_i\}) := 2\lambda_n \sum_{i=0}^{n} \sum_{\substack{j=0 \\ j \neq i}}^{n} V\left(\frac{u_i - u_j}{\lambda_n}\right) + \lambda_n \sum_{i=0}^{n} u_i,$$

which looks more similar to the discrete energies considered so far. The fact that V is even, moreover, allows us to rewrite the energy in the more convenient form

$$E_n(\{u_i\}) := \lambda_n \sum_{k=1}^{n} \sum_{i=0}^{n-k} V\left(\frac{u_{i+k} - u_i}{\lambda_n}\right) + \lambda_n \sum_{i=0}^{n} u_i. \tag{2.4.6}$$

Notice that the energy V is convex, but is not superlinear; in particular it decays exponentially to zero (recall that the Lennard-Jones potential had a polynomial decay).

As in the Lennard-Jones case the lack of the superlinear growth entails that the limit deformations can have discontinuities. In what follows, however, we will only identify the Γ-limit on a subspace of the limit deformations, i.e., on continuous functions in $W^{1,1}(0, 1)$. In this way we will only focus on the effect of long-range interactions. For the general case we refer to [14, 22] (see also [25] for the case of a bounded domain $[0, L]$ for the positions of the walls).

Theorem 2.4.3 *The functionals E_n in* (2.4.6) *Γ-converge with respect to the weak convergence in $W^{1,1}$ to the functional E defined for increasing functions $u \in W^{1,1}(0, 1)$ as*

$$E(u) = \int_0^1 V_{\text{eff}}(u'(s))\, ds + \int_0^1 u(s)\, ds,$$

where $V_{\text{eff}}(t) := \sum_{k=1}^{\infty} V(kt)$ *for every* $t \in \mathbb{R}$.

Proof (*Liminf inequality*) Let $u \in W^{1,1}(0, 1)$ be an increasing function, and let $\{u_i^n\} \subset \mathscr{A}_n(0, 1)$ be a sequence with bounded energy E_n and such that the interpolants $\tilde{u}^n \rightharpoonup u$ weakly in $W^{1,1}$.

We first rewrite the functional E_n in a more convenient way. For every $k \in \mathbb{N}$ we define the function $V^k(t) := V(kt)$. Hence we have

$$E_n(\{u_i^n\}) = \lambda_n \sum_{k=1}^{n} \sum_{j=0}^{n-k} V^k\left(\frac{u_{i+k}^n - u_i^n}{k\lambda_n}\right) + \lambda_n \sum_{i=0}^{n} u_i^n. \tag{2.4.7}$$

The expression in the argument of V^k resembles a gradient, and we make the gradient term appear explicitly using the affine interpolant $\tilde{u}^n$ of the $\{u_i^n\}$. For fixed $k \in \{0,\dots,n-1\}$ and $\ell \in \{0,\dots,n-1\}$ and for $i \le \ell \le k+i-1$, we have

$$\frac{u_{i+k}^n - u_i^n}{k\lambda_n} = \frac{1}{k} \sum_{m=i-\ell}^{i-\ell+k-1} \frac{u_{\ell+m+1}^n - u_{\ell+m}^n}{\lambda_n} = \frac{1}{k} \sum_{m=i-\ell}^{i-\ell+k-1} (\tilde{u}^n)'(s+\lambda_n m)$$

for every $s \in \left(\frac{\ell}{n}, \frac{\ell+1}{n}\right)$. Then

$$\begin{aligned}
\lambda_n V^k\left(\frac{u_{i+k}^n - u_i^n}{k\lambda_n}\right) &= \frac{1}{k} \sum_{\ell=i}^{i+k-1} \int_{\frac{\ell}{n}}^{\frac{\ell+1}{n}} V^k\left(\frac{1}{k} \sum_{m=i-\ell}^{i-\ell+k-1} (\tilde{u}^n)'(s+\lambda_n m)\right) ds \\
&\overset{(j=\ell-i)}{=} \frac{1}{k} \sum_{j=0}^{k-1} \int_{\frac{j+i}{n}}^{\frac{j+i+1}{n}} V^k\left(\frac{1}{k} \sum_{m=-j}^{k-1-j} (\tilde{u}^n)'(s+\lambda_n m)\right) ds.
\end{aligned}$$

Therefore, we can rewrite the first term in (2.4.7) in terms of the functions $\tilde{u}^n$, as

$$\begin{aligned}
\sum_{i=0}^{n-k} \lambda_n V^k\left(\frac{u_{i+k}^n - u_i^n}{k\lambda_n}\right) &= \frac{1}{k} \sum_{i=0}^{n-k} \sum_{j=0}^{k-1} \int_{\frac{j+i}{n}}^{\frac{j+i+1}{n}} V^k\left(\frac{1}{k} \sum_{m=-j}^{k-1-j} (\tilde{u}^n)'(s+\lambda_n m)\right) ds \\
&= \frac{1}{k} \sum_{j=0}^{k-1} \int_{\frac{j}{n}}^{\frac{n-k+j+1}{n}} V^k\left(\frac{1}{k} \sum_{m=-j}^{k-1-j} (\tilde{u}^n)'(s+\lambda_n m)\right) ds \\
&= \frac{1}{k} \sum_{j=0}^{k-1} \int_{\frac{j}{n}}^{1-\frac{k-j-1}{n}} V^k\left(\frac{1}{k} \sum_{m=-j}^{k-1-j} (\tilde{u}^n)'(s+\lambda_n m)\right) ds.
\end{aligned}$$

We claim that for every fixed integer N the following lower bound is satisfied:

$$\liminf_{n\to+\infty} E_n(\{u_i^n\}) \ge \int_0^1 V_{\text{eff}}^N(u'(s))\, ds + \int_0^1 u(s)\, ds, \tag{2.4.8}$$

where the energy density $V_{\rm eff}^N$ is defined as $V_{\rm eff}^N(t) := \sum_{k=1}^N V(kt)$ for every $t \in \mathbb{R}$. This claim implies the lower bound by the arbitrariness of N.

We focus on the first term of the discrete energy E_n; since by assumption $\tilde{u}^n \rightharpoonup u$ in $W^{1,1}$, in particular $\tilde{u}^n \to u$ strongly in L^1, therefore the convergence of the second term of the energy follows directly. (Notice that the second term in the discrete energy is a Riemann sum for the integral of $\tilde{u}^n$.)

Let N be fixed (independent of n). Then, for $n \geq N$,

$$\lambda_n \sum_{k=1}^{n} \sum_{i=0}^{n-k} V^k \left(\frac{u_{i+k}^n - u_i^n}{k\lambda_n} \right)$$

$$\geq \sum_{k=1}^{N} \frac{1}{k} \sum_{j=0}^{k-1} \int_{\frac{j}{n}}^{1-\frac{k-j-1}{n}} V^k \left(\frac{1}{k} \sum_{m=-j}^{k-1-j} (\tilde{u}^n)'(s + \lambda_n m) \right) ds.$$

We note that, since j and k run through a finite set independent of n, for every $\eta > 0$ there exists an integer $\nu(\eta)$ such that, if $n \geq \nu(\eta)$, then

$$\int_{\frac{j}{n}}^{1-\frac{k-j-1}{n}} V^k \left(\frac{1}{k} \sum_{m=-j}^{k-j-1} (\tilde{u}^n)'(s + \lambda_n m) \right) ds$$

$$\geq \int_{\eta}^{1-\eta} V^k \left(\frac{1}{k} \sum_{m=-j}^{k-j-1} (\tilde{u}^n)'(s + \lambda_n m) \right) ds$$

for every j and for every k. Moreover, for every k and j, also the convex combination

$$\tilde{u}_{k,j}^n(s) := \frac{1}{k} \sum_{m=-j}^{k-j-1} \tilde{u}^n (s + \lambda_n m)$$

converges to u weakly in $W^{1,1}(\eta, 1-\eta)$. Therefore, for every k and j

$$\liminf_{n\to\infty} \int_{\eta}^{1-\eta} V^k((\tilde{u}_{k,j}^n)'(s))ds \geq \int_{\eta}^{1-\eta} V^k(u'(s))ds, \tag{2.4.9}$$

since the integral functional in (2.4.9) is lower semicontinuous with respect to the weak convergence in $W^{1,1}$, by the convexity of V^k. In conclusion,

$$\begin{aligned}\liminf_{n\to\infty} E_n(\{u_i^n\}) &\geq \sum_{k=1}^{N}\frac{1}{k}\sum_{j=0}^{k-1}\int_{\eta}^{1-\eta} V^k(u'(s))\,ds + \int_0^1 u(s)\,ds \\ &= \sum_{k=1}^{N}\int_{\eta}^{1-\eta} V^k(u'(s))\,ds + \int_0^1 u(s)\,ds \\ &= \int_{\eta}^{1-\eta} V_{\mathrm{eff}}^N(u'(s))\,ds + \int_0^1 u(s)\,ds,\end{aligned}$$

and the claim (2.4.8) follows by the arbitrariness of N and η.

Limsup inequality. Let $u \in W^{1,1}(0,1)$. The recovery sequence in this case is simply given by the evaluations of the continuous function u on the points of the lattice I_n.

Remark 2.4.2 The discrete energies E_n are coercive in $BV_{\mathrm{loc}}(0,1)$, and the natural convergence to be considered for the Γ-convergence is, also in this case, the strong convergence in L^1.

2.5 Further Directions

We conclude these lecture notes by illustrating a couple of generalisations and interesting ideas.

2.5.1 Asymptotic Expansion via Γ-Convergence

Let us consider the Lennard-Jones interaction energy analysed in Theorem 2.3.7. We noticed that the limit deformations can have discontinuities, which makes the Lennard-Jones potential suitable for dealing with fracture mechanics problems. On the other hand, the Γ-limit energy obtained in Theorem 2.3.7 does not penalise in any way the occurrence of a fracture, since there is no cost associated to discontinuities of the limit deformations. Therefore in principle a deformation could have infinitely many discontinuity points. This is quite unsatisfactory from the mechanical point of view, since the opening of a crack requires energy and it is not energetically convenient for an elastic body to fracture in infinitely many points.

To understand better the limitations of the limit model, let us have a closer look at the limit energy E. For $u \in P\text{-}W^{1,1}(0,1)$, u increasing, and $u(0)=0$, $u(1)=\ell$,

$$E(u) = \int_0^1 J^{**}(u'(s))ds.$$

Let $\ell < \gamma$, where γ is the unique minimum point of the interaction potential J. Then the function $\bar{u}(s) = \ell s$ is the unique minimiser of the Γ-limit. This follows from the Jensen's inequality, since for every admissible u

$$\int_0^1 J^{**}(u'(s))ds \geq J^{**}\left(\int_0^1 u'(s)ds\right) = J^{**}(u(1) - u(0)) = J^{**}(\ell)$$
$$= \int_0^1 J^{**}(\bar{u}'(s))ds.$$

Hence in this case it is energetically unfavourable to create a fracture.

Assume now instead that $\ell > \gamma$. Then the continuous function $\bar{u}(s) = \ell s$ is still a minimiser, but it is not the only one. Also any increasing function u with $u' \geq \gamma$ almost everywhere is a minimiser, with the same energy.

Therefore, while in the case $\ell < \gamma$ the Γ-limit provides us with all the necessary information in order to understand the behaviour of the discrete system, in the case $\ell \geq \gamma$ the information provided by the Γ-limit is very poor, since it cannot distinguish between a continuous deformation and a deformation with infinitely many discontinuity points.

How can we improve the scenario provided by the limit energy E? What we expect from a mechanical reasoning is that if the boundary conditions are in the convex-elastic part of J, i.e., if $\ell < \gamma$, then the best thing to do from the energetic point of view is to deform elastically, rather than creating a fracture. This is exactly what the Γ-limit predicts in the case $\ell < \gamma$. If instead the boundary condition is outside the convex-elastic part of J, i.e., if $\ell \geq \gamma$, then we expect the occurrence of a fracture. This is in contrast with the prediction provided by the energy E, for which also a continuous deformation with slope ℓ is a minimiser.

Hence, we would like to *correct* the energy for $\ell > \gamma$, and penalise continuous deformations with large slope, as well as the number of discontinuity points.

The idea is, as for the Taylor formula for functions, to look at a *higher order approximation* of the discrete energy E_n in the small parameter λ_n (see [4]). In terms of a *Taylor*-expansion, the Γ-limit can be considered as the zero-order term in λ_n. The next term to identify is the first-order one. This is done by looking at the new discrete energy:

$$E_n^{(1)}(\{u_i\}) := \frac{E_n(\{u_i\}) - \min E}{\lambda_n}.$$

Note that, for $\ell > \gamma$, $\min E = J(\gamma)$. Therefore the new energy functional is

$$E_n^{(1)}(u) := \sum_{i=0}^{n-1} \left\{ J\left(\frac{u_{i+1} - u_i}{\lambda_n}\right) - J(\gamma) \right\}. \tag{2.5.1}$$

Heuristically, the energy $E_n^{(1)}$ for large n can be finite only if almost every discrete slope equals γ. In fact in that case the terms in the sum (2.5.1), which are otherwise positive (since γ is the minimum point of J), are zero. Hence, the study of this higher order correction of the energy reduces the set of possible minimisers to the deformations with slope γ.

This result is proved in the next Theorem.

Theorem 2.5.1 *The functionals $E_n^{(1)}$ Γ-converge to the functional $E^{(1)}$ given by*

$$E_n^{(1)}(u) = \begin{cases} -J(\gamma)\#(S(u)) & u \text{ increasing}, u' = \gamma \text{ a.e.} \\ +\infty & \text{otherwise}, \end{cases}$$

on P-$W^{1,1}(0, 1)$.

Proof (Hint for the recovery sequence) Consider a function $u \in P$-$W^{1,1}(0, 1)$, u increasing and such that $u' = \gamma$ a.e. Assume, for simplicity, that u has only one discontinuity point, say $t \in (0, 1)$, and assume that $u^+(t) - u^-(t) = a > 0$. Define $u_i^n := u(\lambda_n i)$.

Let us choose a sequence $(t_n) \subset \mathbb{N}$, such that $t_n \to \infty$, $\lambda_n t_n \leq t$ and $\lambda_n t_n \to t$. Then, clearly,

$$\frac{u^{i+1} - u^i}{\lambda_n} = \begin{cases} \gamma & i \neq t_n, \\ \sim \frac{a}{\lambda_n} & i = t_n. \end{cases}$$

Therefore

$$E^{(1)}(u_i^n) \sim J\left(\frac{a}{\lambda_n}\right) - J(\gamma) \to -J(\gamma).$$

2.5.2 Equivalence by Γ-Convergence

Strictly related to the previous example is the concept of *Γ-equivalence* introduced by Braides and Truskinovsky in [10].

The key observation in [10] is that the first order Γ-expansion obtained in the previous section, namely

$$E(u) + \lambda_n E^{(1)}(u), \tag{2.5.2}$$

regarded as a function of the parameter ℓ (the elongation of the one-dimensional chain) is discontinuous at $\ell = \gamma$, where γ is the minimum point of J. The value $\ell = \gamma$ represents a threshold between the elastic regime ($\ell < \gamma$), and the regime $\ell > \gamma$ in which the chain experiences the opening of a crack.

The idea in [10] is then to construct a uniformly Γ-equivalent approximation that improves the first-order approximation (2.5.2) close to γ.

2.5.3 Empirical Measures Approach

As stated in Remark 2.3.1, it is sometimes convenient to express the discrete energy in terms of the empirical measures μ_n, rather than in terms of the piecewise affine interpolations. To illustrate this alternative formulation, let us consider again the discrete dislocation energy in (2.4.5):

$$\mathscr{E}_n(\{u_i\}) := 2K \sum_{i=0}^{n} \sum_{\substack{j=0 \\ j \neq i}}^{n} V\left(\frac{u_i - u_j}{h}\right) + \sigma \sum_{i=0}^{n} u_i.$$

In Sect. 2.4.2 we analysed the case where the vertical distance h between consecutive dislocations within a wall and the average distance between consecutive dislocations in the horizontal direction is of the same order. We now consider the case where h is of the same order as the *whole* pileup region. In this case, the non-dimensional discrete energy is

$$E_n(\{u_i\}) = 2\frac{1}{n^2} \sum_{i=0}^{n} \sum_{\substack{j=0 \\ j \neq i}}^{n} V\left(u_i - u_j\right) + \frac{1}{n} \sum_{i=0}^{n} u_i,$$

since $\lambda_n = 1/n$. At this point the first term of the energy does not contain a gradient-like term as before, and hence the previous approach does not apply. In terms of the empirical measure $\mu_n = (\sum \delta_{u_i})/n$ instead the energy reads

$$E_n(\{u_i\}) = 2 \iint\limits_{(0,\infty)^2 \setminus D} V(x-y)d\mu_n(x)d\mu_n(y) + \int_0^\infty x d\mu_n(x),$$

where D denotes the diagonal in $(0, \infty)^2$; from this expression is easy to guess that the Γ-limit energy is

$$E(\mu) = 2 \iint\limits_{(0,\infty)^2} V(x-y)d\mu(x)d\mu(y) + \int_0^\infty x d\mu(x).$$

This guess can be made rigorous, and the details of the proof are in [14].

Remark 2.5.1 The measure approach has a more natural extension to higher dimensions, since no underlying lattice is required in this case.

2.5.4 Rate-Independent and Gradient Flow Evolutions

Let us focus again on the discrete dislocation energy (2.4.5). As explained above, one can couple the energy of the system with a suitable *dissipation* and consider the corresponding discrete evolution. Deriving a limit evolution is generally a very subtle issue, and relies on the results [20] (for rate-independent evolutions) and [21] (for gradient flows).

2.6 Exercises

1. Compute the Gamma-limit of the following sequences of functions:

   (a) $f_n : \mathbb{R} \to \mathbb{R}$, $f_n(x) = nxe^{-2n^2x^2}$;
   (b) $f_n = f$, where $f(x) = 1$ if $x < 1/2$ and $f(x) = 0$ if $x \geq 1/2$;
   (c) $f_n = f$, where $f(x) = 1$ if $x \leq 1/2$ and $f(x) = 0$ if $x > 1/2$.
   (d) Prove that it can be

$$\Gamma - \lim_n (f_n + g_n) \neq \Gamma - \lim_n f_n + \Gamma - \lim_n g_n.$$

2. Give an example of a sequence bounded in L^1 but not weakly compact.
3. (a) Let X be a topological space and consider $F_i : X \to \mathbb{R}$ for $i \in I$, with I countable or uncountable. Define

$$F(x) := \sup_{i \in I} F_i(x).$$

   Prove that if for every i the functional F_i is sequentially lower semicontinuous, then F is sequentially lower semicontinuous.
   (b) Under the same assumptions, prove or disprove that

$$G(x) := \inf_{i \in I} F_i(x)$$

   is sequentially lower semicontinuous.
4. (a) Let X be a topological space and let $F : X \to \mathbb{R}$. If F is coercive and lower semicontinuous, then there exists a minimizer of F.

   Hint. Use the following result in general topology. Let X be a topological space and let $(A_n)_n$ be closed sets, $A_n \neq \emptyset$ for every n, and decreasing, i.e.,

$$A_1 \supseteq A_2 \supseteq \cdots \supseteq A_n.$$

Assume, moreover, that there exists a compact set K such that $A_1 \subseteq K$. Then

$$\bigcap_{n=1}^{\infty} A_n \neq \emptyset.$$

(b) Let X be a topological space and let $F : X \to \mathbb{R}$. If F is sequentially coercive and sequentially lower semicontinuous, then there exists a minimizer of F.

5. (a) Does the functional

$$F(u) := \int_0^1 |u'(x)|^2 dx + \int_0^1 \frac{1}{1 + u^2(x)}\, dx$$

have a minimum on $W^{1,2}(0, 1)$?

(b) Does the functional

$$F(u) := \int_{-1}^{1} (1 + x^2)|u'(x)|\, dx$$

have a minimum on $X := \{u \in W^{1,1}(0, 1), u(-1) = -1, u(1) = 1\}$?

(c) Does the functional

$$F(u) := \int_{-1}^{1} (1 + x^2)|u'(x)|^p\, dx, \quad p > 1,$$

have a minimum on $X := \{u \in W^{1,p}(0, 1), u(-1) = -1, u(1) = 1\}$?

(d) Does the functional

$$F(u) := \int_0^1 [((u')^2 - 1)^2 + u^4]\, dx$$

have a minimum on $X := \{u \in W^{1,4}(0, 1), u(0) = 0, u(1) = 0\}$?

6. Let X be a metric space and let $F : X \to \mathbb{R}$; prove that the following are equivalent:

(a) for every $x \in X$ and for every $x_n \to x$: $F(x) \leq \liminf_{n\to\infty} F(x_n)$;
(b) $S_t = \{x : F(x) \leq t\}$ is closed for every t.

7. Let $f : (0, 1) \to [0, \infty]$ be convex, C^1 and Lipschitz. Then the functional

$$F : L^1(0, 1) \to [0, \infty], \quad F(u) = \int_0^1 f(u)dt$$

is weakly lower semicontinuous in L^1.

Acknowledgments The author would like to thank her previous affiliations, the Eindhoven University of Technology and the University of Glasgow, where most of this work was carried out.

References

1. Alicandro, R., Cicalese, M.: Variational analysis of the asymptotics of the XY model. Arch. Ration. Mech. Anal. **192**, 501–536 (2009)
2. Alicandro, R., Cicalese, M., Ponsiglione, M.: Variational equivalence between Ginzburg-Landau, XY spin systems and screw dislocation energies. Indiana Univ. Math. J. **60**, 171–208 (2011)
3. Ambrosio L., Fusco N., Pallara D.: Functions of Bounded Variation and Free Discontinuity Problems. Oxford Mathematical Monographs. Oxford University Press, first edition, 2000
4. Anzellotti, G., Baldo, S.: Asymptotic development by Γ-convergence. Appl. Math. Optim. **27**, 105–123 (1993)
5. Braides, A.: Local Minimisation, Variational Evolution and Γ-Convergence. Springer International Publishing, Switzerland (2014)
6. Braides, A., Cicalese, M.: Surface energies in nonconvex discrete systems. Math. Models Methods Appl. Sci. **17**, 985–1037 (2007)
7. Braides, A., Dal Maso, G., Garroni, A.: Variational formulation of softening phenomena in fracture mechanics: the one-dimensional case. Arch. Rational Mech. Anal. **146**, 23–58 (1999)
8. Braides A., Gelli M.S.: From discrete systems to continuous variational problems: an introduction. In: Topics on Concentration Phenomena and Problems with Multiple Scales, pp. 3–77 (2006)
9. Braides, A., Lew, A., Ortiz, M.: Effective cohesive behavior of layers of interatomic planes. Arch. Rational Mech. Anal. **180**, 151–182 (2006)
10. Braides, A., Truskinovsky, L.: Asymptotic expansions by Gamma-convergence. Cont. Mech. Thermodyn. **20**, 21–62 (2008)
11. Brezis H.: Analyse fonctionnelle. Théorie et Applications. Dunod (2005)
12. Charlotte, M., Truskinovsky, L.: Linear elastic chain with a hyper-pre-stress. J. Mech. Phys. Solids **50**, 217–251 (2002)
13. Dal Maso G.: An introduction to Γ-convergence. Birkhäuser, Boston, 1993
14. Geers, M.G.D., Peerlings, R.H.J., Peletier, M.A., Scardia, L.: Asymptotic behaviour of the dislocation density in a pile-up of edge dislocations. Arch. Rational Mech. Anal. **209**(2), 495–539 (2013)
15. Groma, I.: Derivation of the equations of motion of interacting parallel edge dislocations from a Hamiltonian. Phys. Rev. E **52**(13), 9125–9128 (1995)
16. Groma, I.: Link between the microscopic and mesoscopic length-scale description of the collective behaviour of dislocations. Phys. Rev. B **56**(10), 5807–5813 (1997)
17. Hall, C.: Asymptotic analysis of a pile-up of edge dislocations. Mat. Sci. Eng. A-Struct. **530**, 144–148 (2011)

18. Hudson, T.: Γ-Expansion for a 1D Confined Lennard-Jones Model with point defect. Netw. Heterog. Media **8**(2), 501–527 (2014)
19. Kresse, O., Truskinovsky, L.: Mobility of lattice defects: discrete and continuum approaches. J. Mech. Phys. Solids **51**, 1305–1332 (2003)
20. Mielke, A., Roubíček, T., Stefanelli, U.: Γ-convergence and relaxations for rate-independent evolution problems. Calc. Var. Partial Differ. Equ. **31**, 387–416 (2008)
21. Sandier, E., Serfaty, S.: Gamma-convergence of gradient flows with applications to Ginzburg-Landau, pp. 1627–1672. LVII, Comm. Pure Appl. Math. (2004)
22. Scardia, L., Peerlings, R.H.J., Peletier, M.A., Geers, M.G.D.: Mechanics of dislocation pile-ups: a unification of scaling regimes. J. Mech. Phys. Solids **70**, 42–61 (2014)
23. Scardia, L., Schlömerkemper, A., Zanini, C.: Boundary layer energies for nonconvex discrete energies. Math. Models Methods Appl. Sci. **21**, 777–817 (2011)
24. Schlömerkemper, A., Schmidt, B.: Discrete-to-continuum limits of magnetic forces in dependence on the distance between bodies. Arch. Rational Mech. Anal. **192**, 589–611 (2008)
25. van Meurs, P.J.P., Muntean, A., Peletier, M.A.: Upscaling of dislocation walls in finite domains. Eur. J. Appl. Math. **25**(6), 749–781 (2014)

Chapter 3
On Evolutionary Γ-Convergence for Gradient Systems

Alexander Mielke

In memory of Eduard, Waldemar, and Elli Mielke.

Abstract In these notes we discuss general approaches for rigorously deriving limits of generalized gradient flows. Our point of view is that a generalized gradient system is defined in terms of two functionals, namely the energy functional $\mathcal{E}_\varepsilon$ and the dissipation potential $\mathcal{R}_\varepsilon$ or the associated dissipation distance. We assume that the functionals depend on a small parameter and that the associated gradient systems have solutions u_ε. We investigate the question under which conditions the limits u of (subsequences of) the solutions u_ε are solutions of the gradient system generated by the Γ-limits $\mathcal{E}_0$ and $\mathcal{R}_0$. Here the choice of the right topology will be crucial as well as additional structural conditions. We cover classical gradient systems, where $\mathcal{R}_\varepsilon$ is quadratic, and rate-independent systems as well as the passage from classical gradient to rate-independent systems. Various examples, such as periodic homogenization, are used to illustrate the abstract concepts and results.

3.1 Introduction

This work is devoted to the study of evolutionary systems of the form $\dot{u} = V_\varepsilon(u)$, $u(0) = u_\varepsilon^0$ depending on a small parameter ε, which may arise as a microscopic length scale. The aim is to find effective equations such that any suitable limit u of solutions u_ε satisfies the effective model $\dot{u} = V_0(u)$. We are interested in cases where the dependence on the parameter is singular, i.e. the dependence $\varepsilon \to V_\varepsilon(v)$ is not continuous, but we still hope that the solutions u_ε have a limit. Thus, our results on

A. Mielke (✉)
Weierstraß-Institut für Angewandte Analysis und Stochastik, Berlin, Germany
e-mail: mielke@wias-berlin.de

A. Muntean et al. (eds.), *Macroscopic and Large Scale Phenomena: Coarse Graining, Mean Field Limits and Ergodicity*, Lecture Notes in Applied Mathematics and Mechanics 3, DOI 10.1007/978-3-319-26883-5_3

	microscopic system $\dot{u}_\varepsilon = \boldsymbol{V}_\varepsilon(u_\varepsilon)$	$\varepsilon \to 0$	macroscopic system $\dot{u} = \boldsymbol{V}_0(u)$
initial state	u_ε^0	$\xrightarrow[\mathbb{M}_\varepsilon]{\text{upscaling}}$	u^0
time evolution	$\downarrow$		$\downarrow$
time $t > 0$	$u_\varepsilon(t) = \boldsymbol{S}_\varepsilon(t, u_\varepsilon^0)$	$\xrightarrow[\mathbb{M}_\varepsilon]{\text{upscaling}}$	$u(t) = \boldsymbol{S}_0(t, u^0)$

Fig. 3.1 Illustration of upscaling and semigroup convergence, if $\mathbb{M}_\varepsilon = \text{id}$

evolutionary Γ-convergence can be seen as *singular limits* in the sense of [19]. As in the case of static Γ-convergence, the aim is to derive effective limit equations that still capture the main effects of the original model with $0 < \varepsilon \ll 1$, but are simple in the sense that they do not longer include the small scales, and hence are easier to analyze or to solve numerically.

We emphasize that the justification of such multiscale limits corresponds to showing that it is possible to interchange the time evolution $u_\varepsilon(0) \rightsquigarrow u_\varepsilon(t) = \boldsymbol{S}_\varepsilon(t, u_\varepsilon(0))$ with the limit in $\varepsilon \to 0$. More precisely, we have to show that the diagram in Fig. 3.1 commutes, i.e. we have to prove that

$$\lim_{\varepsilon\to 0} \mathbb{M}_\varepsilon \circ \boldsymbol{S}_\varepsilon(t, \cdot) = \boldsymbol{S}_0(t, \lim_{\varepsilon\to 0} \mathbb{M}_\varepsilon(\cdot)), \tag{3.1.1}$$

where $\mathbb{M}_\varepsilon$ denotes a suitable upscaling operator, which in most of our cases is simply given by id. In the latter case we say that the evolution equations $\dot{u} = \boldsymbol{V}_\varepsilon(u)$ *semigroup converges* to $\dot{u} = \boldsymbol{V}_0(u)$. In principle the justification of this limiting process can be done directly on the (partial) differential equation $\dot{u}_\varepsilon = \boldsymbol{V}_\varepsilon(u_\varepsilon)$.

However, in this work we will concentrate on special evolutionary systems, namely *generalized gradient system* $(\boldsymbol{X}, \mathscr{E}, \mathscr{R})$. This triple consist of a state space $\boldsymbol{X}$ containing the states u, which is usually a Banach space $\boldsymbol{X}$, a closed subset, or even a metric space. The driving functional $\mathscr{E} : \boldsymbol{X} \to \mathbb{R} := \mathbb{R} \cup \{\infty\}$ is usually a (free) energy or the negative of an entropy. The function $\mathscr{R} : \boldsymbol{X} \times \boldsymbol{X} \to [0, \infty]$ is called dissipation potential, which means that $\mathscr{R}(u, \cdot)$ is lower semicontinuous (lsc), proper, and convex and that $\mathscr{R}(u, 0) = 0$. Here $\partial_{\dot{u}}\mathscr{R}(u, \dot{u})$ gives the dissipative force, while $-\mathrm{D}\mathscr{E}(u)$ is the potential restoring force. Using the dual dissipation potential $\mathscr{R}^*(u, \xi)$ (cf. Sect. 3.2.3) we can write the gradient flow induced by $(\boldsymbol{X}, \mathscr{E}, \mathscr{R})$ in three equivalent forms

$$\begin{aligned}
&0 = \partial_{\dot{u}}\mathscr{R}(u, \dot{u}) + \mathrm{D}\mathscr{E}(u) && \text{force balance, Biot's equation}\\
&\dot{u} = \partial_\xi \mathscr{R}^*(u, -\mathrm{D}\mathscr{E}(u)) && \text{rate equation, Onsager system}\\
&\mathscr{R}(u, \dot{u}) + \mathscr{R}^*(u, -\mathrm{D}\mathscr{E}(u)) = -\langle \mathrm{D}\mathscr{E}(u), \dot{u}\rangle && \text{power balance}
\end{aligned}$$

If $\mathscr{R}(u, \cdot)$ is quadratic, i.e. $\mathscr{R}(u, v) = \frac{1}{2}\langle \mathbb{G}(u)v, v\rangle$ and $\mathscr{R}^*(u, \xi) = \frac{1}{2}\langle \xi, \mathbb{K}(u)\xi\rangle$ with $\mathbb{K} = \mathbb{G}^{-1}$, we call $(\boldsymbol{X}, \mathscr{E}, \mathscr{R})$ a *classical gradient system*, and the evolution equations read

$$\dot{u} = \boldsymbol{V}(u) = -\mathbb{K}(u)\mathrm{D}\mathcal{E}(u) \quad \Longleftrightarrow \quad 0 = \mathbb{G}(u)\dot{u} + \mathrm{D}\mathcal{E}(u). \tag{3.1.2}$$

The focus of this work is to derive sufficient conditions for semigroup convergence for the special class of evolutionary systems that are induced by generalized gradient systems. The emphasis is on methods that fully exploit the gradient structure. In particular, it is desirable to derive sufficient conditions on the convergence of the pair $(\mathcal{E}_\varepsilon, \mathcal{R}_\varepsilon)$ to effective functionals $(\mathcal{E}_0, \mathcal{R}_0)$ such that $(\boldsymbol{X}, \mathcal{E}_\varepsilon, \mathcal{R}_\varepsilon)$ semigroup converges to $(\boldsymbol{X}, \mathcal{E}_0, \mathcal{R}_0)$ in the sense of (3.1.1) with $\mathbb{M}_\varepsilon = \mathrm{id}$.

We also emphasize that there is a rich literature on semigroup convergence for general evolutionary systems $\dot{u} + \boldsymbol{A}_\varepsilon(u) = 0$ that often use that $\boldsymbol{A}_\varepsilon$ is defined via a maximal monotone operator on a *fixed* Banach space $\boldsymbol{X}$ (see e.g. [11, Theorem 3.18], [4, Theorem 3.74], and the discussion in [72]), which includes classical gradient systems $\mathbb{G}\dot{u} = -\mathrm{D}\mathcal{E}_\varepsilon(u)$ if the dissipation potential $\mathcal{R}_\varepsilon(u, v) = \frac{1}{2}\langle \mathbb{G}_\varepsilon(u)v, v\rangle$ is independent of u and ε. Note that writing the system in the form $\dot{u} + \boldsymbol{A}_\varepsilon(u) = 0$ with $\boldsymbol{A}_\varepsilon(u) = \mathbb{K}_\varepsilon(u)\mathrm{D}\mathcal{E}_\varepsilon(u)$ the monotonicity for all $\varepsilon \in [0, 1]$ is lost if working with a fixed Hilbert-space norm.

For gradient systems it is also important to keep track of the energies $\mathcal{E}_\varepsilon(u_\varepsilon(t))$ along the solutions and to control their convergence in the limit $\varepsilon \to 0$. In analogy to the notion of static Γ-convergence (cf. [8, 14] or Sect. 3.2.2) we then follow the naming suggested in [69] and speak of *evolutionary Γ-convergence for gradient system.*

More precisely, we introduce two notions of evolutionary Γ-convergence depending of whether we need control of the energy at the initial time or not. We define the statements

$$\mathfrak{S}(t) = \text{“}\big(u_\varepsilon(t) \to u(t)\big)\text{”} \quad \text{and} \quad \mathfrak{E}(t) = \text{“}\big(\mathcal{E}_\varepsilon(u_\varepsilon(t)) \to \mathcal{E}_0(u(t))\big)\text{”}.$$

Then, semigroup convergence simply means “$\mathfrak{S}(0) \Longrightarrow \forall t > 0: \mathfrak{S}(t)$”. For gradient systems we will use the following two definitions for evolutionary Γ-convergence:

$$\text{E-convergence, written } (\boldsymbol{X}, \mathcal{E}_\varepsilon, \mathcal{R}_\varepsilon) \xrightarrow{\mathsf{E}} (\boldsymbol{X}, \mathcal{E}_0, \mathcal{R}_0): \\ \mathfrak{S}(0) \quad \Longrightarrow \quad \forall t > 0: \ \mathfrak{S}(t) \text{ and } \mathfrak{E}(t);$$

$$\text{well-prepared E-convergence, written } (\boldsymbol{X}, \mathcal{E}_\varepsilon, \mathcal{R}_\varepsilon) \xrightarrow{\mathsf{pE}} (\boldsymbol{X}, \mathcal{E}_0, \mathcal{R}_0): \\ \mathfrak{S}(0) \text{ and } \mathfrak{E}(0) \quad \Longrightarrow \quad \forall t > 0: \ \mathfrak{S}(t) \text{ and } \mathfrak{E}(t).$$

Here the letter “E” stands for both ‘E’volutionary as well as ‘E’nergy convergence. In the second definition the assumptions $\mathfrak{S}(0)$ and $\mathfrak{E}(0)$ at $t = 0$ are called *well-‘P’reparedness of the initial conditions.* Clearly, well-prepared E-convergence is implied by the stronger E-convergence, since the latter has a weaker assumptions at $t = 0$, whereas the conclusions for $t > 0$ are the same. We also use the notations $\xrightarrow{\mathsf{E}}$

and $\overset{\mathsf{pE}}{\rightharpoonup}$ if the convergences in $\mathfrak{S}(0)$ and $\mathfrak{S}(t)$ is replaced by the weak convergence in $\boldsymbol{X}$.

In Sect. 3.2 we give preparatory work including the basic modeling ideas of gradient systems (Sect. 3.2.1), basic facts about and a few examples for weak and strong Γ-convergence, Mosco and continuous convergence including the relations with the Legendre transform (Sects. 3.2.2 and 3.2.3), and finally some examples of gradient systems that are used to illustrate the abstract theory in this work or that highlight the development of the general field (Sects. 3.2.5–3.2.8).

The main Sects. 3.3–3.5 are devoted to the question, which types of convergences $\mathscr{E}_\varepsilon \rightsquigarrow \mathscr{E}_0$ and $\mathscr{R}_\varepsilon \rightsquigarrow \mathscr{R}_0$ are sufficient to guarantee that the gradient system $(\boldsymbol{X}, \mathscr{E}_\varepsilon, \mathscr{R}_\varepsilon)$ evolutionary Γ-converges to the limit system $(\boldsymbol{X}, \mathscr{E}_0, \mathscr{R}_0)$. In Sect. 3.3 we will discuss convergence results based on the *energy-dissipation principle* (EDP), also called *De Giorgi's* $(\mathscr{R}, \mathscr{R}^*)$ *formulation*:

$$\mathscr{E}_\varepsilon(u_\varepsilon(T)) + \int_0^T \mathscr{R}_\varepsilon(u_\varepsilon, \dot u_\varepsilon) + \mathscr{R}_\varepsilon^*\big(u_\varepsilon, -\mathrm{D}\mathscr{E}_\varepsilon(u_\varepsilon)\big)\,\mathrm{d}t \;=\; \mathscr{E}_\varepsilon(u_\varepsilon(0)). \tag{3.1.3}$$

This formulation was the starting point of the fundamental paper [69], see also [71], where the crucial conditions for well-prepared E-convergence are given by suitable liminf estimates for $\mathscr{R}_\varepsilon$ and $\mathscr{R}_\varepsilon^*(\cdot, -\mathrm{D}\mathscr{E}_\varepsilon(\cdot))$, respectively. We formulate a corresponding result on evolutionary Γ-convergence in Theorem 3.3.3 and present the more restrictive Theorem 3.3.2, which is based on Mosco convergence and was developed in [72, Theorem 7.2]:

$$\mathscr{E}_\varepsilon \overset{\mathsf{M}}{\to} \mathscr{E}_0 \text{ and } \Psi_\varepsilon \overset{\mathsf{M}}{\to} \Psi_0 \quad \Longrightarrow \quad (\boldsymbol{X}, \mathscr{E}_\varepsilon, \Psi_\varepsilon) \overset{\mathsf{pE}}{\to} (\boldsymbol{X}, \mathscr{E}_0, \Psi_0).$$

We continue to use the notation Ψ_ε for the special case of dissipation potentials $\mathscr{R}_\varepsilon$ that do not depend on the state variable u, i.e. $\mathscr{R}_\varepsilon(u, v) = \Psi_\varepsilon(v)$. Such $\mathscr{R}_\varepsilon$ are also called translation invariant.

The result in [72] is based on a version of the Brézis-Ekeland principle that strongly uses that the translation invariance and the convexity of the energies $\mathscr{E}(t, \cdot)$. We present the approach developed in [51] which is based on the EDP (3.1.3) and works for general dissipation potentials $\mathscr{R}_\varepsilon$.

These abstract convergence results are applied to various examples in Sect. 3.3.5, in particular, to periodic homogenization of parabolic equations, to Tartar's ODE example $\dot u(t, x) = -a(x/\varepsilon)u(t, x)$, and to the passage from a wiggly-energy system with small viscous dissipation to a rate-independent hysteresis model.

In Sect. 3.4 we show that under certain conditions the differential form of gradient systems is equivalent to an *evolutionary variational inequality* (EVI) that is formulated solely in terms of u, $\mathscr{E}$, and $\mathscr{R}$, and does not contain any derivatives, i.e. $\dot u$ and $\partial\mathscr{E}$ do not occur. In the case of convex energies $\mathscr{E}_\varepsilon$ and a translation invariant, quadratic dissipation potential $\mathscr{R}_\varepsilon(u, v) = \Psi_\varepsilon(v) := \frac{1}{2}\langle \mathbb{G}_\varepsilon v, v\rangle$ it takes the simple form of the Integrated Evolutionary Variational Estimate $(\mathrm{IEVE})_{\lambda=0}$:

$$\forall s < t,\ w \in \boldsymbol{X} : \Psi_\varepsilon(u(t) - w) - \Psi_\varepsilon(u(s) - w) \le (t - s)\big(\mathscr{E}_\varepsilon(w) - \mathscr{E}_\varepsilon(u(t))\big). \tag{3.1.4}$$

This formulation is a variant of Bénilan's integral solutions (cf. in [6]) for the Banach-space setting $\mathscr{R}(u, v) = \frac{1}{2}\|v\|_{\boldsymbol{X}}^2$, while the setting of metric spaces $(\mathscr{Q}, \mathscr{D})$ is discussed extensively in [2, 70], see Sect. 3.4.3 for the general form of the IEVE for geodesically λ-convex gradient systems on a geodesic space $(\mathscr{Q}, \mathscr{D})$. Since only the functionals (but not their derivatives) appear, the IEVE formulation is particularly useful for passing to the limit $\varepsilon \to 0$. However, it again relies on a uniform convexity condition and is restricted to the case of classical, but possibly metric gradient systems. A simplified variant of the general result on E-convergence from [70] is given in Theorem 3.4.3:

$$\mathscr{D}_\varepsilon \overset{\mathsf{C}}{\to} \mathscr{D}_0 \text{ and } \mathscr{E}_\varepsilon \overset{\Gamma}{\to} \mathscr{E}_0 \quad \Longrightarrow \quad (\mathscr{Q}, \mathscr{E}_\varepsilon, \mathscr{D}_\varepsilon) \overset{\mathsf{E}}{\to} (\mathscr{Q}, \mathscr{E}_0, \mathscr{D}_0),$$

where $\overset{\mathsf{C}}{\to}$ means continuous convergence, see (3.2.7). We emphasize that this E-convergence does not need well-preparedness of the initial conditions.

The final Sect. 3.5 is devoted to rate-independent systems $(\boldsymbol{X}, \mathscr{E}_\varepsilon, \mathscr{R}_\varepsilon)$, which are a special case of generalized gradient flows where the dissipation potential is positively homogenous of degree 1 in the rate, i.e. $\forall\, \gamma > 0,\ u, v \in \boldsymbol{X}: \quad \mathscr{R}_\varepsilon(u, \gamma v) = \gamma \mathscr{R}_\varepsilon(u, v)$. As in the case of classical gradient systems, the dissipation potential generates a dissipation distance $\mathscr{D} : \boldsymbol{X} \times \boldsymbol{X} \to [0, \infty]$ leading to an energetic rate-independent system $(\boldsymbol{X}, \mathscr{E}, \mathscr{D})$. In fact, one does not need a Banach-space structure and can work on general topological spaces $\mathscr{Q}$ with a suitable metric $\mathscr{D}$. A function $u : [0, T] \to \mathscr{Q}$ is called *energetic solution* for $(\boldsymbol{X}, \mathscr{E}_\varepsilon, \mathscr{D}_\varepsilon)$ if (S) and (E) hold:

$$\text{(S)} \quad \mathscr{E}_\varepsilon(t, u(t)) \le \mathscr{E}_\varepsilon(t, w) + \mathscr{D}_\varepsilon(u(t), w) \text{ for all } t \in [0, T] \text{ and } w \in \mathscr{Q}, \tag{3.1.5}$$

$$\text{(E)} \quad \mathscr{E}_\varepsilon(T, u(T)) + \mathrm{Diss}_{\mathscr{D}_\varepsilon}(u, [0, T]) = \mathscr{E}_\varepsilon(0, u(0)) + \int_0^T \partial_s \mathscr{E}_\varepsilon(s, u(s))\,\mathrm{d}s \tag{3.1.6}$$

with $\mathrm{Diss}_{\mathscr{D}_\varepsilon}(u, [0, T]) = \sup \sum_{j=1}^N \mathscr{D}_\varepsilon(u(t_{j-1}), u(t_j))$, where the supremum is taken over all $N \in \mathbb{N}$ and all partitions $0 = t_0 < t_1 < \cdots < t_{N-1} < t_N = T$. Again, we have a derivative-free formulation such that evolutionary Γ-convergence can be easily connected to Γ-convergence of the energies $\mathscr{E}_\varepsilon$ and of the dissipation distances $\mathscr{D}_\varepsilon$. We present some simplified versions of the results in [37, 55]. Theorem 3.5.1 reads

$$\mathscr{D}_\varepsilon \overset{\mathsf{C}}{\rightharpoonup} \mathscr{D}_0 \text{ and } \mathscr{E}_\varepsilon \overset{\Gamma}{\rightharpoonup} \mathscr{E}_0 \quad \Longrightarrow \quad (\mathscr{Q}, \mathscr{E}_\varepsilon, \mathscr{D}_\varepsilon) \overset{\mathsf{pE}}{\rightharpoonup} (\mathscr{Q}, \mathscr{E}_0, \mathscr{D}_0).$$

If the energy is quadratic, i.e. $\mathscr{E}_\varepsilon(t, u) = \frac{1}{2}\langle A_\varepsilon u, u\rangle - \langle \ell_\varepsilon(t), u\rangle$, and $\mathscr{D}_\varepsilon(u_1, u_2) = \Psi_\varepsilon(u_2 - u_1)$ on a Hilbert space $\boldsymbol{H}$, we can weaken the convergence for Ψ_ε slightly (cf. Theorem 3.5.2):

$$\Psi_\varepsilon \stackrel{\mathsf{C}}{\to} \Psi_0,\ \Psi_\varepsilon \stackrel{\Gamma}{\rightharpoonup} \Psi_0, \text{ and } \mathcal{E}_\varepsilon \stackrel{\Gamma}{\rightharpoonup} \mathcal{E}_0 \quad \Longrightarrow \quad (\boldsymbol{H}, \mathcal{E}_\varepsilon, \Psi_\varepsilon) \stackrel{\mathsf{pE}}{\rightharpoonup} (\boldsymbol{H}, \mathcal{E}_0, \Psi_0).$$

As we see from the above results, it is often necessary to have the rather strong Mosco convergence to establish the desired E-convergence of the whole gradient flow. At first glance, such a result looks quite restrictive; however, we made the experience that a proper understanding of the properties (like microstructure or sharp interfaces) of the solutions of $(\boldsymbol{X}, \mathcal{E}_\varepsilon, \mathcal{R}_\varepsilon)$ is needed even in more direct approaches. Using that knowledge it is often possible to find a suitable blowup or unfolding of the microstructure that turns the usual Γ-convergence into Mosco convergence. A typical example is that of replacing homogenization by two-scale homogenization, as was done in [58] for rate-independent elastoplasticity.

Throughout this survey, we try to give the main themes and to highlight the principal approaches. Hence, we are not always able to give the full details or the optimal results. Moreover, we may not always list all the implicit assumptions, so in any case of doubt the original papers should be consulted.

3.2 Gradient Systems and Γ-Convergence

We first define our notion of gradient systems in the sense of modeling. Then, we recall the main definitions for (static) Γ-convergence and discuss a few properties needed especially for the evolutionary context, such as the relation to Legendre transforms. The definitions of evolutionary Γ-convergence are discussed in Sect. 3.2.4, and finally we give some examples that will be used later to highlight different aspects of the theory.

3.2.1 Gradient Systems from the Modeling Point of View

To begin with, we start from the case that the state space is a smooth (finite-dimensional) manifold $\mathcal{X}$ such that at each state $u \in \mathcal{X}$ the tangent space $\mathrm{T}_u\mathcal{X}$ and its dual space $\mathrm{T}_u^*\mathcal{X} = (\mathrm{T}_u\mathcal{X})^*$, the so-called co-tangent space, are well defined. For the modeling it is important to distinguish the states u, the rates or velocities $v = \dot{u} \in \mathrm{T}_u\mathcal{X}$, and the (thermodynamically conjugate) forces $\xi \in \mathrm{T}_u^*\mathcal{X}$. We denote by $\langle \xi, v\rangle$ the dual pairing between $\mathrm{T}_u^*\mathcal{X}$ and $\mathrm{T}_u\mathcal{X}$. By $\mathrm{T}\mathcal{X}$ we denote the tangent bundle $\cup_{u\in\mathcal{X}}(u, \mathrm{T}_u\mathcal{X})$.

A *gradient system* is a triple $(\mathcal{X}, \mathcal{E}, \mathcal{R})$ with a state space $\mathcal{X}$ as above, a smooth energy functional $\mathcal{E} : [0, T] \times \mathcal{X} \to \mathbb{R}$, and a dissipation potential $\mathcal{R} : \mathrm{T}\mathcal{X} \to [0, \infty]$, i.e. $\mathcal{R}$ depends on the state u and the rate $\dot{u}$. A function $\mathcal{R}$ is called dissipation potential, if for all $u \in \mathcal{X}$, the function $\mathcal{R}(u, \cdot) : \mathrm{T}_u\mathcal{X} \to [0, \infty]$ is convex, lower semicontinuous, and satisfies $\mathcal{R}(u, 0) = 0$.

We consider $\xi = \mathrm{D}_u\mathcal{E}(t, u) \in \mathrm{T}_u\mathcal{X}^*$ as (the negative of) the potential restoring force generated by the energy $\mathcal{E}$, while $\partial_t\mathcal{E}(t, u) \in \mathbb{R}$ will be called the power of the external forces. The partial derivative $\eta = \mathrm{D}_{\dot u}\mathscr{R}(u,\dot u) \in \mathrm{T}^*_u\mathscr{X}$ is the dissipative force induced by the changes of u. The induced evolution of the gradient system $(\mathscr{X},\mathscr{E},\mathscr{R})$ is now given in terms of the force balance also called Biot equation (see e.g. [7, Eq. (2.14)])

$$0 = \mathrm{D}_{\dot u}\mathscr{R}(u(t),\dot u(t)) + \mathrm{D}_u\mathscr{E}(t,u(t)) \quad \in \mathrm{T}^*_u\boldsymbol{X}. \tag{3.2.1}$$

In the simplest case the dissipation potential is a quadratic form $\mathscr{R}(u,v) = \frac{1}{2}\langle \mathbb{G}(u)v, v\rangle$ giving a linear constitutive law $\eta = \mathbb{G}(u)\dot u$, which is usually called viscous dissipation. We call these systems *classical gradient systems* and also write $(\mathscr{X},\mathscr{E},\mathbb{G})$ or $(\mathscr{X},\mathscr{E},\mathbb{K})$. If $\mathscr{R}(u,\cdot)$ is nonquadratic we call $(\mathscr{X},\mathscr{E},\mathscr{R})$ a *generalized gradient system*. See [49] for a natural occurrence of generalized gradient systems obtained via a large-deviation principle and Sect. 3.5 for rate-independent systems, which are characterized by the fact that $\mathscr{R}(u,\cdot)$ is positively homogeneous of degree 1, see (3.5.1).

In the classical case we can invert the operator $\mathbb{G}(u)$ and solve for $\dot u$ to obtain a rate equation, which we also call *Onsager equation* (cf. [62, Eq. (1.11)])

$$\dot u(t) = -\mathbb{K}(u(t))\mathrm{D}_u\mathscr{E}(t,u(t)) \in \mathrm{T}_u\boldsymbol{X}, \quad \text{where } \mathbb{K}(u) = \mathbb{G}(u)^{-1}. \tag{3.2.2}$$

We call $\mathbb{K}(u)$ the Onsager operator to honor Onsager's seminal contributions to the thermodynamics of dissipative processes, cf. [61, 62]. For systems that are always close to local thermodynamic equilibrium, he derived that the relation $\dot u = f(\xi)$ between the thermodynamic driving force ξ and the rate must be linear, i.e. $\dot u = \mathbb{K}\xi$ with a symmetric and positive semidefinite $\mathbb{K}$, where "$\mathbb{K} = \mathbb{K}^\top$" are his famous reciprocal relations. In his setting $\xi = \mathrm{D}\mathscr{S}(u)$ with $\mathscr{S}$ denoting the entropy.

In a more general setting, we use the dual dissipation potential $\mathscr{R}^*(u,\xi)$, which reads $\mathscr{R}^*(u,\xi) = \frac{1}{2}\langle\xi, \mathbb{K}(u)\xi\rangle$ in the quadratic case and is defined in (3.2.13) for the general case, to write the rate equation in the form

$$\dot u(t) = \mathrm{D}_\xi\mathscr{R}^*\big(u(t), -\mathrm{D}_u\mathscr{E}(t,u(t))\big) \in \mathrm{T}_u\boldsymbol{X}. \tag{3.2.3}$$

In principle, Eqs. (3.2.1) and (3.2.3) are equivalent, but depending on the context one may prefer one over the other. For the gradient structures for time-continuous discrete Markov chains (see [40]) or the reaction-diffusion systems (see [28, 44, 46]) it is crucial that $\mathscr{R}^*$ (or the Onsager operator $\mathbb{K}(u)$) is a sum of different dissipative effects. Thus, $\mathbb{K}(u)$ is explicit, whereas $\mathbb{G}(u)$ or $\mathscr{R}(u,\cdot)$ are only implicitly defined.

An important feature in modeling is energy conservation, when taking the dissipation properly into account. For this it is important to realize that the dissipation potential is usually different from the dissipation functions $\Phi(u,\dot u)$ or $\Phi_*(u,\eta)$, where

$$\Phi(u,\dot u) = \langle \mathrm{D}_{\dot u}\mathscr{R}(u,\dot u), \dot u\rangle \quad \text{and} \quad \Phi_*(u,\xi) = \langle \eta, \mathrm{D}_\xi\mathscr{R}^*(u,\xi)\rangle.$$

For quadratic functions we have $\Phi = 2\mathcal{R}$ and $\Phi_* = 2\mathcal{R}^*$, but in general there is no such identity. Along solutions of (3.2.1) or (3.2.3) the chain rule gives

$$\frac{\mathrm{d}}{\mathrm{d}t}\mathcal{E}(t,u(t)) - \partial_t\mathcal{E}(t,u(t)) = \langle \mathrm{D}_u\mathcal{E}(t,u(t)), \dot{u}(t)\rangle = \Phi(u(t),\dot{u}(t)), \tag{3.2.4a}$$

$$\begin{aligned} \text{where } \Phi(u(t),\dot{u}(t)) &= \Phi_*\big(u(t), -\mathrm{D}_u\mathcal{E}(t,u(t))\big) \\ &= \mathcal{R}(u(t),\dot{u}(t)) + \mathcal{R}^*\big(u(t), -\mathrm{D}_u\mathcal{E}(t,u(t))\big). \end{aligned} \tag{3.2.4b}$$

For the latter relation we refer to the Legendre-Fenchel equivalence discussed in Proposition 3.2.2.

Integrating (3.2.4a), with Φ replaced by $\mathcal{R} + \mathcal{R}^*$ as in (3.2.4b), over $t \in [0, T]$ we find a *balance* between the energies and the dissipation, namely

$$\begin{aligned} &\mathcal{E}(T,u(T)) + \int_0^T \mathcal{R}(u(t),\dot{u}(t)) + \mathcal{R}^*\big(u(t), -\mathrm{D}_u\mathcal{E}(t,u(t))\big)\,\mathrm{d}t \\ &\quad = \mathcal{E}(0,u(0)) + \int_0^T \partial_t\mathcal{E}(t,u(t))\,\mathrm{d}t \quad \text{for all } T > 0. \end{aligned} \tag{3.2.5}$$

At first sight it seems that this balance is just simply consequence of (3.2.1) or (3.2.3). Observe that in the quadratic case one always has

$$\mathcal{R}(u(t),\dot{u}(t)) = \mathcal{R}^*\big(u(t), -\mathrm{D}_u\mathcal{E}(t,u(t))\big),$$

hence the balance would be true also when replacing $\mathcal{R} + \mathcal{R}^*$ by $2(1-\theta)\mathcal{R} + 2\theta\mathcal{R}^*$ for any $\theta \in [0, 1]$. However, only when writing the total dissipation in the special form $\mathcal{R} + \mathcal{R}^*$ it can be shown that this balance is equivalent to (3.2.1) and (3.2.3), see Theorem 3.3.1. Hence, we will use the term *energy-dissipation principle (EDP)*, whenever the specific choice $\mathcal{R} + \mathcal{R}^*$ is meant.

We emphasize here that each gradient system $(\mathcal{X}, \mathcal{E}, \mathcal{R})$ has a well-defined associated evolution equation $\dot{u}(t) = \boldsymbol{V}_{\mathcal{E},\mathcal{R}}(t,u(t)) := \mathrm{D}_\xi\mathcal{R}^*\big(u(t), -\mathrm{D}_u\mathcal{E}(t,u(t))\big)$. However, for one given evolution equation $\dot{u}(t) = \boldsymbol{V}_{\mathcal{E},\mathcal{R}}(t,u(t))$ there may exist many different gradient structures. So, the gradient structure is an additional information which contains additional physical information on the model under investigation.

Example 3.2.1 Here we show that ODEs may have several genuinely different gradient structures. For the scalar linear ODE $\dot{u} = -u =: \boldsymbol{V}(u)$ we take any smooth, convex $\phi : \mathbb{R} \to \mathbb{R}$ with $\phi(0) = \phi'(0) = 0$ and $\phi''(0) > 0$. Then, $\mathcal{E}(u) = \phi(u)$ and $\mathbb{K}(u) = \frac{u}{\phi'(u)} > 0$ defines a gradient structure $(\mathbb{R}, \mathcal{E}, \mathbb{K})$ for $\dot{u} = -u$. A similar construction also works for systems of ODEs. For example, consider

$$\dot{u}_1 = u_2 - u_1, \quad \dot{u}_2 = -3u_2,$$

choose a uniformly convex ϕ and set $\mathscr{E}(u_1,u_2) = \phi(u_1) + \phi(u_2) + \phi(-u_1-u_2)$ and $\mathscr{R}^*(u_1,u_2,\xi_1,\xi_2) =$

$$\frac{u_2-u_1}{\phi'(u_2)-\phi'(u_1)}(\xi_2-\xi_1)^2 + \frac{2u_1+u_2}{\phi'(u_1)-\phi'(-u_1-u_2)}\xi_1^2 + \frac{u_1+2u_2}{\phi'(u_2)-\phi'(-u_1-u_2)}\xi_2^2.$$

See [40] for similar gradient structures for time-continuous Markov chains.

Of course, in general thermodynamical modeling we are interested in partial differential equations, where the above theory has to be generalized suitably. In particular, the underlying space $\mathscr{X}$ may not be a smooth manifold but is usually a subset of a Banach space $\boldsymbol{X}$. Moreover, the functionals $\mathscr{E}$ and $\mathscr{R}$ will no longer be smooth but may attain the value $+\infty$. Hence, the derivatives $\mathrm{D}_u\mathscr{E}(t,u)$, $\mathrm{D}_v\mathscr{R}(u,v)$, and $\mathrm{D}_\xi\mathscr{R}^*(u,\xi)$ must be generalized in a suitable way. We refer to Sect. 3.3.1 for a mathematical precise setup. The main point to be remembered is, that the state u, the rates $v=\dot{u}$, and the forces ξ lie in different spaces. Only for the rates and the forces we have linear spaces and a duality pairing $\langle \xi, \dot{u}\rangle$, which denotes a power.

Working with partial differential equations, we may proceed formally without specifying a function space $\boldsymbol{X}$, which is the typical approach in thermomechanical modeling, see e.g. [44, 46, 57]. Functional derivatives can be interpreted as variational derivatives by assuming that all functions are sufficiently smooth. As an example consider the PDE

$$\dot{u} = \Delta u \text{ in } \Omega \subset \mathbb{R}^d, \quad \nabla u\cdot \nu = 0 \text{ on } \partial\Omega, \tag{3.2.6}$$

where Ω is a smooth bounded domain in $\mathbb{R}^d$. There are two simple quadratic gradient structures, namely $(\mathrm{L}^2(\Omega), \mathscr{E}_0, \mathbb{K}_0)$ and $(\mathrm{H}^1(\Omega), \mathscr{E}_1, \mathbb{K}_1)$ with

$$\mathscr{E}_0(u) = \int_\Omega \frac{u^2}{2}\,\mathrm{d}x, \quad \mathbb{K}_0\xi = -\Delta\xi, \qquad \mathscr{E}_1(u) = \int_\Omega \frac{1}{2}|\nabla u|^2\,\mathrm{d}x, \quad \mathbb{K}_1\xi = \xi.$$

However, if we interpret (3.2.6) as a heat equation for the absolute temperature u of a gas, we choose the entropy $\mathscr{S}$ for $\dot{u} = +\mathbb{K}_{\mathrm{heat}}(u)\mathrm{D}\mathscr{S}(u)$ with

$$\mathscr{S}(u) = \int_\Omega \log(u(x))\,\mathrm{d}x \quad \text{and} \quad \mathbb{K}_{\mathrm{heat}}(u)\xi = -\operatorname{div}\left(u^2\nabla\xi\right),$$

see [44, 46]. If (3.2.6) describes the diffusion of a density u, then the Wasserstein gradient structure for the relative entropy $\mathscr{H}$, found in [32, 63], is the right choice:

$$\mathscr{H}(u) = \int_\Omega u\log u\,\mathrm{d}x \quad \text{and} \quad \mathbb{K}_{\mathrm{Wass}}(u)\xi = -\operatorname{div}(u\nabla\xi).$$

Of course, the different choices of the gradient structures may influence the theory of evolutionary Γ-convergence. In Sect. 3.3.5.2 we discuss the simple model $\dot{u}(t,x) = -a(\frac{1}{\varepsilon}x)u(t,x)$ and show that different gradient structures lead to different evolutionary Γ-limits, see Corollary 3.3.1.

3.2.2 *Γ-Convergence for (Static) Functionals*

We consider a reflexive Banach space $\boldsymbol{X}$ and functionals $\mathscr{J}_\varepsilon : \boldsymbol{X} \to \mathbb{R}_\infty$. Strong and weak convergence in $\boldsymbol{X}$ will be denoted by $u_k \to u$ and $v_k \rightharpoonup v$, respectively. We first introduce more classical notions of convergence of functionals, namely the pointwise convergence $\mathscr{J}_\varepsilon \xrightarrow{\text{pw}} \mathscr{J}_0$ and the strong or weak continuous convergence defined via

$$\mathscr{J}_\varepsilon \xrightarrow{\mathsf{C}} \mathscr{J}_0, \quad \text{if } u_\varepsilon \to u \;\Rightarrow\; \mathscr{J}_\varepsilon(u_\varepsilon) \to \mathscr{J}_0(u); \tag{3.2.7a}$$

$$\mathscr{J}_\varepsilon \overset{\mathsf{C}}{\rightharpoonup} \mathscr{J}_0, \quad \text{if } u_\varepsilon \rightharpoonup u \;\Rightarrow\; \mathscr{J}_\varepsilon(u_\varepsilon) \to \mathscr{J}_0(u). \tag{3.2.7b}$$

In the context of minimization of functionals, the concept of *Γ-convergence* is more natural, see Theorem 3.2.1. This convergence was originally called variational convergence or epi-graph convergence (cf. [4, 16]), but nowadays the term *Γ-convergence* is more common and we refer to [8–10, 14] for further details.

Definition 3.2.1 (*Γ* and *Mosco convergence*) Let $\boldsymbol{X}$ be a reflexive Banach space. We say that $\mathscr{J}_\varepsilon$ weakly Γ-converges to $\mathscr{J}_0$ and write $\mathscr{J}_\varepsilon \overset{\Gamma}{\rightharpoonup} \mathscr{J}_0$, if (G1w) and (G2w) hold:

(G1w) $u_\varepsilon \rightharpoonup u \implies \mathscr{J}_0(u) \le \liminf_{\varepsilon\to 0} \mathscr{J}_\varepsilon(u_\varepsilon)$ (liminf estimate)

(G2w) $\forall \widehat{u}\, \exists\, (\widehat{u}_\varepsilon)_\varepsilon \colon \widehat{u}_\varepsilon \rightharpoonup \widehat{u}$ and $\mathscr{J}_0(\widehat{u}) = \lim_{\varepsilon\to 0} \mathscr{J}_\varepsilon(\widehat{u}_\varepsilon)$ (recovery seq. exist)

We say that $\mathscr{J}_\varepsilon$ strongly Γ-converges to $\mathscr{J}_0$ and write $\mathscr{J}_\varepsilon \xrightarrow{\Gamma} \mathscr{J}_0$, if (G1s) and (G2s) hold:

(G1s) $u_\varepsilon \to u \implies \mathscr{J}_0(u) \le \liminf_{\varepsilon\to 0} \mathscr{J}_\varepsilon(u_\varepsilon)$ (liminf estimate)

(G2s) $\forall \widehat{u}\, \exists\, (\widehat{u}_\varepsilon)_\varepsilon \colon \widehat{u}_\varepsilon \to \widehat{u}$ and $\mathscr{J}_0(\widehat{u}) = \lim_{\varepsilon\to 0} \mathscr{J}_\varepsilon(\widehat{u}_\varepsilon)$ (recovery seq. exist)

We say that $\mathscr{J}_\varepsilon$ *Mosco converges* to $\mathscr{J}_0$ and write $\mathscr{J}_\varepsilon \xrightarrow{\mathsf{M}} \mathscr{J}_0$, if $\mathscr{J}_\varepsilon \xrightarrow{\Gamma} \mathscr{J}_0$ and $\mathscr{J}_\varepsilon \overset{\Gamma}{\rightharpoonup} \mathscr{J}_0$.

Since all strongly converging sequences are also weakly converging we have the implications (G1w) ⇒ (G1s) and (G2s) ⇒ (G2w). Hence, for Mosco convergence one needs to check only (G1w) and (G2s). We will see in Lemma 3.2.2 that there are simple quadratic functionals for which weak and strong Γ-limits exist, but they are different.

The conditions (G2w) and (G2s) on the existence of recovery sequences is also called "limsup estimate". The recovery sequences are crucial since they capture the correct microscopic behavior that is needed to recover the correct (namely the lowest possible) macroscopic energy $\mathcal{J}_0(\widehat{u})$.

Clearly, *Continuous convergence* is much stronger than Γ-convergence. We have the following relations.

Lemma 3.2.1 *Assume that* $\mathbf{X}$ *is a reflexive Banach space and* $\mathcal{J}_\varepsilon, \mathcal{K}_\varepsilon : \mathbf{X} \to \mathbb{R}_\infty$.

(a) $\mathcal{J}_\varepsilon \overset{\Gamma}{\rightharpoonup} \mathcal{J}_0$ *and* $\mathcal{K}_\varepsilon \overset{\mathsf{C}}{\rightharpoonup} \mathcal{K}_0 \implies \mathcal{J}_\varepsilon + \mathcal{K}_\varepsilon \overset{\Gamma}{\rightharpoonup} \mathcal{J}_0 + \mathcal{K}_0$

(b) $\mathcal{J}_\varepsilon \overset{\Gamma}{\to} \mathcal{J}_0$ *and* $\mathcal{K}_\varepsilon \overset{\mathsf{C}}{\to} \mathcal{K}_0 \implies \mathcal{J}_\varepsilon + \mathcal{K}_\varepsilon \overset{\Gamma}{\to} \mathcal{J}_0 + \mathcal{K}_0$

(c) $\mathcal{J}_\varepsilon \overset{\mathsf{C}}{\rightharpoonup} \mathcal{J}_0 \implies \mathcal{J}_\varepsilon \overset{\mathsf{M}}{\to} \mathcal{J}_0$

(d) *If* $(\mathcal{J}_\varepsilon)_{\varepsilon\in[0,1]}$ *is strongly equicontinuous, then* $\mathcal{J}_\varepsilon \overset{\Gamma}{\to} \mathcal{J}_0$ *implies* $\mathcal{J}_\varepsilon \overset{\mathsf{C}}{\to} \mathcal{J}_0$.

The origin for the definition of Γ-convergence, which is clearer in the original name "variational convergence", is the following convergence of minimizers, see [8, 14].

Theorem 3.2.1 (Convergence of minimizers) *Assume* $\mathcal{J}_\varepsilon \overset{\Gamma}{\rightharpoonup} \mathcal{J}_0$ *in* $\mathbf{X}$ *and that* $\inf \mathcal{J}_0 =: \alpha_0 > -\infty$. *Moreover, assume that there exists a closed bounded set* $B \subset \mathbf{X}$ *such that* $\mathcal{J}_\varepsilon(u) \le \alpha_0 + 1$ *implies* $u \in B$, *then for every sequence* $(u_{\varepsilon_k})_{k\in\mathbb{N}}$ *with* $\varepsilon_k \to 0$ *we have:*

$$\text{if } u_\varepsilon \rightharpoonup \widetilde{u} \text{ and } \mathcal{J}_{\varepsilon_k}(u_{\varepsilon_k}) \to \alpha_0, \quad \text{then } \mathcal{J}_0(\widetilde{u}) = \alpha_0.$$

In particular, if u_ε *is a minimizer of* $\mathcal{J}_\varepsilon$ *(i.e.* $\mathcal{J}_\varepsilon(u_\varepsilon) = \inf_X \mathcal{J}_\varepsilon$*), then any accumulation point of* $(u_\varepsilon)_{\varepsilon\in]0,1[}$ *is a minimizer of* $\mathcal{J}_0$.

The following useful result seems to be folklore, but is not easy to find.

Proposition 3.2.1 (Γ-versus Mosco convergence) *Assume that* $\mathbf{X}$ *and* $\mathbf{Z}$ *are reflexive Banach spaces such that* $\mathbf{Z}$ *is compactly embedded in* $\mathbf{X}$, *written* $\mathbf{Z} \Subset \mathbf{X}$. *Moreover, assume that the functionals* $\mathcal{J}_\varepsilon$ *are equicoercive in* $\mathbf{Z}$, *i.e.*

$$\forall J > 0\, \exists R > 0\, \forall \varepsilon > 0,\ u \in \mathbf{X}: \quad \mathcal{J}_\varepsilon(u) \le J \Rightarrow \|u\|_Z \le R: \tag{3.2.8}$$

Then, $\mathcal{J}_\varepsilon \overset{\mathsf{M}}{\to} \mathcal{J}_0$ *in* $\mathbf{X}$ *is equivalent to* $\mathcal{J}_\varepsilon \overset{\Gamma}{\rightharpoonup} \mathcal{J}_0$ *in* $\mathbf{Z}$.

Proof The equicoercivity is meant such that all $\mathcal{J}_\varepsilon$ take the value $+\infty$ on $\mathbf{X} \setminus \mathbf{Z}$.

"$\Rightarrow$" We start from $\mathcal{J}_\varepsilon \overset{\mathsf{M}}{\to} \mathcal{J}_0$ in $\mathbf{X}$. If $u_\varepsilon \rightharpoonup u$ in $\mathbf{Z}$, then this also holds in $\mathbf{X}$. Hence, the liminf estimate follows. To construct a recovery sequence $\widehat{u}_\varepsilon \rightharpoonup \widehat{u}$ in $\mathbf{Z}$ for arbitrary $\widehat{u} \in \mathbf{Z}$, we first assume $\mathcal{J}_0(\widehat{u}) < \infty$. We choose the recovery sequence $\widehat{u}_\varepsilon$ guaranteed by $\mathcal{J}_\varepsilon \overset{\mathsf{M}}{\to} \mathcal{J}_0$ in $\mathbf{X}$, i.e. we know $\widehat{u}_\varepsilon \to \widehat{u}$ in $\mathbf{X}$. The equicoercivity (3.2.8) and $\mathcal{J}_\varepsilon(\widehat{u}_\varepsilon) \to \mathcal{J}_0(\widehat{u}) < \infty$ imply $\|\widehat{u}_\varepsilon\|_Z \le R$. Hence, $\widehat{u}_\varepsilon \rightharpoonup \widehat{u}$ in $\mathbf{Z}$ by reflexivity of $\mathbf{Z}$. If $\mathcal{J}_0(\widehat{u}) = \infty$, we choose $\widehat{u}_\varepsilon = \widehat{u}$ giving $\widehat{u}_\varepsilon \to \widehat{u}$ in $\mathbf{Z}$. Hence, the liminf estimate yields $\infty = \mathcal{J}_0(\widehat{u}) \le \liminf_{\varepsilon\to 0} \mathcal{J}_\varepsilon(\widehat{u})$, which shows that we have a recovery sequence in $\mathbf{Z}$.

"$\Leftarrow$" Given $\mathcal{J}_\varepsilon \xrightharpoonup{\Gamma} \mathcal{J}_0$ in $\boldsymbol{Z}$, we take any sequence $u_\varepsilon \rightharpoonup u$ in $\boldsymbol{X}$. If we have $\liminf_{\varepsilon\to 0} \|u_\varepsilon\|_{\mathrm{Z}} = \infty$, then the equicoercivity implies $\mathcal{J}_\varepsilon(u_\varepsilon) \to \infty$ and the liminf estimate holds. If for some subsequence $\|u_{\varepsilon_k}\|_{\mathrm{Z}} \leq C$, then $u_{\varepsilon_k} \rightharpoonup u$ in $\boldsymbol{Z}$, and the liminf estimate follows from that of $\mathcal{J}_\varepsilon \xrightharpoonup{\Gamma} \mathcal{J}_0$ in $\boldsymbol{Z}$. For the construction of recovery sequences, we can choose $\widehat{u}_\varepsilon = \widehat{u}$ if $\widehat{u} \in \boldsymbol{X} \setminus \boldsymbol{Z}$. If $\widehat{u} \in \boldsymbol{Z}$ we choose a recovery sequence $\widehat{u}_\varepsilon \rightharpoonup \widehat{u}$ in $\boldsymbol{Z}$. By the compact embedding we have $\widehat{u}_\varepsilon \to \widehat{u}$ in $\boldsymbol{X}$ and the proof is finished. □

The following lemma presents a simple quadratic example in which the weak and the strong Γ-limits exist but they are different. We define

$$\mathcal{F}_\varepsilon(w) = \int_\Omega \frac{1}{2} w(x) \cdot \mathbb{A}\big(\frac{1}{\varepsilon}x\big) w(x)\,\mathrm{d}x \quad \text{for } w \in \boldsymbol{X} = \mathrm{L}^2(\Omega;\mathbb{R}^m),$$

where $\Omega \subset \mathbb{R}^d$ is a bounded Lipschitz domain and $\mathbb{A} \in \mathrm{L}^\infty(\mathbb{R}^d;\mathbb{R}^{m\times m}_{\mathrm{sym}})$ is 1-periodic, i.e. $\mathbb{A}(y+n) = \mathbb{A}(y)$ for all $y \in \mathbb{R}^d$ and all $n \in \mathbb{Z}^d$. Moreover, we assume that $\mathbb{A}$ is uniformly positive definite, i.e. $\underline{a}|w|^2 \leq w \cdot \mathbb{A}(y) w \leq \overline{a}|w|^2$ for $\overline{a} > \underline{a} > 0$.

Lemma 3.2.2 *Define the arithmetic and harmonic mean of* $\mathbb{A}$ *via*

$$\mathbb{A}_{arith} := \int_{[0,1]^d} \mathbb{A}(y)\,\mathrm{d}y \quad \text{and} \quad \mathbb{A}_{harm} := \Big(\int_{[0,1]^d} \mathbb{A}(y)^{-1}\,\mathrm{d}y\Big)^{-1}$$

and the two functionals

$$\mathcal{F}_{arith}(w) = \int_\Omega \frac{1}{2} w(x) \cdot \mathbb{A}_{arith} w(x)\,\mathrm{d}x \quad \text{and} \quad \mathcal{F}_{harm}(w) = \int_\Omega \frac{1}{2} w(x) \cdot \mathbb{A}_{harm} w(x)\,\mathrm{d}x.$$

In $\boldsymbol{X} = \mathrm{L}^2(\Omega;\mathbb{R}^m)$ *we have* $\mathcal{F}_\varepsilon \xrightharpoonup{\Gamma} \mathcal{F}_{harm}$ *and* $\mathcal{F}_\varepsilon \xrightarrow{\mathrm{C}} \mathcal{F}_{arith}$, *which implies* $\mathcal{F}_\varepsilon \xrightarrow{\Gamma} \mathcal{F}_{arith}$.

Proof We first prove $\mathcal{F}_\varepsilon \xrightharpoonup{\Gamma} \mathcal{F}_{\mathrm{harm}}$. For the liminf estimate assume $w_\varepsilon \rightharpoonup w$ in $\mathrm{L}^2(\Omega)$. Writing $\mathbb{A}_\varepsilon(x) = \mathbb{A}(\frac{1}{\varepsilon}x)$ we have

$$\mathcal{F}_\varepsilon(w_\varepsilon) = \frac{1}{2}\int_\Omega w_\varepsilon \cdot \mathbb{A}_\varepsilon w_\varepsilon\,\mathrm{d}x = \tag{3.2.9}$$

$$= \frac{1}{2}\int_\Omega \underbrace{(w_\varepsilon - \mathbb{A}_\varepsilon^{-1}\mathbb{A}_{\mathrm{harm}}w)\cdot\mathbb{A}_\varepsilon(w_\varepsilon - \mathbb{A}_\varepsilon^{-1}\mathbb{A}_{\mathrm{harm}}w)}_{\geq 0} \tag{3.2.10}$$

$$+ 2 \underbrace{w_\varepsilon}_{\rightharpoonup w} \cdot\mathbb{A}_{\mathrm{harm}}w - \mathbb{A}_{\mathrm{harm}}w \cdot \underbrace{\mathbb{A}_\varepsilon^{-1}}_{\overset{*}{\rightharpoonup}\mathbb{A}_{\mathrm{harm}}^{-1}} \mathbb{A}_{\mathrm{harm}}w\,\mathrm{d}x.$$

Dropping the nonnegative term and taking the limit $\varepsilon \to 0$ give the desired lower estimate $\liminf_\varepsilon \mathscr{F}_\varepsilon(w_\varepsilon) \geq \frac{1}{2}\int_\Omega 0 + 2w\cdot\mathbb{A}_{\mathrm{harm}}w - w\cdot\mathbb{A}_{\mathrm{harm}}w\,\mathrm{d}x = \mathscr{F}_0(w)$.

For the limsup-estimate we use the same reformulation of $\mathscr{F}_\varepsilon$ as in (3.2.9). For a given $\widehat{w}$ we choose $\widehat{w}_\varepsilon = \mathbb{A}_\varepsilon^{-1}\mathbb{A}_{\mathrm{harm}}\widehat{w}$. Since by construction the first term in the integral is 0 we find $\mathscr{F}_\varepsilon(\widehat{w}_\varepsilon) = \frac{1}{2}\int_\Omega 0 + 2\mathbb{A}_\varepsilon^{-1}\mathbb{A}_{\mathrm{harm}}\widehat{w}\cdot\mathbb{A}_{\mathrm{harm}}\widehat{w} - \mathbb{A}_{\mathrm{harm}}\widehat{w}\cdot\mathbb{A}_\varepsilon^{-1}\mathbb{A}_{\mathrm{harm}}\widehat{w}\,\mathrm{d}x \to \mathscr{F}_{\mathrm{harm}}(\widehat{w})$.

For strong continuous convergence take any $w_\varepsilon \to w$ in $\mathrm{L}^2(\Omega)$ and write

$$\mathscr{F}_\varepsilon(w_\varepsilon) = \frac{1}{2}\int_\Omega w\cdot \underbrace{\mathbb{A}_\varepsilon w}_{\to \mathbb{A}_{\mathrm{arith}} w} - 2w\cdot\mathbb{A}_\varepsilon \underbrace{(w - w_\varepsilon)}_{\to 0} + \underbrace{(w - w_\varepsilon)}_{\to 0}\cdot\mathbb{A}_\varepsilon(w - w_\varepsilon)\,\mathrm{d}x \tag{3.2.11}$$

$$\to \mathscr{F}_{\mathrm{arith}}(w). \tag{3.2.12}$$

This proves the strong continuous and hence the strong Γ-convergence. □

3.2.3 Prerequisites from Convex Analysis

For each $u \in \boldsymbol{X}$, the dissipation potentials $\mathscr{R}_\varepsilon(u, \cdot) : \boldsymbol{X} \to [0, \infty]$ are always convex, and lower semicontinuous (lsc). So we can apply the Legendre-Fenchel theory for convex functionals $\Psi : \boldsymbol{X} \to \mathbb{R}_\infty$, where we always assume that $\boldsymbol{X}$ is a reflexive Banach space. The *Legendre-Fenchel transform* $\Psi^* = \mathrm{Leg}(\Psi) : \boldsymbol{X}^* \to \mathbb{R}_\infty$ is defined via

$$\Psi^*(\xi) := \sup\{\langle \xi, v\rangle - \Psi(v) \mid v \in \boldsymbol{X}\}, \tag{3.2.13}$$

where $\langle\cdot,\cdot\rangle$ is the natural dual pairing of $\boldsymbol{X}^*$ and $\boldsymbol{X}$, see [18, 66]. Clearly, Ψ^* is again convex, lsc, and satisfies $\Psi^*(0) = 0$. In particular, the dual dissipation potential $\mathscr{R}_\varepsilon^*$ is defined via $\mathscr{R}_\varepsilon^*(u, \cdot) = \mathrm{Leg}\big(\mathscr{R}_\varepsilon(u, \cdot)\big)$. Elementary examples are

$$\Psi(v) = \frac{1}{2}\langle \mathbb{G}v, v\rangle \iff \Psi^*(\xi) = \frac{1}{2}\langle \xi, \mathbb{G}^{-1}\xi\rangle$$

$$\Psi(v) = \frac{1}{p}\|v\|_X^p \iff \Psi^*(\xi) = \frac{1}{p^*}\|\xi\|_{X^*}^{p^*}, \text{ where } 1 < p < \infty \text{ and } p^* = \frac{p}{p-1}.$$

The fundamental properties of the Legendre-Fenchel transform are the duality relation $\mathrm{Leg}\big(\mathrm{Leg}\Psi\big) = \Psi$ or $\Psi^{**} = \Psi$ and the *Young-Fenchel estimate*

$$\forall v \in \boldsymbol{X}\ \forall \xi \in \boldsymbol{X}^* :\ \Psi(v) + \Psi^*(\xi) \geq \langle \xi, v\rangle. \tag{3.2.14}$$

To discuss the case of equality in this estimate we need the *subdifferential of the convex functional* Ψ, which is defined via

$$\partial\Psi(v) := \{\, \eta \in \boldsymbol{X}^* \mid \forall\, w \in \boldsymbol{X} : \ \Psi(w) \geq \Psi(v) + \langle \eta, w - v\rangle \,\} \subset \boldsymbol{X}^*.$$

Note that $\partial\Psi$ is set-valued and we will write $\partial\Psi : \boldsymbol{X} \rightrightarrows \boldsymbol{X}^*$. However, if Ψ is differentiable with Gateaux derivative $\mathrm{D}\Psi(v)$, then $\partial\Psi(v) = \{\mathrm{D}\Psi(v)\}$.

The Fenchel equivalence characterizes equality in the Young-Fenchel estimate (3.2.14).

Proposition 3.2.2 (Fenchel equivalence [20]) *Let* $\boldsymbol{X}$ *be a reflexive Banach space and* $\Psi : \boldsymbol{X} \to \mathbb{R}_\infty$ *be proper, convex, and lsc. Then, we have*

$$(i)\ \xi \in \partial\Psi(v) \quad \Longleftrightarrow \quad (ii)\ v \in \partial\Psi^*(\xi) \quad \Longleftrightarrow \quad (iii)\ \Psi(v) + \Psi^*(\xi) = \langle \xi, v\rangle.$$

For a proof we refer to [18]. We emphasize that the relation (i) is a relation in dual space $\boldsymbol{X}^*$, (ii) is a relation in $\boldsymbol{X}$, and (iii) is a relation in $\mathbb{R}$.

A further fundamental property of the Legendre transform is related to its continuity with respect to weak or strong Γ-convergence. This result will be important for studying E-convergence in Sect. 3.3.

Theorem 3.2.2 ([4, p. 271]) *Let* $\boldsymbol{X}$ *be a reflexive Banach space and assume that all* $\mathscr{F}_\varepsilon : \boldsymbol{X} \to [0, \infty]$ *are convex, lsc, and satisfy* $\mathscr{F}_\varepsilon(0) = 0$*. Then,*

$$\mathscr{F}_\varepsilon \overset{\Gamma}{\rightharpoonup} \mathscr{F} \quad \Longleftrightarrow \quad \mathscr{F}_\varepsilon^* \overset{\Gamma}{\to} \mathscr{F}^*.$$

The duality and the switch between the weak and strong convergence appears natural, because the definition of the Legendre transform involves the duality product $\langle \xi, v\rangle$. A well-known result from linear functional analysis states that $\langle \xi_\varepsilon, v_\varepsilon\rangle \to \langle \xi, v\rangle$ if either $v_\varepsilon \rightharpoonup v$ and $\xi_\varepsilon \to \xi$ or vice versa $v_\varepsilon \to v$ and $\xi_\varepsilon \rightharpoonup \xi$.

Under the assumptions of the above theorem the definition of Mosco convergence gives the equivalences

$$\mathscr{F}_\varepsilon \overset{\mathsf{M}}{\to} \mathscr{F} \quad \Longleftrightarrow \quad \mathscr{F}_\varepsilon^* \overset{\mathsf{M}}{\to} \mathscr{F}^* \quad \Longleftrightarrow \quad \Big(\mathscr{F}_\varepsilon \overset{\Gamma}{\rightharpoonup} \mathscr{F} \text{ and } \mathscr{F}_\varepsilon^* \overset{\Gamma}{\rightharpoonup} \mathscr{F}^* \Big). \tag{3.2.15}$$

Lemma 3.2.2 provides an interesting example for the application of Theorem 3.2.2. In fact, we have $\mathscr{F}_\varepsilon^*(\xi) = \frac{1}{2}\int_\Omega \xi \cdot \mathbb{A}_\varepsilon^{-1}\xi \,\mathrm{d}x$. Thus, the strong convergence for $\mathscr{F}_\varepsilon^*$ leads to an effective matrix $\mathrm{arith}(\mathbb{A}^{-1}) = \mathrm{harm}(\mathbb{A})^{-1}$.

Another important tool of convex analysis is the weak-strong closedness of the graphs of the subdifferentials $\partial\mathscr{E}_\varepsilon : \boldsymbol{X} \rightrightarrows \boldsymbol{X}^*$ (which is in fact equivalent to Mosco convergence).

Proposition 3.2.3 (Strong weak-closedness, [4, Theorem 3.66]) *Assume that all* $\mathscr{E}_\varepsilon : \boldsymbol{X} \to [0, \infty]$ *are lsc and convex and that* $\mathscr{E}_\varepsilon \overset{\mathsf{M}}{\to} \mathscr{E}_0$*. Then, we have*

$$\left.\begin{array}{l} u_\varepsilon \to u,\ \mathcal{E}_\varepsilon(u_\varepsilon) \to e_0 \in \mathbb{R}, \\ \partial\mathcal{E}_\varepsilon(u_\varepsilon) \ni \xi_\varepsilon \rightharpoonup \xi \end{array}\right\} \implies e_0 = \mathcal{E}_0(u) \text{ and } \xi \in \partial\mathcal{E}_0(u). \tag{3.2.16}$$

The same conclusion holds under the assumptions $u_\varepsilon \rightharpoonup u$ *and* $\partial\mathcal{E}_\varepsilon(u_\varepsilon) \ni \xi_\varepsilon \to \xi$.

Proof The convexity of $\mathcal{E}_\varepsilon$ gives $\mathcal{E}_\varepsilon(w) \geq \mathcal{E}_\varepsilon(u_\varepsilon) + \langle \xi_\varepsilon, w - u_\varepsilon\rangle$ of all $w \in \boldsymbol{X}$. For fixed $\widehat{u}$ the Mosco convergence provides a recovery sequence $\widehat{u}_\varepsilon$ with $\widehat{u}_\varepsilon \to \widehat{u}$, $\mathcal{E}_\varepsilon(\widehat{u}_\varepsilon) \to \mathcal{E}_0(\widehat{u})$. Hence, inserting $w = \widehat{u}_\varepsilon$ yields $\mathcal{E}_\varepsilon(\widehat{u}_\varepsilon) \geq \mathcal{E}_\varepsilon(u_\varepsilon) + \langle \xi_\varepsilon, \widehat{u}_\varepsilon - u_\varepsilon\rangle$.

Taking the limit $\varepsilon \to 0$, all three terms converge (where we use weak-strong continuity of the duality product $\langle \cdot, \cdot\rangle$), and we obtain the relation $\mathcal{E}_0(\widehat{u}) \geq e_0 + \langle \xi, \widehat{u} - u\rangle$ and choosing $\widehat{u} = u$ yields $\mathcal{E}_0(u) \geq e_0$. However, the liminf estimate for $u_\varepsilon \rightharpoonup u$ gives $e_0 \geq \mathcal{E}_0(u)$. Thus, $e_0 = \mathcal{E}_0(u)$ and we conclude $\xi \in \partial\mathcal{E}_0(u)$ as desired. □

3.2.4 Definitions for Γ-Convergence of Evolutionary Systems

According to Theorem 3.2.1 the definition of (static) Γ-convergence implies the convergence of minimizers. If we interpret minimizers as "solutions of a static variational problem", we may state that variational convergence implies that "the solutions of variational problems converge". We now define several versions of evolutionary convergence in a similar way.

We start with *semiflow convergence* as was defined in (3.1.1). Assume that for all $\varepsilon \in [0, 1]$ the evolutionary systems $\dot{u}_\varepsilon = \boldsymbol{V}_\varepsilon(u_\varepsilon)$ have for each $u^0_\varepsilon \in \boldsymbol{X}$ at least one solution $u_\varepsilon : [0, T] \to \boldsymbol{X}$ with $u_\varepsilon(0) = u^0_\varepsilon$. We say that $(\boldsymbol{X}, \boldsymbol{V}_\varepsilon)$ *strongly (or weakly) semiflow-converge* to $(\boldsymbol{X}, \boldsymbol{V}_0)$, if $u_\varepsilon(0) \to u^0$ (or $u_\varepsilon(0) \rightharpoonup u^0$) implies that there exist a sequence $\varepsilon_k \to 0$ and a solution $u : [0, T] \to \boldsymbol{X}$ of $\dot{u} = \boldsymbol{V}_0(u)$ with $u(0) = u^0$ such that $u_{\varepsilon_k}(t) \to u(t)$ (or $u_{\varepsilon_k}(t) \rightharpoonup u(t)$) for all $t \in\,]0, T]$.

For gradient systems $(\boldsymbol{X}, \mathcal{E}_\varepsilon, \mathcal{R}_\varepsilon)$ we will use a special types of semiflow convergence that includes the additional condition on energy convergence, which is the reason why we include the symbol Γ in the name. Throughout this work we call $u_\varepsilon : [0, T] \to \boldsymbol{X}$ a solution for the gradient system $(\boldsymbol{X}, \mathcal{E}_\varepsilon, \mathcal{R}_\varepsilon)$ if it satisfies the gradient evolution $0 \in \partial_{\dot{u}}\mathcal{R}_\varepsilon(u, \dot{u}) + \mathrm{D}\mathcal{E}_\varepsilon(t, u(t))$ in a suitable sense.

Definition 3.2.2 (*Evolutionary Γ-convergence*) For $\varepsilon \in [0, 1]$ consider gradient systems $(\boldsymbol{X}, \mathcal{E}_\varepsilon, \mathcal{R}_\varepsilon)$.

(A) We say $(\boldsymbol{X}, \mathcal{E}_\varepsilon, \mathcal{R}_\varepsilon)$ *E-converges to* $(\boldsymbol{X}, \mathcal{E}_0, \mathcal{R}_0)$ and write $(\boldsymbol{X}, \mathcal{E}_\varepsilon, \mathcal{R}_\varepsilon) \xrightarrow{\mathsf{E}} (\boldsymbol{X}, \mathcal{E}_0, \mathcal{R}_0)$, if

$$\left.\begin{array}{c} u_\varepsilon : [0, T] \to \boldsymbol{X} \\ \text{is sol. of } (\boldsymbol{X}, \mathcal{E}_\varepsilon, \mathcal{R}_\varepsilon) \\ \text{and } u_\varepsilon(0) \to u^0 \end{array}\right\} \implies \left\{\begin{array}{c} \exists\, u \text{ sol. of } (\boldsymbol{X}, \mathcal{E}_0, \mathcal{R}_0) \text{with } u(0) = u^0 \\ \text{and a subsequence } \varepsilon_k \to 0 : \\ \forall\, t \in\,]0, T] :\ u_{\varepsilon_k}(t) \to u(t) \text{ and} \\ \qquad \mathcal{E}_\varepsilon(u_{\varepsilon_k}(t)) \to \mathcal{E}_0(u(t)). \end{array}\right. \tag{3.2.17}$$

(B) We say that $(\boldsymbol{X}, \mathscr{E}_\varepsilon, \mathscr{R}_\varepsilon)$ *E-converges with well-prepared initial conditions to* $(\boldsymbol{X}, \mathscr{E}_0, \mathscr{R}_0)$, and shortly write $(\boldsymbol{X}, \mathscr{E}_\varepsilon, \mathscr{R}_\varepsilon) \overset{\mathsf{pE}}{\to} (\boldsymbol{X}, \mathscr{E}_0, \mathscr{R}_0)$, if

$$\left.\begin{array}{c} u_\varepsilon : [0,T] \to \boldsymbol{X} \\ \text{is sol. of } (\boldsymbol{X}, \mathscr{E}_\varepsilon, \mathscr{R}_\varepsilon), \\ u_\varepsilon(0) \to u^0, \text{ and} \\ \mathscr{E}_\varepsilon(u_\varepsilon(0)) \to \mathscr{E}_0(u^0) < \infty \end{array}\right\} \Longrightarrow \left\{\begin{array}{l} \exists\, u \text{ sol. of } (\boldsymbol{X}, \mathscr{E}_0, \mathscr{R}_0) \text{ with } u(0) = u^0 \\ \qquad \text{and a subsequence } \varepsilon_k \to 0 : \\ \quad \forall\, t \in \,]0,T] : \; u_{\varepsilon_k}(t) \to u(t) \text{ and} \\ \qquad\qquad\qquad \mathscr{E}_\varepsilon(u_{\varepsilon_k}(t)) \to \mathscr{E}_0(u(t)). \end{array}\right. \tag{3.2.18}$$

(C) If all the strong convergences in (A) or (B) are replaced by weak convergences, then we write $\overset{\mathsf{E}}{\rightharpoonup}$ or $\overset{\mathsf{pE}}{\rightharpoonup}$, respectively. Any of these four types of convergence is called *evolutionary Γ -convergence*.

As in [69] we use the symbol Γ in the name "evolutionary Γ-convergence" to indicate the relation to Γ-convergence for the energy. In fact, we typically expect that $\mathscr{E}_\varepsilon \overset{\Gamma}{\to} \mathscr{E}_0$ is a necessary condition for evolutionary Γ-convergence. Then, the statement that the energies $\mathscr{E}_\varepsilon(u_\varepsilon(t))$ converge to $\mathscr{E}_0(u(t))$ means that $u_\varepsilon(t)$ is a recovery sequence for $u(t)$. If one asks this condition at the initial time $t = 0$, one speaks of *well-preparedness of the initial conditions*, i.e. the initial conditions must be a recovery sequence as well. This explains our name "well-prepared E-convergence".

Note that E-convergence implies well-prepared E-convergence, since in the later case the assumptions are stronger while the conclusions are the same. Often the energy convergence improves the convergence $u_\varepsilon(t) \rightharpoonup u_0(t)$ into strong convergence or weak convergence in a better space.

The aim of these notes is to provide conditions for suitable convergences for $(\mathscr{E}_\varepsilon, \mathscr{R}_\varepsilon)$ towards a limiting pair $(\mathscr{E}_0, \mathscr{R}_0)$ on a suitable space $\boldsymbol{X}$ such that evolutionary Γ-convergence is guaranteed. We emphasize that separate convergence of the two functionals will not be sufficient. It is important to realize that there has to be some *compatibility* between the convergences $\mathscr{E}_\varepsilon \rightsquigarrow \mathscr{E}_0$ and $\mathscr{R}_\varepsilon \rightsquigarrow \mathscr{R}_0$. This will be the topic of Sects. 3.3 and 3.4, while the rest of the present section introduces a few examples that will be used later for applying the abstract theory.

3.2.5 An ODE Example

In the first example we discuss a linear ODE. *Since the solutions and the Γ-limits can be explicitly constructed we can check the validity of evolutionary Γ-convergence without any abstract theory. In particular, we will show that in general it is not enough to have separate Mosco convergences $\mathscr{E}_\varepsilon \overset{\mathsf{M}}{\to} \mathscr{E}_0$ and $\Psi_\varepsilon \overset{\mathsf{M}}{\to} \Psi_0$ to conclude evolutionary Γ-convergence.*

For $\boldsymbol{X} = \mathbb{R}^2$ we consider the gradient system $(\mathbb{R}^2, \mathscr{E}_\varepsilon, \Psi_\varepsilon)$ with

$$\mathscr{E}_\varepsilon(u) = \frac{1}{2}u \cdot \mathbb{A}_\varepsilon u = \frac{1}{2}u_1^2 + \frac{1}{2\varepsilon^2}(u_2 - \varepsilon u_1)^2,$$
$$\Psi_\varepsilon(v) = \frac{1}{2}\dot{u} \cdot \mathbb{G}_\varepsilon \dot{u} = \frac{1}{2}\dot{u}_1^2 + \frac{1}{2\varepsilon^\beta}\dot{u}_2^2,$$
$$\text{where } \mathbb{A}_\varepsilon = \begin{pmatrix} 2 & -\frac{1}{\varepsilon} \\ -\frac{1}{\varepsilon} & \frac{1}{\varepsilon^2} \end{pmatrix} \text{ and } \mathbb{G}_\varepsilon = \begin{pmatrix} 1 & 0 \\ 0 & \frac{1}{\varepsilon^\beta} \end{pmatrix}.$$

The solution of the gradient evolution $\mathbb{G}_\varepsilon \dot{u}_\varepsilon = -\mathbb{A}_\varepsilon u_\varepsilon$ with general initial conditions $u_\varepsilon^0 = (a_\varepsilon, b_\varepsilon)^\top$ can be calculated explicitly for all $\varepsilon > 0$ and all $\beta > 0$.

Before discussing the question of the types of evolutionary convergence, we study the convergence properties of the functionals $\mathscr{E}_\varepsilon$ and Ψ_ε. Because the weak and the strong topology coincide on $\mathbb{R}^2$, Γ-convergence equals Mosco convergence. In fact, we have

$$\mathscr{E}_\varepsilon \xrightarrow{\mathsf{M}} \mathscr{E}_0 : u \mapsto \begin{cases} \frac{1}{2}u_1^2 & \text{for } u_2 = 0, \\ \infty & \text{otherwise,} \end{cases} \quad \text{and} \quad \Psi_\varepsilon \xrightarrow{\mathsf{M}} \Psi_0 : v \mapsto \begin{cases} \frac{1}{2}v_1^2 & \text{for } v_2 = 0, \\ \infty & \text{otherwise.} \end{cases}$$

We also mention that $\mathscr{E}_\varepsilon \xrightarrow{\text{pw}} \mathscr{E}_{\text{pw}} = 2\mathscr{E}_0 \gneqq \mathscr{E}_0$, but neither $\mathscr{E}_\varepsilon$ nor Ψ_ε converge continuously. Finally we consider well-preparedness of initial conditions, namely $(a_\varepsilon, b_\varepsilon) \to (a, b)$ and $\mathscr{E}_\varepsilon(a_\varepsilon, b_\varepsilon) \to \mathscr{E}_0(a, b) < \infty$, which implies

$$b = 0 \quad \text{and} \quad b_\varepsilon = \varepsilon a + o(\varepsilon) \text{ for } \varepsilon \to 0. \tag{3.2.19}$$

The general solutions can be calculated explicitly. With $\delta_\varepsilon = \varepsilon^{\beta-2}/2$ the matrix $\mathbb{G}_\varepsilon^{-1}\mathbb{A}_\varepsilon$ has the eigenvalues $\mu_1 = 1 + \delta_\varepsilon - \left(1 + \delta_\varepsilon^2\right)^{1/2}$ and $\mu_2 = 1 + \delta_\varepsilon + \left(1 + \delta_\varepsilon^2\right)^{1/2}$, and we obtain

$$u_\varepsilon(t) = \tfrac{\varepsilon(\mu_2-2)a+b}{\varepsilon(\mu_2-\mu_1)}\,\mathrm{e}^{-\mu_1 t}\begin{pmatrix} 1 \\ \varepsilon(2-\mu_1) \end{pmatrix} + \tfrac{b+\varepsilon(\mu_1-2)a}{\varepsilon(\mu_2-\mu_1)}\,\mathrm{e}^{-\mu_2 t}\begin{pmatrix} -1 \\ \varepsilon(\mu_2-2) \end{pmatrix}.$$

In the case $\beta \in \,]0, 2[$ we have $\delta_\varepsilon \to \infty$, $\mu_1 \to 1$, $\mu_2 \to \infty$, and hence for $t > 0$ we have $\mathrm{e}^{-\mu_2 t} \to 0$ faster than $O(\varepsilon^k)$ for all $k \in \mathbb{N}$. This fact leads to the convergences

$$u_\varepsilon(t) \to a\,\mathrm{e}^{-t}\begin{pmatrix} 1 \\ 0 \end{pmatrix} \quad \text{and} \quad \mathscr{E}_\varepsilon(u_\varepsilon(t)) \to \frac{a^2}{2}\,\mathrm{e}^{-2t}.$$

Thus, we conclude $(\mathbb{R}^2, \mathscr{E}_\varepsilon, \Psi_\varepsilon) \xrightarrow{\mathsf{E}} (\mathbb{R}^2, \mathscr{E}_0, \Psi_0)$ for $\beta \in \,]0, 2[$.

For $\beta > 2$ we have $\delta_\varepsilon \to 0$, $\mu_1 \to 0$, and $\mu_2 \to 2$. For $b_\varepsilon \to b \neq 0$ the solutions are unbounded and $\mathscr{E}_\varepsilon(u_\varepsilon(t)) \to \infty$ for $t \in [0, T_b[$ for some $T(b) > 0$. So we consider the case $b_\varepsilon/\varepsilon \to \widehat{b}$ for some $\widehat{b} \in \mathbb{R}$ and find the limits

$$u_\varepsilon(t) \to \Big(\frac{\widehat{b}}{2} + \big(a - \frac{\widehat{b}}{2}\big)\,\mathrm{e}^{-2t}\Big)\begin{pmatrix} 1 \\ 0 \end{pmatrix}, \qquad \mathscr{E}_\varepsilon(u_\varepsilon(t)) \to \frac{\widehat{b}^2}{4} + \big(a - \frac{\widehat{b}}{2}\big)^2 \mathrm{e}^{-4t}.$$

Note that well-prepared initial conditions require $\widehat{b} = a$, while the case $\widehat{b} = 0$ implies semigroup convergence to the solutions of the gradient system $(\mathbb{R}^2, \mathcal{E}_{\text{pw}}, \Psi_0)$, but we do not have evolutionary Γ-convergence.

For the crossover case $\beta = 2$ the assumption $b_\varepsilon/\varepsilon \to \widehat{b}$ leads to

$$u_\varepsilon(t) \to \begin{pmatrix} w(t) \\ 0 \end{pmatrix} \text{ with } w(t) = \frac{1}{2\sqrt{5}}\Big(\big((\sqrt{5}-1)a + 2\widehat{b}\big)\mathrm{e}^{-\overline{\mu}_1 t} + \big((\sqrt{5}+1)a - 2\widehat{b}\big)\mathrm{e}^{-\overline{\mu}_2 t}\Big),$$

where $\overline{\mu}_{1,2} = (3 \pm \sqrt{5})/2$. Assuming $\widehat{b} = 0$ we obtain the linear evolution $w(t) = B(t)w(0)$, where $\dot{B}(t)/B(t)$ is not constant. Hence, the limit evolution cannot be described by an autonomous equation $\dot{w} = \boldsymbol{V}(w)$, so we do not have semigroup convergence.

Remark 3.2.1 (Scaling changes evolutionary convergence) We note that a rescaling of the variables may change the convergence behavior dramatically. E.g. using $z = (z_1, z_2) = (u_1, u_2/\varepsilon)$, we obtain

$$\widetilde{\mathcal{E}}_\varepsilon(z) = \mathcal{F}(z) := \frac{1}{2}z_1^2 + \frac{1}{2}(z_2 - z_1)^2 \quad \text{and} \quad \widetilde{\Psi}_\varepsilon(\dot{z}) = \frac{1}{2}\dot{z}_1^2 + \frac{\varepsilon^{2-\beta}}{2}\dot{z}_2^2,$$

where $\widetilde{\mathcal{E}}_\varepsilon$ is independent of ε and hence continuously converging to $\mathcal{F}$. For $\beta > 2$ we obtain $(\mathbb{R}^2, \mathcal{F}, \Psi_\varepsilon) \xrightarrow{\mathsf{E}} (\mathbb{R}^2, \mathcal{F}, \Psi_0)$.

The alternative scaling $w = (u_1, \varepsilon^{-\beta/2}u_2)$ makes the dissipation distance ε-independent:

$$\overline{\mathcal{E}}_\varepsilon(w) = \frac{1}{2}w_1^2 + \frac{1}{2}\big(w_1 - \varepsilon^{\beta/2-1}w_2\big)^2 \quad \text{and} \quad \overline{\Psi}(\dot{w}) = \frac{1}{2}|\dot{w}|^2.$$

For $\beta \geq 2$ we have the continuous convergence $\overline{\mathcal{E}}_\varepsilon \xrightarrow{\mathsf{C}} \overline{\mathcal{E}}_0 : w \mapsto w_1^2$, while for $\beta \in \,]0, 2[$ we have $\overline{\mathcal{E}}_\varepsilon \xrightarrow{\mathsf{M}} \mathcal{E}_0$. In all cases we have $(\mathbb{R}^2, \overline{\mathcal{E}}_\varepsilon, \overline{\Psi}) \xrightarrow{\mathsf{E}} (\mathbb{R}^2, \overline{\mathcal{E}}_0, \overline{\Psi})$ cf. Theorem 3.3.2.

An important conclusion of this remark is that linear transformations, here scalings, may change the convergence behavior dramatically.

3.2.6 Homogenization of a 1D Parabolic Equation

Our main guiding example for the different levels of the theory will be the *homogenization* of a simple parabolic equation, since for this case it is possible to do all the calculation by hand. We emphasize that the theory works equally well for any space dimension, of course then the calculations are no longer explicit.

On the domain $\Omega = \,]0, \ell[$ for $t > 0$ we consider the parabolic equation

$$c(\frac{x}{\varepsilon})\dot{u}(t,x) = \big(a(\frac{x}{\varepsilon})u_x(t,x)\big)_x - b(\frac{x}{\varepsilon})u(t,x), \qquad u_x(t,0) = 0 = u_x(t,\ell). \tag{3.2.20}$$

Here the coefficient functions $a, b, c \in \mathrm{L}^\infty(\mathbb{R})$ are 1-periodic and are bounded from below by a positive constant $c_0 > 0$. Choosing $\boldsymbol{X} = \mathrm{L}^2(\Omega)$ this equation is induced by the gradient system $(\boldsymbol{X}, \mathcal{E}_\varepsilon, \Psi_\varepsilon)$ with

$$\mathcal{E}_\varepsilon(u) = \frac{1}{2}\int_\Omega a(\frac{x}{\varepsilon})u_x(x)^2 + b(\frac{x}{\varepsilon})u(x)^2 \mathrm{d}x \quad \text{and} \quad \Psi_\varepsilon(v) = \frac{1}{2}\int_\Omega c(\frac{x}{\varepsilon})v(x)^2 \mathrm{d}x, \tag{3.2.21}$$

if the derivatives of Ψ_ε and $\mathcal{E}_\varepsilon$ are calculated in the duality pairing of $\boldsymbol{X} = \mathrm{L}^2(\Omega)$.

We are interested in the question of evolutionary Γ-convergence and expect that in the limit $\varepsilon \to 0$ we find effective functionals

$$\mathcal{E}_{\mathrm{eff}}(u) = \frac{1}{2}\int_\Omega a_{\mathrm{eff}} u_x(x)^2 + b_{\mathrm{eff}} u^2 \mathrm{d}x \quad \text{and} \quad \Psi_{\mathrm{eff}}(v) = \frac{1}{2}\int_\Omega c_{\mathrm{eff}} v^2 \mathrm{d}x,$$

where the main task remains to determine the effective coefficients a_{eff}, b_{eff}, and c_{eff}. Moreover, when taking the Γ-limits of $\mathcal{E}_\varepsilon$ and Ψ_ε in (3.2.21) there is major issue in choosing a suitable topology.

The natural function space for the dissipation functionals Ψ_ε is $\boldsymbol{X} = \mathrm{L}^2(\Omega)$, which we will call the *dissipation space*. According to Lemma 3.2.2 we have $\Psi_\varepsilon \xrightharpoonup{\Gamma} \Psi_{\mathrm{harm}}$ and $\Psi_\varepsilon \xrightarrow{\Gamma} \Psi_{\mathrm{arith}}$ in $\mathrm{L}^2(\Omega)$, where $c_{\mathrm{eff}} = c_{\mathrm{harm}}$ and $c_{\mathrm{eff}} = c_{\mathrm{arith}}$, respectively.

Similarly, we can study the energy functionals in their natural space $\boldsymbol{Z} = \mathrm{H}^1(\Omega)$, i.e. the *energy space*. Using the compact embedding of $\mathrm{H}^1(\Omega) \Subset \mathrm{L}^2(\Omega)$, we find

$$\mathcal{E}_\varepsilon \xrightharpoonup{\Gamma} \mathcal{E}_{\mathrm{harm}} : u \mapsto \frac{1}{2}\int_\Omega a_{\mathrm{harm}} u_x^2 + b_{\mathrm{arith}} u^2 \mathrm{d}x \quad \text{in } \boldsymbol{Z} = \mathrm{H}^1(\Omega),$$

$$\mathcal{E}_\varepsilon \xrightarrow{\Gamma} \mathcal{E}_{\mathrm{arith}} : u \mapsto \frac{1}{2}\int_\Omega a_{\mathrm{arith}} u_x^2 + b_{\mathrm{arith}} u^2 \mathrm{d}x \quad \text{in } \boldsymbol{Z} = \mathrm{H}^1(\Omega).$$

For later use we prepare the following result which has an obvious extension to general elliptic homogenization problems, see [14, Chaps. 24, 25] and [9, Sect. 5].

Proposition 3.2.4 *Consider the functionals $\mathcal{E}_\varepsilon : \mathrm{L}^2(\Omega) \to [0, \infty]$ as in (3.2.21), which are set to ∞ for $u \in \mathrm{L}^2(\Omega) \backslash \mathrm{H}^1(\Omega)$. Then, $\mathcal{E}_\varepsilon \xrightarrow{\mathrm{M}} \mathcal{E}_{harm}$ in $\mathrm{L}^2(\Omega)$.*

Proof For the liminf estimate in the weak L^2 topology consider $u_\varepsilon \rightharpoonup u$ in $\mathrm{L}^2(\Omega)$ with $\alpha := \liminf \mathcal{E}_\varepsilon(u_\varepsilon)$. For $\alpha = \infty$ there is nothing to be shown. For $\alpha < \infty$ the uniform coercivity of $\mathcal{E}_\varepsilon$ in $\mathrm{H}^1(\Omega)$ implies $u_\varepsilon \rightharpoonup u$ in H^1. Now using $u_\varepsilon \to u$ and $\nabla u_\varepsilon \rightharpoonup \nabla u$ in $\mathrm{L}^2(\Omega)$ together with Lemma 3.2.2 give $\mathcal{E}_{\mathrm{harm}}(u) \le \liminf \mathcal{E}_\varepsilon(u_\varepsilon)$ as desired.

Table 3.1 Possible effective coefficients in 1D homogenization

Topology	a_{eff}	b_{eff}	c_{eff}
L^2 weak	a_{harm}	b_{arith}	c_{harm}
L^2 strong	a_{harm}	b_{arith}	c_{arith}
H^1 weak	a_{harm}	b_{arith}	c_{arith}
H^1 strong	a_{arith}	b_{arith}	c_{arith}

For the limsup estimate in the strong L^2 topology we construct a recovery sequence that converges weakly in $\text{H}^1(\Omega)$ and hence strongly in $\text{L}^2(\Omega)$. Here we use that the domain Ω is one-dimensional: for a given $\widehat{u}$ we simply define $\widehat{u}_\varepsilon$ such that $\nabla\widehat{u}_\varepsilon(x) = a_{\text{harm}}\nabla\widehat{u}(x)/a(\frac{1}{\varepsilon}x)$ and $\widehat{u}_\varepsilon(0) = \widehat{u}(0)$. □

We leave the question concerning the evolutionary Γ-convergence with the proper effective coefficients open until Sects. 3.3.2 and 3.4.2. At this moment we observe that we have several choices for the topology that lead to different Γ-limits, see Table 3.1. *This fact is our motivation to derive mathematical theories giving us guidance for choosing the correct topology via suitable compatibility conditions of the two convergences.*

3.2.7 Tartar's Model Equation

In [73, 74] a more general version of following example is considered:

$$\dot{u}_\varepsilon(t,x) = -a_\varepsilon(x)u_\varepsilon(t,x) + f(t,x), \ t > 0, \ x \in \Omega, \ u_\varepsilon(0,x) = u_\varepsilon^0(x). \quad (3.2.22)$$

It is shown there that for $u_\varepsilon^0 = 0$ and $(a_\varepsilon)_{\varepsilon\in(0,1)}$ bounded in $\text{L}^\infty(\Omega)$ one can pass to the nonlocal limit equation

$$\dot{u}_0(t,x) = -a_0(x)u_0(t,x) + f(t,x) + \int_0^t K^0(x,t-s)u_0(s,x)\,\mathrm{d}s,$$

where the memory kernel K^0 is determined from the sequence a_ε alone (via its Young measure).

This model serves three purposes: it shows that (i) simple Γ-convergence in the weak or the strong topology is not sufficient for evolutionary convergence, (ii) that there can be quite different gradient structures, and (iii) that we may obtain different effective equations when doing pE-convergence for different gradient structures, see Corollary 3.3.1.

(i) The simplest gradient structure is the one in $\text{L}^2(\Omega)$, namely $(\text{L}^2(\Omega), \mathcal{E}_\varepsilon, \Psi_\varepsilon)$ with

$$\mathscr{E}_\varepsilon(u) = \int_\Omega \frac{a_\varepsilon}{2} u^2 \,\mathrm{d}x \quad \text{and} \quad \Psi_\varepsilon(v) = \int_\Omega \frac{1}{2} v^2 \,\mathrm{d}x,$$

Assuming that $a_\varepsilon(x) = a(\frac{1}{\varepsilon}x)$ with a nontrivial 1-periodic $a \in \mathrm{L}^\infty(\mathbb{R})$, we have $\mathscr{E}_\varepsilon \xrightarrow{\mathsf{E}} \mathscr{E}_{\text{arith}}$ and $\mathscr{E}_\varepsilon \xrightharpoonup{\mathsf{E}} \mathscr{E}_{\text{harm}}$ but no Mosco convergence, see Lemma 3.2.2. Since the explicit solution is given by $u_\varepsilon(t, x) = \mathrm{e}^{-a_\varepsilon(x)t} u_\varepsilon(0, x)$ we see that evolutionary Γ-convergence does not hold (not even semigroup convergence).

In [10, Exercise 7.2.5] this gradient structure for Tartar's model was studied via time-incremental minimization, i.e. for a fixed time step τ one solves iteratively

$$u_k^\tau = \operatorname{arg\,min}\{ \mathscr{E}_\varepsilon(u) + \frac{1}{\tau} \Psi_\varepsilon(u - u_{k-1}^\tau) \mid u \in \mathrm{L}^2(\Omega) \}.$$

Taking first the limit $\varepsilon \to 0$ for fixed τ and then the limit $\tau \to 0$, the convergence to the effective equation $\dot{u} = -a_{\text{harm}} u$.

(ii) Under the assumption $u^0(x) > 0$, we can introduce further gradient structures by considering u as the density of a measure on Ω. Choosing $\boldsymbol{X} := \boldsymbol{M}_+(\Omega)$, the set of nonnegative measures on Ω, we define the functionals

$$\widetilde{\mathscr{E}}_\varepsilon(u) = \int_\Omega b_\varepsilon(x) \mathrm{d}u(x) \quad \text{and} \quad \widetilde{\mathscr{R}}_\varepsilon^*(u, \xi) = \frac{1}{2} \int_\Omega \frac{a_\varepsilon}{b_\varepsilon} \xi^2 \mathrm{d}u(x),$$

where now a_ε and b_ε are assumed to be positive continuous functions on $\overline{\Omega}$. The dual dissipation potential $\widetilde{\mathscr{R}}_\varepsilon^*$ defines the state-dependent Onsager operator

$$\widetilde{\mathbb{K}}_\varepsilon(u)\xi := \mathrm{D}_\xi \widetilde{\mathscr{R}}_\varepsilon^*(u, \xi) = \frac{a_\varepsilon}{b_\varepsilon} u\,\xi.$$

Using $\mathrm{D}\widetilde{\mathscr{E}}_\varepsilon(u) = b_\varepsilon$ we easily find the linear gradient flow $\dot{u} = -\widetilde{\mathbb{K}}_\varepsilon(u) \mathrm{D}\widetilde{\mathscr{E}}_\varepsilon(u) = -a_\varepsilon u$.

(iii) In Sect. 3.3.5.2 we will discuss pE-convergence and highlight that different choices of b_ε, which do not change the gradient flow equation, lead to different Γ-limits. In particular, we will emphasize the role of the well-preparedness $\widetilde{\mathscr{E}}_\varepsilon(u_\varepsilon(0)) \to \widetilde{\mathscr{E}}_0(u(0))$.

3.2.8 Further Examples of Gradient Systems

Here we discuss some additional gradient systems, which highlight that generally both, the energy and the dissipation potential, may depend on the small parameter ε.

3.2.8.1 Interfaces as Limits of Thin Bulk Layers

This example shows that E-convergence can be used for dimension reduction, in particular for deriving new interface models. We consider a bulk material in $\Omega =]-1,1[$ where a layer of width ε around 0 is filled with a material of having quite different material coefficients, that scale suitably with ε. The parabolic equation is given by

$$\gamma_\varepsilon(x)\dot{u} = \big(A_\varepsilon(x)u_x\big)_x - \partial_u F(x,u) \text{ in } \Omega, \quad u_x(t,-1) = 0 = u_x(t,1).$$

For the coefficients we assume the scalings

$$(\gamma_\varepsilon(x), A_\varepsilon(x)) = \begin{cases} (c\,,\alpha) & \text{for } |x| > \varepsilon/2, \\ (\rho/\varepsilon\,,\beta\varepsilon) & \text{for } |x| < \varepsilon/2. \end{cases}$$

For $\mathcal{E}_\varepsilon(u) = \int_\Omega \frac{A_\varepsilon}{2}u_x^2 + F_\varepsilon(x,u)\mathrm{d}x$ and $\Psi_\varepsilon(v) = \int_\Omega \frac{\gamma_\varepsilon}{2}v^2\mathrm{d}x$ we find the Γ-limits

$$\mathcal{E}_0(u) = \int_{\Omega_*} \frac{\alpha}{2}u_x^2 + F(x,u)\mathrm{d}x + \frac{\beta}{2}\big(u(0^-) - u(0^+)\big)^2 \text{ and}$$

$$\Psi_0(v) = \int_{\Omega_*} \frac{c}{2}v^2\mathrm{d}x + \frac{\rho}{2}v(0)^2,$$

where $\Omega_* =]-1,0[\,\cup\,]0,1[$. Hence, the limit functionals consist of two integral terms as well as a point contribution at the interface $x = 0$. These extra terms at $x = 0$ determine the effective properties of the limiting sharp interface. We refer to [28, 35, 36] and Sect. 3.3.5.1 for more details in this direction.

3.2.8.2 Ginzburg-Landau Vortices

This gradient system is included for historical reasons: the fundamental paper [69] *is the first that develops a method for using the gradient structure for establishing evolutionary Γ-convergence*, see Sect. 3.3.3 for the abstract approach and [10, Chap. 10] for a short survey. The gradient system $(\mathrm{L}^2(\Omega), \mathcal{E}_\varepsilon, \Psi_\varepsilon)$ is given by Ginzburg-Landau functional for the order parameter $\psi : \Omega \to \mathbb{C}$ in the form

$$\mathcal{E}_\varepsilon(\psi) = \int_\Omega \frac{1}{2}|\nabla\psi|^2 + \frac{1}{\varepsilon^2}(1 - |\psi|^2)^2\mathrm{d}x - n\pi\,|\log\varepsilon|$$

and $\Psi_\varepsilon(v) = \frac{1}{2|\log\varepsilon|}\|v\|_2^2$. Well-prepared solutions $\psi_\varepsilon(t)$ are then well approximated by simple vortices with positions $x_i(t) \in \mathbb{R}^2$, $i = 1, .., n$. The evolutionary Γ-limit induces an ODE for the evolution of the vortex positions $x_i(t)$.

3.2.8.3 Slow and Fast Chemical Reactions

This example describes a gradient system where the energy $\mathcal{E}$ is independent of ε, but *$\mathcal{R}_\varepsilon$ is state-dependent and strongly dependent on ε*. In [44] it is shown that general reaction systems for concentrations $c_1, ..., c_I > 0$ with R reactions of mass-action type can be written as a gradient system on $\mathcal{Q} =]0, \infty[^I$ if the reactions satisfy the detailed-balance condition for some concentration vector $\boldsymbol{w} = (w_1, \dots, w_I) \in \mathcal{Q}$. We then have

$$\dot{\boldsymbol{c}} = -\sum_{r=1}^{R} \kappa_r \Big(\frac{\boldsymbol{c}^{\boldsymbol{\alpha}^r}}{\boldsymbol{w}^{\boldsymbol{\alpha}^r}} - \frac{\boldsymbol{c}^{\boldsymbol{\beta}^r}}{\boldsymbol{w}^{\boldsymbol{\beta}^r}}\Big)\Big(\boldsymbol{\alpha}^r - \boldsymbol{\beta}^r\Big) = -\mathbb{K}(\boldsymbol{c})\mathrm{D}\mathcal{E}(\boldsymbol{c}),$$

where the vectors $\boldsymbol{\alpha}^r, \boldsymbol{\beta}^r \in \mathbb{N}_0^I$ contain the stoichiometric coefficients for the rth reaction and $\boldsymbol{c}^\gamma$ denotes the monomial $\prod_1^I c_i^{\gamma_i}$. The gradient structure $(\mathcal{Q}, \mathcal{E}, \mathcal{R}_\varepsilon)$ is given via $\Lambda(a, b) = (a - b)/(\log a - \log b) > 0$ in the form

$$\mathcal{E}(\boldsymbol{c}) = \sum_{i=1}^{I} \big(c_i \log(c_i/w_i) - c_i + w_i\big) \quad \text{and}$$

$$\mathcal{R}_\varepsilon^*(\boldsymbol{c}, \boldsymbol{\xi}) = \frac{1}{2}\sum_{r=1}^{R} \kappa_r \Lambda\big(\tfrac{\boldsymbol{c}^{\boldsymbol{\alpha}^r}}{\boldsymbol{w}^{\boldsymbol{\alpha}^r}}, \tfrac{\boldsymbol{c}^{\boldsymbol{\beta}^r}}{\boldsymbol{w}^{\boldsymbol{\beta}^r}}\big)\Big(\big(\boldsymbol{\alpha}^r - \boldsymbol{\beta}^r\big)\cdot \boldsymbol{\xi}\Big)^2.$$

This defines $\mathbb{K}_\varepsilon$ via $\mathcal{R}_\varepsilon^*(\boldsymbol{c}, \boldsymbol{\xi}) = \frac{1}{2}\langle \mathbb{K}_\varepsilon(\boldsymbol{c})\boldsymbol{\xi}, \boldsymbol{\xi}\rangle$.

Often the reaction coefficients κ_r have quite different magnitudes, i.e. some reactions are very fast while others are slow. Assuming $k_r = \rho_r/\varepsilon$ for $r = 1, \dots, R_{\text{fast}}$ and $\kappa_r = O(1)$ for $r > R_{\text{fast}}$ leads to a decomposition $\mathcal{R}_\varepsilon = \frac{1}{\varepsilon}\mathcal{R}_{\text{fast}} + \mathcal{R}_{\text{slow}}$ and a Mosco limit $\mathcal{R}_0^*(\boldsymbol{c}, \boldsymbol{\xi}) = \mathcal{R}_{\text{slow}}(\boldsymbol{c}, \boldsymbol{\xi})$ if $\mathcal{R}_{\text{fast}}^*(\boldsymbol{c}, \boldsymbol{\xi}) = 0$ and $+\infty$ otherwise. In particular, this is a case where the limit gradient system is only defined on a nonlinear submanifold of $\mathcal{Q}$, namely the one defined by $\frac{\boldsymbol{c}^{\boldsymbol{\alpha}^r}}{\boldsymbol{w}^{\boldsymbol{\alpha}^r}} = \frac{\boldsymbol{c}^{\boldsymbol{\beta}^r}}{\boldsymbol{w}^{\boldsymbol{\beta}^r}}$ for $r = 1, \dots, R_{\text{fast}}$.

3.3 pE-Convergence via the Energy-Dissipation Principle

In this section we discuss formulations that are based on the differential formulations involving derivatives such as $\dot{u}(t)$, $\mathrm{D}\mathcal{E}$, $\mathrm{D}_{\dot{u}}\mathcal{R}(u, \dot{u})$, or $\mathrm{D}_\xi\mathcal{R}^*(u, -\mathrm{D}\mathcal{E}(u))$. In Sect. 3.4 we will see that under certain structural assumptions such relations can be replaced by evolutionary variational inequalities only involving $u(t)$, $\mathcal{E}(w)$, and a dissipation distance $\mathcal{D}(u(t), w)$, i.e. derivatives are not required.

3.3.1 Four Equivalent Formulations via Legendre Transform

We now convert the formal modeling ideas in Sect. 3.2.1 into exact mathematical statements by working in a reflexive Banach space $\boldsymbol{X}$. We consider generalized gradient systems $(\boldsymbol{X}, \mathscr{E}, \mathscr{R})$ with time-dependent energy functionals $\mathscr{E} : [0,T]\times\boldsymbol{X} \to \mathbb{R}_\infty$. Then, by Fenchel's equivalence in Proposition 3.2.2 the following three formulations are equivalent, but have different physical interpretations. They are formulated in terms of the dissipation potential $\mathscr{R}$, the dual dissipation potential $\mathscr{R}^*$, or in terms of an extremum principle involving both:

Force balance in $\boldsymbol{X}^*$ *Biot equation* [7]

(FB) $\quad 0 \in \partial_{\dot u}\mathscr{R}(u(t), \dot u(t)) + \mathrm{D}\mathscr{E}(t, u(t)) \in \boldsymbol{X}^*$ for a.a. $t \in [0, T]$.

Rate equation in $\boldsymbol{X}$ *Onsager equation* [62]

(RE) $\quad \dot u(t) \in \partial_\xi\mathscr{R}^*(u(t), -\mathrm{D}\mathscr{E}(t, u(t))) \in \boldsymbol{X}$ for a.a. $t \in [0, T]$.

Power balance in $\mathbb{R}$ *De Giorgi's $(\mathscr{R}, \mathscr{R}^*)$ formulation* [17]

(PB) $\quad \mathscr{R}(u(t), \dot u(t)) + \mathscr{R}^*(u(t), -\mathrm{D}\mathscr{E}(t, u(t))) = -\langle \mathrm{D}\mathscr{E}(t, u(t)), \dot u(t)\rangle$.

Before returning to the general situation, we highlight the three different cases for the classical viscous dissipation, i.e. $\mathscr{R}(u, v) = \frac{1}{2}\langle \mathbb{G}v, v\rangle$ and $\mathscr{R}^*(u, \xi) = \frac{1}{2}\langle \xi, \mathbb{K}\xi\rangle$ with $\mathbb{K} = \mathbb{G}^{-1}$. Then, we have

$$\text{(FB)} \quad \mathbb{G}\dot u = -\mathrm{D}\mathscr{E}(u)$$

$$\text{(RE)} \quad \dot u = -\mathbb{K}\mathrm{D}\mathscr{E}(u) = -\nabla_{\mathbb{G}}\mathscr{E}(u)$$

$$\text{(PB)} \quad \frac{1}{2}\langle \mathbb{G}\dot u, \dot u\rangle + \frac{1}{2}\langle \mathrm{D}\mathscr{E}(u), \mathbb{K}\mathrm{D}\mathscr{E}(u)\rangle = -\langle \mathrm{D}\mathscr{E}(u), \dot u\rangle,$$

where (RE) can be seen as a "gradient evolution", because $\nabla_{\mathbb{G}}$ is the gradient operator.

The Young-Fenchel estimate (3.2.14) states that $\geq$ in (PB) always holds. Hence, in limit passages it will be sufficient to control the opposite estimate. Moreover, it is advantageous to use the integrated form. For this we employ a version of the chain rule. Indeed, it will be sufficient to have a chain-rule estimate in terms of a suitable notion of a set-valued subdifferential $\widetilde\partial\mathscr{E}$ for the functional $\mathscr{E}$, see (3.3.2) for the Fréchet subdifferential or [67] for a general theory. As before $\widetilde\partial\mathscr{E}(t, u)$ denotes a differential for the function $\mathscr{E}(t, \cdot) : \boldsymbol{X} \to \mathbb{R}_\infty$ for fixed t, while $\partial_t\mathscr{E}(t, u)$ denotes the partial derivative $\lim_{h\to 0} \frac{1}{h}\big(\mathscr{E}(t + h, u) - \mathscr{E}(t, u)\big)$, which is always assumed to exist.

Definition 3.3.1 (*Abstract chain rule*) We say that the triple $(\boldsymbol{X}, \mathscr{E}, \widetilde\partial\mathscr{E})$ satisfies the *chain rule* if for all $p \geq 1$ the following holds. If $u \in \mathrm{W}^{1,p}([0, T]; \boldsymbol{X})$ and

$\xi \in \mathrm{L}^{p^*}([0,T];\boldsymbol{X}^*)$ with $\xi(t) \in \widetilde{\partial}\mathscr{E}(t,u(t))$ a.e. in $[0,T]$, then $t \mapsto \mathscr{E}(t,u(t))$ is absolutely continuous and

$$\frac{\mathrm{d}}{\mathrm{d}t}\mathscr{E}(t,u(t)) = \langle \xi(t), \dot{u}(t)\rangle + \partial_t\mathscr{E}(t,u(t)) \text{ a.e. in } [0,T]. \tag{3.3.1}$$

We refer to [51, 67] for a general treatments of such abstract chain rule. In particular, the chain rule holds for λ-convex functionals, i.e. $u \mapsto \mathscr{E}(t,u) - \frac{\lambda}{2}\|u\|_X^2$ is convex, and $\widetilde{\partial}\mathscr{E}$ denotes the Fréchet subdifferential

$$\widetilde{\partial}\mathscr{E}(u) := \{\, \xi \in \boldsymbol{X}^* \mid \mathscr{E}(u+w) \geq \mathscr{E}(u) + \langle \xi, w\rangle + o(\|w\|) \text{ for } w \to 0\,\}. \tag{3.3.2}$$

Theorem 3.3.1 (Upper energy-dissipation estimate) *Assume that $(\boldsymbol{X}, \mathscr{E}, \widetilde{\partial}\mathscr{E})$ satisfies the chain rule (3.3.1). Furthermore, assume that $u \in \mathrm{W}^{1,p}([0,T];\boldsymbol{X})$ and $\xi \in \mathrm{L}^{p^*}([0,T];\boldsymbol{X}^*)$ with $\xi(t) \in \widetilde{\partial}\mathscr{E}(t,u(t))$ a.e. in $[0,T]$. Then, u solves (FB), (RE), or (PB) if and only if the Upper Energy-Dissipation Estimate (UEDE) holds:*

$$\textit{UEDE:}\quad \begin{cases} \mathscr{E}(T,u(T)) + \displaystyle\int_0^T \mathscr{R}(u(t),\dot{u}(t)) + \mathscr{R}^*(u(t),-\xi(t))\,\mathrm{d}t \\ \qquad\qquad \leq \mathscr{E}(0,u(0)) + \displaystyle\int_0^T \partial_t\mathscr{E}(t,u(t))\,\mathrm{d}t. \end{cases}$$

Then, the UEDE is equivalent to the energy-dissipation principle (EDP)

$$\textit{EDP:}\quad \begin{cases} \mathscr{E}(T,u(T)) + \displaystyle\int_0^T \big(\mathscr{R}(u(t),\dot{u}(t)) + \mathscr{R}^*(u(t),-\xi(t))\,\mathrm{d}t \\ \qquad\qquad = \mathscr{E}(0,u(0)) + \displaystyle\int_0^T \partial_t\mathscr{E}(t,u(t))\,\mathrm{d}t. \end{cases}$$

Proof By the Fenchel equivalence (FB), (RE), and (PB) are equivalent, where now $\mathrm{D}\mathscr{E}(t,u(t))$ is replaced by $\xi(t)$. Integrating (PB) and using the chain rule (3.3.1) provide the EDP and hence the UEDE. For the opposite implication we start from the Young-Fenchel estimate (3.2.14) and obtain the following chain of estimates:

$$\int_0^T -\langle \xi(t), \dot{u}(t)\rangle\,\mathrm{d}t \overset{\mathrm{YF}}{\leq} \int_0^T \mathscr{R}(u,\dot{u}) + \mathscr{R}^*(u,-\xi)\,\mathrm{d}t$$

$$\overset{\text{UEDE}}{\leq} \mathscr{E}(0,u(0)) + \int_0^T \partial_t \mathscr{E}(t,u)\,\mathrm{d}t - \mathscr{E}(T,u(T)) = \int_0^T \partial_t \mathscr{E}(t,u) - \frac{\mathrm{d}}{\mathrm{d}t}\mathscr{E}(t,u)\,\mathrm{d}t$$

$$\overset{\text{CR}}{=} \int_0^T -\langle \xi(t), \dot{u}(t)\rangle\,\mathrm{d}t.$$

Thus, we conclude that all the above estimates must be equalities. Moreover, using the Young-Fenchel estimate again with the first estimate $\overset{\text{YF}}{\leq}$ being an equality, we see that (PB) has to hold a.e. in $[0,T]$. For the last conclusion we simply use that $g(t) \leq h(t)$ and $\int_0^T g\,\mathrm{d}t = \int_0^T h\,\mathrm{d}t$ imply that $g(t) = h(t)$ a.e. Hence, the proof is complete. □

We remark that the UEDE relates the final energy $\mathscr{E}(T,u(T))$ plus the dissipated energy $\int_0^T \mathscr{R} + \mathscr{R}^*\,\mathrm{d}t$ to the initial energy $\mathscr{E}(0,u(0))$ plus the external work $\int_0^T \partial_s \mathscr{E}(s,u)\,\mathrm{d}s$, which arises via time-dependence of the system. The main importance of the UEDE is that the final and the dissipated energies only need to have a good upper bound. Hence, in passing to a Γ-limit it will be sufficient to have good liminf estimates for these terms, while the right-hand side can be controlled by the well-preparedness of the initial conditions and proper assumptions on the work of the external forces. In fact, we will mostly assume $\mathscr{E}(t,u) = \mathscr{F}(u) - \langle \ell(t), u\rangle$, so it will be easy to control the linear term $-\int_0^T \langle \dot{\ell}, u\rangle\,\mathrm{d}t = \int_0^T \partial_t \mathscr{E}(t,u(t))\,\mathrm{d}t$.

3.3.2 pE-Convergence Obtained from Mosco Convergence

We now provide a first general result on pE-convergence for an ε-dependent family of gradient systems $(\boldsymbol{X}, \mathscr{E}_\varepsilon, \mathscr{R}_\varepsilon)$ that is based on the UEDE

$$\mathscr{E}_\varepsilon(u_\varepsilon(T)) + \int_0^T \mathscr{R}_\varepsilon(u_\varepsilon, \dot{u}_\varepsilon) + \mathscr{R}^*_\varepsilon(u_\varepsilon, -\mathrm{D}\mathscr{E}_\varepsilon(u_\varepsilon))\,\mathrm{d}t \leq \mathscr{E}_\varepsilon(u_\varepsilon(0)).$$

The proof presented here is a simplified version of the Mosco-convergence theory developed in [51]. In fact, the results there (cf. Theorem 4.8) are more general, while we present a version which displays the general structure more clearly by restricting to the case that $\mathscr{E}_\varepsilon$ is independent of time and that $\mathscr{R}_\varepsilon$ is translation invariant, i.e. $\mathscr{R}_\varepsilon(u,v) = \Psi_\varepsilon(v)$. Thus, the restricted results corresponds to [72, Theorem 7.2] for variational doubly nonlinear equations with convex energies $\mathscr{E}(t,\cdot)$, which can be reformulated via a Brézis-Ekeland principle, i.e. u solves $0 \in \partial\Psi(\dot{u}) + \mathrm{D}\mathscr{F}(u) - \ell(t)$ if and only if there exists $\xi(t) \in \partial\mathscr{E}(t,u(t)$ such that (u,ξ) minimizes $\mathscr{I}$ with $\mathscr{I}(u,\xi) = 0$, where $\mathscr{I}$ is given by

$$\mathscr{I}(u,\xi) = \left(\int_0^T \Psi(\dot u) + \Psi^*(-\xi) - \partial_t \mathscr{E}(t,u)\,\mathrm{d}t + \mathscr{E}(T,u(t)) - \mathscr{E}(0,u_0) \right)^+ \\ + \int_0^T \mathscr{E}(t,u) + \mathscr{E}^*(t,\xi) - \langle \xi, u\rangle\,\mathrm{d}t + \|u(0)-u_0\|_X^2,$$

where $(a)^+ = \max\{a,0\}$ and $\mathscr{E}^*(,\cdot)$ is the Fenchel-Legendre transform of the convex function $\mathscr{E}(t,\cdot)$. A more restrictive result for Hilbert spaces $\boldsymbol{X}$ with the constant dissipation potential $\Psi_\varepsilon(v) = \frac{1}{2}\|v\|_X^2$ has already been obtained in [4, Theorem 3.74].

Theorem 3.3.2 (Mosco convergence implies pE-convergence) *Let $\boldsymbol{X}$ be a reflexive Banach space and let $c, C, \lambda_{cx} > 0,\ p > 1$ be such that $\mathscr{E}_\varepsilon(\cdot) + \lambda_{cx}\|\cdot\|_X^2$ is convex and that the dissipation potentials satisfy the coercivity assumption*

$$\forall v \in \boldsymbol{X},\ \xi \in \boldsymbol{X}^*: \quad \Psi_\varepsilon(v) \geq c\|v\|_X^p - C,\ \Psi_\varepsilon^*(\xi) \geq c\|\xi\|_{X^*}^{p^*} - C. \tag{3.3.3}$$

Moreover, assume that $\mathbf{Z}$ is compactly embedded into $\boldsymbol{X}$ such that the energies are uniformly $\mathbf{Z}$-coercive, cf. (3.2.8). Then, we have

$$\big(\mathscr{E}_\varepsilon \overset{\mathsf{M}}{\to} \mathscr{E}_0 \text{ and } \Psi_\varepsilon \overset{\mathsf{M}}{\to} \Psi_0 \text{ in } \boldsymbol{X}\big) \quad \Longrightarrow \quad (\boldsymbol{X}, \mathscr{E}_\varepsilon, \Psi_\varepsilon) \overset{\mathsf{pE}}{\to} (\boldsymbol{X}, \mathscr{E}_0, \Psi_0).$$

In this result the convergence of the two functionals is rather strong, namely Mosco convergence. At first glance, there seems to be no compatibility condition between the two convergences. However, the necessary compatibility reduces to the fact that the Mosco convergences $\mathscr{E}_\varepsilon \overset{\mathsf{M}}{\to} \mathscr{E}_0$ and $\Psi_\varepsilon \overset{\mathsf{M}}{\to} \Psi_0$ occur in the same topology, namely in the dynamic function space $\boldsymbol{X}$. Here the uniform coercivity of Ψ_ε and Ψ_ε^* is crucial.

The proof of Theorem 3.3.2 relies on two important arguments, namely a lower semicontinuity result of Ioffe type (cf. [30]) and the strong-weak closedness of the graphs of $\widetilde{\partial}\mathscr{E}_\varepsilon$.

Proposition 3.3.1 (Lower semicontinuity, e.g. [51, Theorem A.2]) *Assume that all $\mathscr{J}_\varepsilon : \boldsymbol{X} \to \mathbb{R}_\infty$ are proper, lsc, and convex and that $\mathscr{J}_\varepsilon \overset{\Gamma}{\rightharpoonup} \mathscr{J}_0$ in $\boldsymbol{X}$, then*

$$w_\varepsilon \rightharpoonup w \in \mathrm{L}^p([0,T],\boldsymbol{X}) \quad \Longrightarrow \quad \int_0^T \mathscr{J}_0(w(t))\,\mathrm{d}t \leq \liminf_{\varepsilon\to 0} \int_0^T \mathscr{J}_\varepsilon(w_\varepsilon(t))\,\mathrm{d}t.$$

Definition 3.3.2 (*Strong-weak closedness*) The triples $(\boldsymbol{X}, \mathscr{E}_\varepsilon, \widetilde{\partial}\mathscr{E}_\varepsilon)_{\varepsilon\in[0,1]}$ satisfies the *strong-weak closedness of the graph of* $\widetilde{\partial}\mathscr{E}$, if the following holds:

$$\text{If } \left\{ \begin{array}{c} u_\varepsilon \to u \text{ in } \boldsymbol{X},\ \mathscr{E}_\varepsilon(u_\varepsilon) \to e_0 \in \mathbb{R}, \\ \xi_\varepsilon \in \widetilde{\partial}\mathscr{E}_\varepsilon(u_\varepsilon),\ \xi_\varepsilon \rightharpoonup \xi \text{ in } \boldsymbol{X}^*, \end{array} \right\} \text{ then } \xi \in \widetilde{\partial}\mathscr{E}_0(u) \text{ and } \mathscr{E}_0(u) = e_0. \tag{3.3.4}$$

Proposition 3.2.3 shows that strong-weak closedness holds if all $\mathcal{E}_\varepsilon$ are lsc and convex and the convex subdifferential is used. The assumption of convexity of $\mathcal{E}_\varepsilon(\cdot) + \lambda_{\mathrm{cx}}\|\cdot\|_X^2$, which is also called uniform λ_{cx} convexity, provides the same result for the Fréchet subdifferential defined in (3.3.2), see [51, 67].

Sketch of proof of Theorem 3.3.2 We only give the main arguments of the proof and refer to [51, Theorem 4.8] for the details. We start from a family of solutions $u_\varepsilon : [0, T] \to \boldsymbol{X}$ of $(\boldsymbol{X}, \mathcal{E}_\varepsilon, \Psi_\varepsilon)$ and want to show that an accumulation point u is a solution of the limit system $(\boldsymbol{X}, \mathcal{E}_0, \Psi_0)$.

Step 1 The well-preparedness of the initial condition gives $\mathcal{E}_\varepsilon(u_\varepsilon(0)) \leq C$. Since the solutions u_ε satisfy the EDP, we obtain the uniform bounds $\mathcal{E}_\varepsilon(u_\varepsilon(t)) \leq C$, $\int_0^T \Psi_\varepsilon(\dot{u}_\varepsilon)\mathrm{d}t \leq C$, and $\int_0^T \Psi_\varepsilon^*(-\xi_\varepsilon)\mathrm{d}t \leq C$, where $\xi_\varepsilon(t) \in \widetilde{\partial}\mathcal{E}(u_\varepsilon(t))$ a.e. The coercivity of $\mathcal{E}_\varepsilon$, Ψ_ε, and Ψ_ε^* implies the uniform a priori bounds:

$$\|u_\varepsilon\|_{\mathrm{L}^\infty([0,T];\mathrm{Z})} + \|u_\varepsilon\|_{\mathrm{W}^{1,p}([0,T];X)} + \|\xi_\varepsilon\|_{\mathrm{L}^p([0,T];X^*)} \leq C.$$

Step 2 Using the reflexivity of $\boldsymbol{X}$ and the compact embedding of $\boldsymbol{Z}$ into $\boldsymbol{X}$, we can extract a convergent subsequence (not relabeled) giving

$$\begin{aligned} &u_\varepsilon \rightharpoonup u \text{ in } \mathrm{W}^{1,p}([0,T]; \boldsymbol{X}),\\ &\xi_\varepsilon \rightharpoonup \xi \text{ in } \mathrm{L}^p([0,T]; \boldsymbol{X}^*),\\ &\forall t \in [0,T]: \; u_\varepsilon(t) \rightharpoonup u(t) \text{ in } \boldsymbol{Z}. \end{aligned}$$

Step 3 Following (3.2.15), $\Psi_\varepsilon \xrightarrow{\mathrm{M}} \Psi_0$ provides the weak Γ-convergences $\Psi_\varepsilon \xrightharpoonup{\Gamma} \Psi_0$ and $\Psi_\varepsilon^* \xrightharpoonup{\Gamma} \Psi_0^*$. Hence we can apply Proposition 3.3.1 to obtain lower semicontinuity of the dissipation:

$$\int_0^T \Psi_0(\dot{u}) + \Psi_0^*(\xi)\mathrm{d}t \leq \liminf_{\varepsilon\to 0} \int_0^T \Psi_\varepsilon(\dot{u}_\varepsilon) + \Psi_\varepsilon^*(\xi_\varepsilon)\mathrm{d}t.$$

Step 4 Next we use the strong-weak closedness to conclude that the two limit functions u and ξ are related. Using a Banach-space valued Young measure ν generated by a suitable subsequence of (ξ_ε), the strong-weak closedness implies that for a.a. $t \in [0, T]$ the support of $\nu(t)$ is concentrated in $\widetilde{\partial}\mathcal{E}_0(u(t))$ a.e. in $[0, T]$. Assuming additionally that $\widetilde{\partial}\mathcal{E}_0$ is single-valued (see the proof of [51, Theorem 4.4] for the general case) we conclude $\xi_\varepsilon(t) \to \xi(t)$ and hence $\xi(t) \in \widetilde{\partial}\mathcal{E}_0(u(t))$ a.e.

Step 5 We now pass to the limit $\varepsilon \to 0$ in the EDP. Note that the right-hand side converges because of the well-preparedness of the initial data. From $u_\varepsilon(T) \to u(T)$ in $\boldsymbol{X}$ and $\mathcal{E}_\varepsilon \xrightarrow{\mathrm{M}} \mathcal{E}_0$ we find $\mathcal{E}_0(u(T)) \leq \liminf_{\varepsilon\to 0} \mathcal{E}_\varepsilon(u_\varepsilon(T))$. Now using the weak convergence of u_ε and ξ_ε we can apply the lsc result of Proposition 3.3.1 to conclude the UEDE

$$\mathcal{E}_0(u(T)) + \int_0^T \Psi_0(\dot{u}) + \Psi_0^*(\xi)\,\mathrm{d}t \;\le\; \mathcal{E}_0(u(0)) \quad \text{with } \xi(t) \in \widetilde{\partial}\mathcal{E}_0(u(t)),$$

where we used the well-preparedness of the initial data to obtain $\mathcal{E}_0(u(0))$ on the left-hand side. Since the uniform λ-convexity is preserved under Mosco convergence, the limit system $(\boldsymbol{X}, \mathcal{E}_0, \Psi_0)$ satisfies the chain rule, and Theorem 3.3.1 is applicable. Thus, u is a solution of $(\boldsymbol{X}, \mathcal{E}_0, \Psi_0)$.

Step 6 It remains to show the energy convergence $\mathcal{E}_\varepsilon(u_\varepsilon(t)) \to \mathcal{E}_0(u(t))$ for all t. For this we can repeat the derivation of the above UEDE on the interval $[0, t]$ where the energy term and the dissipation term satisfy a liminf estimate separately. However, using the chain rule estimate on this interval, we know that the UEDE is in fact an equality and we conclude that in both liminf estimates we must have a limit:

$$\mathcal{E}_\varepsilon(u_\varepsilon(t)) \to \mathcal{E}_0(u(t)) \text{ and } \int_0^t \Psi_\varepsilon(\dot{u}_\varepsilon(t)) + \Psi_\varepsilon^*(\xi_\varepsilon(t))\mathrm{d}t \to \int_0^t \Psi_0(\dot{u}(t)) + \Psi_0^*(\xi(t))\mathrm{d}t$$

for all t. Thus, pE-convergence is established. □

We now discuss the applicability of the above result to our major examples. Note that the dissipation potentials Ψ_ε determine the choice of the dynamic space $\boldsymbol{X}$, while the energy space $\boldsymbol{Z}$ will be determined by $\mathcal{E}_\varepsilon$.

ODE model on $X = \mathbf{R}^2$ from Sect. 3.2.5

Here $\boldsymbol{X} = \mathbb{R}^2$ is the only possible choice, and because in finite dimensions weak and strong convergence coincide we have $\mathcal{E}_\varepsilon \xrightarrow{\mathrm{M}} \mathcal{E}_0$ and $\Psi_\varepsilon \xrightarrow{\mathrm{M}} \Psi_0$. However, $\Psi_\varepsilon(v) = \frac{1}{2}(v_1^2 + v_2^2/\varepsilon^\beta)$ gives the dual dissipation potential $\Psi_\varepsilon^*(\xi) = \frac{1}{2}(\xi_1^2 + \varepsilon^\beta \xi_2^2)$. Thus, we have uniform coercivity of Ψ_ε and Ψ_ε^* only for the case $\beta = 0$, where Theorem 3.3.2 is applicable.

Moreover, the uniform-coercivity assumption guides us to a proper choice for the rescaling. Using $w = (u_1, \varepsilon^{-\beta/2}u_2)$ we obtain the gradient system $(\mathbb{R}^2, \overline{\mathcal{E}}_\varepsilon, \overline{\Psi})$ with

$$\overline{\mathcal{E}}_\varepsilon(w) = \frac{1}{2}w_1^2 + \frac{1}{2}\big(w_1 - \varepsilon^{\beta/2-1}w_2\big)^2 \quad \text{and} \quad \overline{\Psi}(\dot{w}) = \frac{1}{2}|\dot{w}|^2.$$

Obviously, we have uniform coercivity of $\overline{\Psi}$ and $\overline{\Psi}^*$, while uniform coercivity of $\overline{\mathcal{E}}_\varepsilon$ only holds for $\beta \le 2$. Thus, Theorem 3.3.2 is only applicable to $(\mathbb{R}^2, \overline{\mathcal{E}}_\varepsilon, \overline{\Psi})$ for $\beta \ge 2$. However, the same strategy of proof shows that pE-convergence also holds for $\beta \in [0, 1[$.

1D homogenization from Sect. 3.2.6

We have $\Omega =]0, \ell[$, $a_\varepsilon(x) = a(x/\varepsilon)$, $b_\varepsilon(x) = b(x/\varepsilon)$, and $c_\varepsilon(x) = c(x/\varepsilon)$ with upper and lower positive bounds. Since the dissipation potentials are given via $\Psi_\varepsilon(v) = \frac{1}{2}\int_\Omega c_\varepsilon v^2\,\mathrm{d}x$ and $\Psi_\varepsilon^*(\xi) = \frac{1}{2}\int_\Omega 1/c_\varepsilon\, \xi^2\,\mathrm{d}x$, we have uniform coercivity of the dissipation in $X = \mathrm{L}^2(\Omega)$. The energies $\mathcal{E}_\varepsilon(u) = \frac{1}{2}\int_0^\ell a_\varepsilon u_x^2 + b_\varepsilon u^2 \mathrm{d}x$ are convex and uniformly coercive in $Z = \mathrm{H}^1(\Omega)$, which is compactly embedded in X.

By Proposition 3.2.4 we know $\mathcal{E}_\varepsilon \xrightarrow{\mathsf{M}} \mathcal{E}_{\mathrm{harm}}$ in X. Lemma 3.2.2 gives $\Psi_\varepsilon \overset{\Gamma}{\rightharpoonup} \Psi_{\mathrm{harm}}$ and $\Psi_\varepsilon \xrightarrow{\Gamma} \Psi_{\mathrm{arith}}$. Thus, we have Mosco convergence if and only if c_ε is constant. In that case, Theorem 3.3.2 applies, and we conclude pE-convergence, i.e. the solution of $c\dot{u} = (a_\varepsilon u_x)_x - b_\varepsilon u$ converge to the solutions of the effective equation $c\dot{u} = (a_{\mathrm{harm}} u_x)_x - b_{\mathrm{arith}} u$.

The case of an oscillating c_ε will be covered in Sect. 3.4.2.

Tartar's model from Sect. 3.2.7

We emphasize that even in the case of constant $\Psi_\varepsilon = \Psi_0$ it is not sufficient to have $\mathcal{E}_\varepsilon \overset{\Gamma}{\rightharpoonup} \mathcal{E}_0$ or $\mathcal{E}_\varepsilon \xrightarrow{\Gamma} \mathcal{E}_0$ in X. This is seen by considering Tartar's example of Sect. 3.2.7 on the Hilbert space $\mathrm{L}^2(\Omega)$ with $\Psi_\varepsilon = \Psi_0 : v \mapsto \frac{1}{2}\|v\|_2^2$ and the energy functionals $\mathcal{E}_\varepsilon(u) = \int_\Omega \frac{a_\varepsilon}{2} u^2 \mathrm{d}x$. Then, according to Lemma 3.2.2 the weak and the strong Γ-limit exist and are different. However, none of them characterizes the limit dynamics.

3.3.3 The Sandier-Serfaty Approach to pE-Convergence

The *approach of Sandier and Serfaty* (cf. [69, 71]) is more general than the previous result using Mosco convergence, because it directly imposes assumptions on the liminf estimates for the two dissipation terms in the EDP. For this, we now even allow that the dissipation potentials depend on u (in a mild way), but restrict, for simplicity, to the case that the subdifferential is at most single-valued, i.e. $\partial\mathcal{E}_\varepsilon(u) = \{\mathrm{D}\mathcal{E}_\varepsilon(u)\}$ or $\emptyset$.

$$\text{If } u_\varepsilon \to u \text{ in } X, \quad v_\varepsilon \rightharpoonup v \text{ in } X, \quad \text{and } \mathcal{E}_\varepsilon(u_\varepsilon) \le C, \text{ then}$$
$$\mathcal{R}_0(u, v) \le \liminf_{\varepsilon\to 0} \mathcal{R}_\varepsilon(u_\varepsilon, v_\varepsilon) \text{ and} \tag{3.3.5a}$$
$$\mathcal{R}_0^*(u, -\mathrm{D}\mathcal{E}_0(u)) \le \liminf_{\varepsilon\to 0} \mathcal{R}_\varepsilon^*(u_\varepsilon, -\mathrm{D}\mathcal{E}_\varepsilon(u_\varepsilon)). \tag{3.3.5b}$$

Clearly, the strong-weak closedness of the graph of $(\partial\mathcal{E}_\varepsilon)_{\varepsilon\in[0,1]}$ and the Mosco convergence $\Psi_\varepsilon \xrightarrow{\mathsf{M}} \Psi_0$ imposed in Theorem 3.3.2 imply these two assumptions. In particular, condition (3.3.5b) implicitly imposes a closedness of the graph of the differential $\mathrm{D}\mathcal{E}_\varepsilon$.

Theorem 3.3.3 (Sandier-Serfaty [69, 71]) *Let X and Z be reflexive Banach spaces with compact embedding $Z \Subset X$. Assume that $(X, \mathcal{E}_\varepsilon, \mathcal{R}_\varepsilon)$ satisfy the chain rule (3.3.1) and that there are $c, C > 0$ and $p > 1$ such that*

$$\forall\, u, v \in X\ \forall\, \varepsilon \in [0, 1]: \quad \mathcal{R}_\varepsilon(u, v) \ge c\|v\|_X^p - C, \quad \mathcal{E}_\varepsilon(u) \ge c\|u\|_Z - C.$$

Then, $\mathcal{E}_\varepsilon \xrightarrow{\Gamma} \mathcal{E}_0$ in X and (3.3.5) imply $(X, \mathcal{E}_\varepsilon, \mathcal{R}_\varepsilon) \xrightarrow{\mathsf{pE}} (X, \mathcal{E}_0, \mathcal{R}_0)$.

Proof We follow the steps in the proof of Theorem 3.3.2. Consider a family u_ε of solutions with well-prepared initial conditions. The EDP and the well-preparedness of the initial conditions provide the a priori bounds $\|u_\varepsilon\|_{\mathrm{L}^\infty([0,T];\boldsymbol{Z})} + \|u_\varepsilon\|_{\mathrm{W}^{1,p}([0,T];\boldsymbol{X})} \leq C$. The compact embedding $\boldsymbol{Z} \Subset \boldsymbol{X}$ allows us to extract a subsequence (not relabeled) such that

$$u_\varepsilon \rightharpoonup u \text{ in } \mathrm{W}^{1,p}([0,T];\boldsymbol{X}) \quad \text{and } \forall\, t \in [0,T] : u_\varepsilon(t) \rightharpoonup u(t) \text{ in } \boldsymbol{Z}.$$

Combining a generalization of Ioffe's lsc result in Proposition 3.3.1, assumption (3.3.5a), the strong convergence $u_\varepsilon(t) \to u(t)$ in $\boldsymbol{X}$, and the weak convergence $\dot u_\varepsilon \rightharpoonup u$ in $\mathrm{L}^p([0,T];\boldsymbol{X})$ yields

$$\int_0^T \mathscr{R}_0(u(t),\dot u(t))\,\mathrm{d}t \leq \liminf_{\varepsilon\to 0} \int_0^T \mathscr{R}_\varepsilon(u_\varepsilon(t),\dot u_\varepsilon(t))\,\mathrm{d}t.$$

Similarly, assumption (3.3.5b) and Fatou's lemma give

$$\int_0^T \mathscr{R}_0^*(u(t),-\mathrm{D}\mathscr{E}_0(u(t)))\,\mathrm{d}t \leq \liminf_{\varepsilon\to 0} \int_0^T \mathscr{R}_\varepsilon(u_\varepsilon(t),-\mathrm{D}\mathscr{E}_\varepsilon(u_\varepsilon(t)))\,\mathrm{d}t.$$

Using $u_\varepsilon(T) \to u(T)$, $\mathscr{E}_\varepsilon \xrightarrow{\Gamma} \mathscr{E}_0$ in $\boldsymbol{X}$, and the well-preparedness of the initial data yield

$$\mathscr{E}_0(u(T)) + \int_0^T \mathscr{R}_0(u(t),\dot u(t)) + \mathscr{R}_0^*(u(t),-\mathrm{D}\mathscr{E}_0(u(t)))\,\mathrm{d}t \;\leq\; \mathscr{E}_0(u(0)).$$

Since the gradient system $(\boldsymbol{X},\mathscr{E}_0,\mathscr{R}_0)$ satisfies the chain-rule estimate (3.3.1), the proof can be completed as for Theorem 3.3.2. □

For applications of this result in a situation where the Mosco convergence of Theorem 3.3.2 does not hold we refer to [36, 47, 69]. In [5] this method is used to show that the solutions of $\dot u = \big(W'(u) - \varepsilon^2 u_{xx}\big)_{xx}$ converge to the solutions of the Stefan problem, which is the H^{-1}-gradient flow for the functional $\mathscr{E}_0(u) = \int_\Omega W^{**}(u)\mathrm{d}x$, where W^{**} is the lsc convex envelope of W.

In [34] a family of Cahn-Hilliard systems are considered as the gradient systems $(\boldsymbol{X},\mathscr{E}_\varepsilon,\mathscr{R})$ with $\boldsymbol{X} = (H^1_{\mathrm{av}}(\Omega))^*$,

$$\mathscr{E}_\varepsilon(u) = \int_\Omega \frac{\varepsilon}{2}|\nabla u|^2 + \frac{1}{\varepsilon}W(u)\,\mathrm{d}x, \quad \text{and} \quad \mathscr{R}^*(\xi) = \int_\Omega \frac{1}{2}|\nabla \xi|^2\,\mathrm{d}x,$$

where W is a double-well potential, e.g. $W(u) = \frac{1}{4}(u^2 - 1)^2$. Using the well-known Γ-convergence of $\mathcal{E}_\varepsilon$ to the perimeter functional $\mathcal{E}_0$ it is shown in [34, Theorem 1.3] that under suitable assumptions on the initial data there exists a time $T > 0$ such that the solutions $u_\varepsilon : [0, T] \to \boldsymbol{X}$ converge to the solutions $u : [0, T] \to \boldsymbol{X}$ of the Mullins-Sekerka system, which is given by the limitng gradient system $(\boldsymbol{X}, \mathcal{E}_\varepsilon, \mathcal{R})$.

3.3.4 More General Results on pE-Convergence Using the EDP

In fact, the energy-dissipation formulation via $(\mathcal{R}, \mathcal{R}^*)$ is even more flexible. We do not even need an independent bound for each of the terms $\int_0^T \mathcal{R}_\varepsilon \,\mathrm{d}t$ and $\int_0^T \mathcal{R}_\varepsilon^* \,\mathrm{d}t$. Moreover, in passing to the liminf for the total dissipation $\int_0^T \mathcal{R}_\varepsilon + \mathcal{R}_\varepsilon^* \,\mathrm{d}t$ we may even give up the special dual form $\mathcal{R} + \mathcal{R}^*$ of the integrand. This idea, which was applied successfully in [3, 38, 45, 49], can be summarized as follows.

We define the functional $\mathcal{J}_\varepsilon : \mathrm{W}^{1,1}([0, T]; \boldsymbol{X}) \to [0, \infty]$ via

$$\mathcal{J}_\varepsilon(u) := \int_0^T \mathcal{R}_\varepsilon(u, \dot{u}) + \mathcal{R}_\varepsilon^*(u, -\mathrm{D}\mathcal{E}_\varepsilon(u))\,\mathrm{d}t,$$

and we have to find a sufficiently good lower bound for the Γ-liminf, namely

$$\text{(i)}\quad u_\varepsilon(\cdot) \rightharpoonup u(\cdot) \text{ in } \mathrm{W}^{1,1}([0, T]; \boldsymbol{X}) \implies \int_0^T \mathcal{M}_0(u(t), \dot{u}(t))\,\mathrm{d}t \le \liminf_{\varepsilon \to 0} \mathcal{J}_\varepsilon(u_\varepsilon),$$

where the integrand $\mathcal{M}_0$ does not need to be of the form $\mathcal{R}_0 + \mathcal{R}_0^*$. Hence, finding the best (i.e. largest) $\mathcal{M}_0$ is nothing else than finding the (static) Γ-limit of the functionals $\mathcal{J}_\varepsilon$. It suffices to find $(\boldsymbol{X}, \mathcal{E}_0, \mathcal{R}_0)$ and $\mathcal{M}_0$ such that

(ii) $\mathcal{E}_\varepsilon \xrightharpoonup{\Gamma} \mathcal{E}_0$;
(iii) $\mathcal{M}_0(u, v) \ge -\langle \mathrm{D}\mathcal{E}_0(u), v\rangle$;
(iv) the chain rule (3.1) holds for $(\boldsymbol{X}, \mathcal{E}_0, \mathcal{R}_0)$;
(v) $\mathcal{M}_0(u, v) = -\langle \mathrm{D}\mathcal{E}_0(u), v\rangle \implies \mathcal{R}_0(u, v) + \mathcal{R}_0^*(u, -\mathrm{D}\mathcal{E}_0(u)) = -\langle \mathrm{D}\mathcal{E}_0(u), v\rangle$.

As before, we can start from the EDP $\mathcal{E}_\varepsilon(u_\varepsilon(T)) + \mathcal{J}_\varepsilon(u_\varepsilon) = \mathcal{E}_\varepsilon(u_\varepsilon(0))$. Using the well-preparedness of the initial datum, (i), and (ii) we pass to the limit and obtain the UEDE

$$\mathcal{E}_0(u(T)) + \int_0^T \mathcal{M}_0(u(t), \dot{u}(t))\,\mathrm{d}t \le \mathcal{E}_0(u(0)).$$

Now using the (iii) and the chain rule (iv) we find

$$\mathcal{E}_0(u(0)) \overset{\text{(iv)}}{=} \mathcal{E}_0(u(T)) - \int_0^T \langle \mathrm{D}\mathcal{E}(u(t)), \dot{u}(t)\rangle \,\mathrm{d}t$$

$$\overset{\text{(iii)}}{\leq} \mathcal{E}_0(u(T)) + \int_0^T \mathcal{M}_0(u(t), \dot{u}(t))\,\mathrm{d}t \leq \mathcal{E}_0(u(0)).$$

Thus, we conclude that we must have equality in (iii) for almost all $t \in [0, T]$, such that we can use (v) to conclude that u is a solution for $(\boldsymbol{X}, \mathcal{E}_0, \mathcal{R}_0)$.

Section 3.3.5.3 summarizes the results of [45], which show that the above strategy can even be applied to justify the passage from small viscous dissipation (i.e. $\mathcal{R}_\varepsilon(u, \cdot)$ is quadratic) to a limit problem with large rate-independent dissipation (i.e. $\mathcal{R}_0(u, \cdot)$ is positively homogeneous of degree 1, see Sect. 3.5.1), like dry friction.

In fact, under a slight and natural strengthening of the conditions (i) to (v), it is possible to contruct $\mathcal{R}_0$ directly from $\mathcal{M}_0$. Indeed, assume that $\mathcal{M}_0(u, \cdot)$ is additionally even, convex, and lsc, then it is immeadiate that $\mathcal{R}_{\mathcal{M}}$ defined via

$$\mathcal{R}_{\mathcal{M}}(u, v) := \mathcal{M}_0(u, v) - \mathcal{M}_0(u, 0)$$

is a dissipation potential. Moreover, using property (iii) we find the estimate

$$\mathcal{R}^*_{\mathcal{M}}(u, -\mathrm{D}\mathcal{E}_0(u)) = \sup_{v\in \boldsymbol{X}} \Big(\langle -\mathrm{D}\mathcal{E}_0(u), v\rangle - \mathcal{M}_0(u, v) + \mathcal{M}_0(u, 0)\Big) \leq \mathcal{M}_0(u, 0).$$

Thus, we find the desired UEDE $\mathcal{E}_0(u(T)) + \int_0^T \mathcal{R}_{\mathcal{M}} + \mathcal{R}^*_{\mathcal{M}}\,\mathrm{d}t \leq \mathcal{E}_0(u(0))$. We emphasize that the choice $\mathcal{R}_{\mathcal{M}}$ in (iv) and (v) is admissible, but not unique. In particular, it may be possible to find simpler $\mathcal{R}_0$ as is the case in the application discussed in Sect. 3.3.5.3.

3.3.5 Applications for pE-Convergence Based on the EDP

We treat three different applications of the abstract approaches described above, which rely on calculating the Γ-limit $\mathcal{J}_0$ of the functionals $\mathcal{J}_\varepsilon(u) = \int_0^T \mathcal{R}(u, \dot{u}) + \mathcal{R}(u, -\mathrm{D}\mathcal{E}(u))\,\mathrm{d}t$. The first two are based on the Sandier-Serfaty approach of Sect. 3.3.3, while the third uses the generalization of Sect. 3.3.4.

3.3.5.1 Dimension Reduction

The theory of evolutionary Γ-convergence was applied to cases of *dimension reduction* in [35, 36]. There, the passage from a thin bulk layer to hypersurface was considered. Dimension reduction is also relevant for the modeling of delamination via thin damage layers (cf. [54]), for elastoplastic behavior of plates (cf. [37]), or for the modeling of electronically active interfaces in thin-film photovoltaics (cf. [27, 28]).

We treat the following simplified diffusion model with $\Omega =]-1, 1[\subset \mathbb{R}^1$:

$$u_t = (A_\varepsilon(x)u_x)_x \quad \text{for } t > 0,\ x \in \Omega, \qquad u_x(t, \pm 1) = 0, \tag{3.3.6}$$

where the diffusion coefficient A_ε equals 1 for $|x| > \varepsilon/2$ and $\alpha\varepsilon$ for $|x| \leq \varepsilon/2$. While the system can be written as a L^2-gradient flow with quadratic energy (see Sect. 3.2.8.1), here we use the logarithmic entropy that is physically more relevant for diffusion, see [32, 44, 63]. Thus, we consider the gradient system $(\mathcal{Q}, \mathcal{E}, \mathcal{R}_\varepsilon)$ with state space $\mathcal{Q} = \{ u \in \mathrm{L}^1(\Omega) \mid u > 0,\ \int_\Omega u(x)\mathrm{d}x = 1 \}$, ε-independent energy functional $\mathcal{E}(u) = \int_\Omega u \log u - u \,\mathrm{d}x$, and the dual dissipation potential $\mathcal{R}^*_\varepsilon(u, \xi) = \frac{1}{2}\int_\Omega A_\varepsilon u \xi_x^2 \mathrm{d}x$. Using

$$\mathbb{K}_\varepsilon(u)\xi = \mathrm{D}_\xi \mathcal{R}^*_\varepsilon(u, \xi) = -\big(u A_\varepsilon \xi_x\big)_x \quad \text{and} \quad \mathrm{D}\mathcal{E}(u) = \log u,$$

we see that $\dot{u} = -\mathbb{K}_\varepsilon(u)\mathrm{D}\mathcal{E}(u) = \big(u A_\varepsilon (\log u)_x\big)_x = (A_\varepsilon u_x)_x$ is the diffusion equation (3.3.6).

The EDP takes the form $\mathcal{E}(u_\varepsilon(T)) + \mathcal{J}_\varepsilon(u_\varepsilon) = \mathcal{E}(u_\varepsilon(0))$ with $\mathcal{J}_\varepsilon(u) =$

$$\int_0^T \mathcal{R}_\varepsilon(u, \dot{u}) + \mathcal{R}^*_\varepsilon(u, -\mathrm{D}\mathcal{E}_\varepsilon(u))\mathrm{d}t = \int_0^T\!\!\int_\Omega \Big(\frac{1}{2u A_\varepsilon}\Big(\int_{-1}^x \dot{u}\,\mathrm{d}y\Big)^2 + \frac{A_\varepsilon}{2u}u_x^2\Big)\mathrm{d}x\,\mathrm{d}t.$$

We note that $\mathcal{J}_\varepsilon$ is a convex functional over the space-time domain $\Omega \times [0, T]$ depending on ε via the coefficient A_ε. The calculation of the Γ-limit of $\mathcal{J}_\varepsilon$ can be done with classical methods, see [35, Sect. 3.2], and for $u_\varepsilon \rightharpoonup u$ one finds

$$\mathcal{J}_0(u) := \int_0^T \mathcal{R}_0(u, \dot{u}) + M(u)\mathrm{d}t \leq \liminf_{\varepsilon \to 0} \mathcal{J}_\varepsilon(u_\varepsilon), \quad \text{where}$$

$$\mathcal{R}_0(u, v) = \int_{-1}^0 \frac{1}{2u}\left(\int_{-1}^x v\,\mathrm{d}y\right)^2 \mathrm{d}x + \frac{(\int_0^1 v\,\mathrm{d}y)^2}{2\alpha \Lambda(u(0^-), u(0^+))} + \int_0^1 \frac{1}{2u}\left(\int_x^1 v\,\mathrm{d}y\right)^2 \mathrm{d}x,$$

$$M(u) = \int_{-1}^{0} \frac{u_x^2}{2u}\,\mathrm{d}x + \frac{\alpha(u(0^-) - u(0^+))^2}{2\Lambda(u(0^-), u(0^+))} + \int_{0}^{1} \frac{u_x^2}{2u}\,\mathrm{d}x$$

with $\Lambda(a, b) = \frac{a-b}{\log a - \log b} > 0$, see Sect. 3.2.8.3.

Since $\mathscr{E}$ does not depend on ε, we have $\mathscr{E}_0 = \mathscr{E}$, and one can check $\mathscr{R}_0(u, v) + M(u) \geq -\langle \mathrm{D}\mathscr{E}(u), v\rangle = -\int_\Omega (\log u)v\ \mathrm{d}x$, which is condition (iii) for $\mathscr{M}_0$ in Sect. 3.3.4. Since v was arbitrary we have $\mathscr{R}_=^*(u, -\mathrm{D}\mathscr{E}(u)) \leq M(u)$ and conclude

$$\mathscr{E}(u(T)) + \int_0^T \mathscr{R}_0(u, \dot{u}) + \mathscr{R}_0^*(u, -\mathrm{D}\mathscr{E}(u))\mathrm{d}t \ \leq\ \mathscr{E}(u(T)) + \mathscr{J}_0(u)$$

$$\underset{u_\varepsilon(\cdot)\rightharpoonup u(\cdot)}{\overset{\mathscr{J}_\varepsilon \xrightarrow{\Gamma} \mathscr{J}_0}{\leq}} \liminf_{\varepsilon\to 0} \mathscr{E}(u_\varepsilon(T)) + \mathscr{J}_\varepsilon(u_\varepsilon) \overset{\mathrm{EDP}_\varepsilon}{=} \lim_{\varepsilon\to 0} \mathscr{E}_\varepsilon(u_\varepsilon(0)) \overset{\text{wellpr.}}{=} \mathscr{E}(u(0)).$$

Invoking the chain rule again, we find that u is a solution of the gradient system $(\mathscr{Q}, \mathscr{E}, \mathscr{R}_0)$, cf. Theorem 3.3.1. The associated PDE is given by classical diffusion in the two bulk parts $]-1, 0[$ and $]0, 1[$ and a transmission condition at the interface $x = 0$:

$$\begin{aligned}
\dot{u} &= u_{xx} && \text{for } x \in\ ]0, 1[,\\
0 &= u_x(0^+) + \alpha(u(0^+) - u(0^-)) && \text{for } x = 0^+,\\
0 &= u_x(0^-) + \alpha(u(0^-) - u(0^+)) && \text{for } x = 0^-,\\
\dot{u} &= u_{xx} && \text{for } x \in\ ]-1, 0[.
\end{aligned}$$

We refer to [28, 35, 36] for more details concerning gradient structures for reaction-diffusion systems with nontrivial interface conditions.

3.3.5.2 pE-Convergence = Homogenization for Tartar's Model

This example highlights the fact that a gradient structure is an additional information for a given evolution equation. In particular, choosing different gradient structures we may obtain different limiting evolution equations via evolutionary Γ-convergence.

We consider a bounded open domain $\Omega \subset \mathbb{R}^d$ and the damped parametrized ODE

$$\dot{u}(t, x) = -a_\varepsilon(x)u(t, x), \quad u(0, x) = u^0(x) \geq 0. \tag{3.3.7}$$

Here we will interprete $u(t, \cdot) \in \mathrm{L}^1_+(\Omega)$ as a density of a measure and assume that the absorption coefficient a_ε has the form $a_\varepsilon(x) = \mathbb{A}(x, \frac{1}{\varepsilon}x)$, where $\mathbb{A}$ is continuous on $\overline{\Omega} \times \mathbb{R}^d$ and 1-periodic in the second argument, i.e. $\mathbb{A}(x, y + k) = \mathbb{A}(x, y)$ for all $k \in \mathbb{Z}^d$.

As indicated in Sect. 3.2.7 we use the gradient structure $(\boldsymbol{M}_+(\overline{\Omega}), \mathscr{E}_\varepsilon, \mathscr{R}_\varepsilon^{\mathrm{H}})$ with

$$\mathscr{E}_\varepsilon(x) = \int_{\partial\Omega} b_\varepsilon(x)\mathrm{d}u(x) \quad \text{and} \quad \mathscr{R}_\varepsilon^{\mathrm{H}*}(u,\xi) = \frac{1}{2}\int_{\overline{\Omega}} \frac{a_\varepsilon(x)}{b_\varepsilon(x)}\xi(x)^2\,\mathrm{d}u(x),$$

where b_ε is again assumed in the form $b_\varepsilon(x) = \mathbb{B}(x, \frac{1}{\varepsilon}x)$. In particular, the energy functional is linear, but the dual dissipation potential depends on u, so we are *not* using the Hilbert-space structure of $\mathrm{L}^2(\Omega)$ as in [73, 74]. As in [39, 48] we call $\mathscr{R}_\varepsilon^{\mathrm{H}*}$ the dual Hellinger-dissipation potential, because it gives rise to a weighted Hellinger distance between nonnegative measures.

We equip $\boldsymbol{M}_+(\overline{\Omega})$ with the weak* convergence and find the Γ-limit of $\mathscr{E}_\varepsilon$ via

$$\mathscr{E}_\varepsilon \overset{\Gamma*}{\rightharpoonup} \mathscr{E}_0 : u \mapsto \int_{\overline{\Omega}} b_0(x)\mathrm{d}u(x) \text{ with } b_0(x) := \min\{\, \mathbb{B}(x,y) \mid y \in \mathbb{R}^d \,\},$$

where $u_\varepsilon \overset{*}{\rightharpoonup} u$ denotes weak* convergence in $\boldsymbol{M}_+$, i.e. $\int \phi \mathrm{d}u_\varepsilon \to \int \phi \mathrm{d}u$ for all $\phi \in \mathrm{C}(\overline{\Omega})$. For the proof of the liminf estimate simply note $b_\varepsilon(x) \geq b_0(x)$, which implies $\mathscr{E}_\varepsilon(u) \leq \mathscr{E}_0(u)$ for all u. Since $\mathscr{E}_0$ is weakly* continuous the liminf estimate follows.

For a given $\widehat{u}$ we have to find a recovery sequence $\widehat{u}_\varepsilon$ such that $\widehat{u}_\varepsilon \overset{*}{\rightharpoonup} \widehat{u}$ and $\mathscr{E}_\varepsilon(\widehat{u}_\varepsilon) \to \mathscr{E}_0(u)$. We do the construction as follows. For positive ε decompose Ω into finitely many sets of the form $Q_{k,\varepsilon} = \varepsilon(k + [0,1[^d) \cap \Omega$ with $k \in K_\varepsilon := \{\, k \in \mathbb{Z}^d \mid Q_{k,\varepsilon} \neq \emptyset \,\}$. Denote by $y_{k,\varepsilon}$ the point in $Q_{k,\varepsilon}$ which minimizes b_ε. Then, define the functions

$$\widehat{u}_\varepsilon = \sum_{k\in K_\varepsilon} \int_{Q_{k,\varepsilon}} \mathrm{d}\widehat{u}(y)\ \delta_{y_{k,\varepsilon}} \quad \text{and obtain } \widehat{u}_\varepsilon \overset{*}{\rightharpoonup} \widehat{u}.$$

Moreover, using $b_\varepsilon(y_{k,\varepsilon}) \leq b_0(y_{k,\varepsilon})$ we have $\mathscr{E}_\varepsilon(\widehat{u}_\varepsilon) \leq \mathscr{E}_0(\widehat{u}_\varepsilon)$ and conclude the limsup estimate. Finally we prove the pE-convergence using the EDP. For $\varepsilon > 0$ the solutions satisfy

$$\mathscr{E}_\varepsilon(u_\varepsilon(T)) + \int_0^T \int_{\overline{\Omega}} \Big(\frac{b_\varepsilon}{2a_\varepsilon} \frac{\dot{u}_\varepsilon^2}{u_\varepsilon} + \frac{1}{2} a_\varepsilon b_\varepsilon u_\varepsilon \Big) \mathrm{d}x\ \mathrm{d}t = \mathscr{E}_\varepsilon(u_\varepsilon(0)).$$

To pass to the limit $\varepsilon \to 0$ we define the two functions

$$\begin{aligned} g_0(x) &:= \min\{\, \mathbb{B}(x,y)/\mathbb{A}(x,y) \mid y \in \mathbb{R}^d \,\} \text{ and} \\ h_0(x) &:= \min\{\, \mathbb{B}(x,y)\mathbb{A}(x,y) \mid y \in \mathbb{R}^d \,\}. \end{aligned} \tag{3.3.8}$$

Using the well-preparedness of the initial condition on the right-hand side and the liminf estimates any pointwise weak* limit $u(t)$ of $u_\varepsilon(t)$ satisfies

$$\mathscr{E}_0(u(T)) + \int_0^T \mathscr{M}_0(u(t), \dot{u}(t))\mathrm{d}t \leq \mathscr{E}_0(u(0)) \quad \text{with}\ \mathscr{M}_0(u, v) = \int_{\overline{\Omega}} \frac{g_0}{2}\frac{v^2}{u} + \frac{h_0}{2}\, u\mathrm{d}x.$$

To show this, we first estimate $b_\varepsilon/a_\varepsilon \geq g_0$ and $a_\varepsilon b_\varepsilon \geq h_0$ and then use that the integrand is jointly convex in $(u, \dot{u})$ such that weak lsc follows.

To complete the proof of pE-convergence we have to prove that $\mathscr{M}_0(u, v) \geq -\langle \mathrm{D}\mathscr{E}_0(u), v\rangle$ (cf. (iii) in Sect. 3.3.4), which means

$$\frac{g_0 v^2}{2u} + \frac{h_0 u}{2} \geq -b_0 v \quad \text{for all}\ u > 0\ \text{and}\ v \in \mathbb{R}.$$

Since this condition is equivalent to $g_0 h_0 \geq b_0^2$ and since the chain rule holds, we obtain the following result.

Proposition 3.3.2 *Consider the gradient systems $(\boldsymbol{M}_+(\overline{\Omega}), \mathscr{E}_\varepsilon, \mathscr{R}_\varepsilon^{\mathrm{H}})$ for $\varepsilon > 0$ and the functions b_0, g_0, and h_0 as given in (3.3.8). Define*

$$\mathscr{E}_0(u) := \int_{\overline{\Omega}} b_0(x)\mathrm{d}u(x) \quad \textit{and} \quad \mathscr{R}_0^*(u, \xi) := \int_{\overline{\Omega}} \frac{h_0(x)}{2b_0(x)^2}\,\xi(x)^2\mathrm{d}u(x)$$

and assume $g_0 h_0 \geq b_0^2$, then $(\boldsymbol{M}_+(\overline{\Omega}), \mathscr{E}_\varepsilon, \mathscr{R}_\varepsilon) \xrightarrow{\mathrm{pE}}{}^{\!*} (\boldsymbol{M}_+(\overline{\Omega}), \mathscr{E}_0, \mathscr{R}_0)$ and the limiting equation reads $\dot{u} = -\frac{k_0}{b_0}u = -\frac{b_0}{g_0}u$.

We emphasize that this result is in contrast, but not in contradiction, to the results in [73, 74]. There, the solutions $u_\varepsilon(t, \cdot)$ converge weakly in $\mathrm{L}^2(\Omega)$. However, this does not fit to our well-preparedness asking $u_\varepsilon(t) \overset{*}{\rightharpoonup} u(t)$ in $\boldsymbol{M}_+(\overline{\Omega})$ and $\mathscr{E}_\varepsilon(u_\varepsilon(t)) \to \mathscr{E}_0(u(t))$, which requires a concentration at the minimizing points of $\mathbb{B}(x, \cdot)$.

More importantly, this provides an example where for the same equation $\dot{u} = -a_\varepsilon u$ we have two different gradient structures that produce two different evolutionary Γ-limits having different effective equations.

Corollary 3.3.1 (Different gradient structures lead to different limit equations) *Equation (3.3.7) has the two gradient structures $(\boldsymbol{M}_+(\overline{\Omega}), \mathscr{E}_\varepsilon^{(j)}, \mathscr{R}_\varepsilon^{(j)})$, $j = 1$ and 2, with*

$$\mathscr{E}_\varepsilon^{(1)}(u) = \int_{\overline{\Omega}} a_\varepsilon\,\mathrm{d}u, \qquad \mathscr{R}_\varepsilon^{(1)*}(u, \xi) = \int_{\overline{\Omega}} \frac{1}{2}\xi^2\mathrm{d}u,$$

$$\mathscr{E}_\varepsilon^{(2)}(u) = \int_{\overline{\Omega}} \frac{1}{a_\varepsilon}\mathrm{d}u, \qquad \mathscr{R}_\varepsilon^{(2)*}(u, \xi) = \int_{\overline{\Omega}} \frac{a_\varepsilon^2}{2}\xi^2\mathrm{d}u.$$

Setting $a_{min}(x) := \min\{\, \mathbb{A}(x,y) \mid y \in \mathbb{R}^d \,\}$ *and* $a_{max}(x) := \max\{\, \mathbb{A}(x,y) \mid y \in \mathbb{R}^d \,\}$,

$$\mathscr{E}_0^{(1)}(u) = \int_{\overline{\Omega}} a_{min}\,\mathrm{d}u, \qquad \mathscr{R}_0^{(1)*}(u,\xi) = \int_{\overline{\Omega}} \frac{1}{2}\xi^2\,\mathrm{d}u,$$

$$\mathscr{E}_0^{(2)}(u) = \int_{\overline{\Omega}} \frac{1}{a_{max}}\,\mathrm{d}u, \qquad \mathscr{R}_0^{(2)*}(u,\xi) = \int_{\overline{\Omega}} \frac{a_{max}^2}{2}\xi^2\,\mathrm{d}u,$$

we obtain the evolutionary Γ*-convergences*

$$(\boldsymbol{M}_+(\overline{\Omega}), \mathscr{E}_\varepsilon^{(j)}, \mathscr{R}_\varepsilon^{(j)}) \overset{\mathsf{pE}}{\rightharpoonup^*} (\boldsymbol{M}_+(\overline{\Omega}), \mathscr{E}_0^{(j)}, \mathscr{R}_0^{(j)}) \text{ for } j = 1 \text{ and } 2,$$

which give rise to the two different effective limit equations:

$$\begin{aligned} (\boldsymbol{M}_+(\overline{\Omega}), \mathscr{E}_0^{(1)}, \mathscr{R}_0^{(1)}) &: \ \dot u(t,x) = -a_{min}(x)u(t,x) \\ (\boldsymbol{M}_+(\overline{\Omega}), \mathscr{E}_0^{(2)}, \mathscr{R}_0^{(2)}) &: \ \dot u(t,x) = -a_{max}(x)u(t,x). \end{aligned} \tag{3.3.9}$$

Proof We simply apply Proposition 3.3.2. For $j = 1$ we choose $b_\varepsilon = a_\varepsilon$ (or $\mathbb{B} = \mathbb{A}$), then $b_0(x) = a_{\min}(x) := \min\{\, \mathbb{A}(x,y) \mid y \in \mathbb{R}^d \,\}$, $g_0 = 1$, and $h_0 = a_{\min}^2$, such that condition $g_0 h_0 \geq b_0^2$ holds. Similarly, for $j = 2$ we choose $b_\varepsilon = 1/a_\varepsilon$, then $b_0 = 1/a_{\max}$, $g_0(x) = 1/a_{\max}^2$, and $h_0 = 1$. □

Again, there is no contradiction between the two different limit equations. However, the choice of the energy functionals induces different conditions for the convergence of well-prepared initial conditions. Since this well-preparedness is preserved during the full evolution, we obtain truely different effective evolution equations.

The two different limit equations in (3.3.9) *for Tartar's model* (3.3.7) *show clearly, that the choice of the gradient structure for given equation is an additional modeling information that cannot be extracted from the equation alone.*

3.3.5.3 From Viscous to Rate-Independent Friction

In this section we discuss evolutionary Γ*-convergence, where we start form classical gradient systems* $(\boldsymbol{X}, \mathscr{E}_\varepsilon, \Psi_\varepsilon)$ *with quadratic* Ψ_ε *and obtain a rate-independent generalized gradient system* $(\boldsymbol{X}, \mathscr{E}_0, \Psi_0)$ with $\Psi_0(v) = \rho|v|$, cf. Sect. 3.5.1. *In particular, we emphasize that in this case* $\Psi_\varepsilon \overset{\mathsf{C}}{\to} 0$ while $\Psi_0 \neq 0$. To analyze the evolutionary Γ-limit via the EDP we use that the functional $\mathscr{J}_\varepsilon(u) = \int_0^T \Psi_\varepsilon(\dot u) + \Psi_\varepsilon^*(-\mathrm{D}\mathscr{E}_\varepsilon(u))\mathrm{d}t$ may lead to a more general $\mathscr{J}_0$, as was discussed in Sect. 3.3.4.

First investigations concerning the derivation of dry friction from viscous models can be found in [64], see Fig. 3.2. The work was re-initiated in [31] and further investigated by [59, 65] to understand hysteresis in materials. Here we summarize the analytical approach in [45] based on the EDP. The main interest is to establish

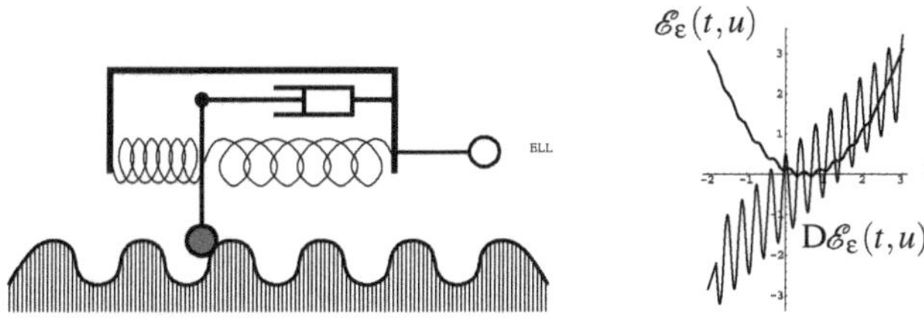

Fig. 3.2 *Left* Prandtl's Gedankenmodell [64]: a microscopically wiggly energy landscape gives rise to macroscopic dry friction. *Right* Wiggly energy and its derivative

the evolutionary Γ-convergence $(\mathbb{R}, \mathcal{E}_\varepsilon, \Psi_\varepsilon) \xrightarrow{\mathsf{pE}} (\mathbb{R}, \mathcal{E}_0, \Psi_0)$ where $\Psi_\varepsilon(v) = \frac{\varepsilon^\alpha}{2}v^2$ is a small viscous (i.e. quadratic) dissipation potential while the limit friction $\Psi_0(v) = \rho|v|$ is homogeneous of degree 1, giving Coulomb's rate-independent friction law $\xi = \rho\mathrm{Sign}(v)$. Moreover, the friction coefficient $\rho > 0$ will depend on the size of the wiggles in the wiggly energy landscape $\mathcal{E}_\varepsilon(t, \cdot)$.

We consider the time-dependent gradient system $(\mathbb{R}, \mathcal{E}_\varepsilon, \Psi_\varepsilon)$ with

$$\begin{aligned}
&\mathcal{E}_\varepsilon(t, u) = \mathcal{E}_0(t, u) + W_\varepsilon(u) \;\text{ with} \\
&\mathcal{E}_0(t, u) := \frac{1}{2}u^2 - \ell(t)u \text{ and } W_\varepsilon(u) := \varepsilon\rho\cos(u/\varepsilon).
\end{aligned}$$

Here the macroscopic part $\mathcal{E}_0$ is independent of ε and will be the Γ-limit of $\mathcal{E}_\varepsilon$. For $t \in \mathbb{R}$ we have continuous and Mosco convergence: $\mathcal{E}_\varepsilon \xrightarrow{\mathsf{C}} \mathcal{E}_0$ and $\mathcal{E}_\varepsilon \xrightarrow{\mathsf{M}} \mathcal{E}_0$. However, the derivative

$$\xi_\varepsilon = \mathrm{D}_u\mathcal{E}_\varepsilon(t, u) = \mathrm{D}_u\mathcal{E}_0(t, u) + \mathrm{D}W_\varepsilon(u) = u - \ell(t) - \rho\sin(u/\varepsilon)$$

has fluctuations of fixed size $\rho > 0$ and period ε. Moreover, for each $t \in \mathbb{R}$ there are many steady states, namely approximately ρ/ε, see Fig. 3.2. This explains the name *wiggly energy landscape.*

The gradient system induces the ODE

$$\varepsilon^\alpha \dot{u} = -\mathrm{D}_u\mathcal{E}_\varepsilon(t, u) = -\big(u - \ell(t)\big) + \rho\sin(u/\varepsilon). \tag{3.3.10}$$

A plot of a numerical simulation is depicted in Fig. 3.3, which shows that the solutions u_ε converge for $\varepsilon \to 0$ to a limit solution u that does not solve the naïve limit equation $0 = -\mathrm{D}_u\mathcal{E}_0(t, u(t))$, which might be guessed from the limits $\mathcal{E}_\varepsilon \xrightarrow{\mathsf{C}} \mathcal{E}_0$ and $\Psi_\varepsilon \xrightarrow{\mathsf{C}} 0$.

The limit passage can be achieved by using the EDP in the form

$$\mathcal{E}_\varepsilon(T, u_\varepsilon(T)) + \int_0^T \frac{\varepsilon^\alpha}{2}\dot{u}_\varepsilon^2 + \frac{1}{2\varepsilon^\alpha}|\mathrm{D}_u\mathcal{E}_\varepsilon(t, u_\varepsilon(t))|^2\,\mathrm{d}t = \mathcal{E}_\varepsilon(0, u_\varepsilon(0)) - \int_0^T \dot{\ell}(t)u_\varepsilon(t)\,\mathrm{d}t,$$

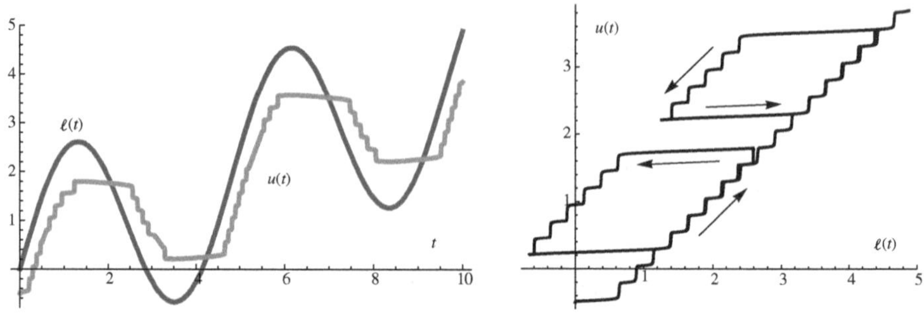

Fig. 3.3 Simulation of (3.3.10) for $\ell(t) = 2.1\sin(1.3t) + 0.4\,t$, $\rho = \alpha = 1$, and $\varepsilon = 0.04$

where the last term denotes the work of the external forces. The main difficulty in the proof of evolutionary Γ-convergence is to find a liminf estimate for the dissipation integral.

Proposition 3.3.3 ([45, Proposition 3.1]) *If $u_\varepsilon(t) \to u(t)$ and $|u_\varepsilon(t) - u_\varepsilon(s)| \leq C(\varepsilon^\delta + |t-s|)$ for some $\delta > 0$. Then,*

$$\liminf_{\varepsilon\to 0} \int_0^T \frac{\varepsilon^\alpha}{2}\dot{u}_\varepsilon^2 + \frac{1}{2\varepsilon^\alpha}|\mathrm{D}_u\mathscr{E}_\varepsilon(t,u_\varepsilon(t))|^2\,\mathrm{d}t \geq \int_0^T \mathscr{M}_0(u(t),\dot{u}(t),t)\,\mathrm{d}t$$

with $\mathscr{M}_0(u,v,t) = |v|K(\ell(t)-u) + \chi_{[-\rho,\rho]}(\ell(t)-u)$ and $K(\xi) = \frac{1}{2\pi}\int_0^{2\pi}|\xi + \rho\cos y|\,\mathrm{d}y$, where $\chi_{[-\rho,\rho]}(\xi) = 0$ for $|\xi| \leq \rho$ and ∞ otherwise.

Since $K(\xi) = |\xi|$ for $|\xi| \geq \rho$ and $K(\xi) \gneqq |\xi|$ for $|\xi| < \rho$ we find the desired relation (iii) in Sect. 3.3.4, viz.

$$\mathscr{M}_0(u,v,t) \geq |v|\,|\ell(t)-u| \geq -v\,\mathrm{D}\mathscr{E}_0(t,u).$$

Moreover, by the chain rule (here in $\mathbb{R}^1$) the limit functions will realize equality which holds if and only if $|\mathrm{D}\mathscr{E}_0(t,u)| \leq \rho$ and $\rho|v| + v\,\mathrm{D}\mathscr{E}_0(t,u) = 0$. Thus, using Lemma 3.5.1 the limit u of solutions u_ε satisfies the differential inclusion

$$0 \in \partial\Psi_0(\dot{u}) + \mathrm{D}\mathscr{E}_0(t,u) \quad \text{with } \Psi_0(v) = \rho|v|.$$

Thus, we have established the following evolutionary Γ-convergence result that connects quadratic Ψ_ε to a one-homogeneous Ψ_0 in the limit.

Theorem 3.3.4 ([65], [45, Theorem 3.2] *For the above gradient structure for (3.3.10) we have $(\mathbb{R},\mathscr{E}_\varepsilon,\Psi_\varepsilon) \overset{\mathsf{pE}}{\to} (\mathbb{R},\mathscr{E}_0,\Psi_0)$. Moreover, along well-prepared solutions we have $\int_0^T 2\Psi_\varepsilon(\dot{u}_\varepsilon(t))\,\mathrm{d}t \to \int_0^T \Psi_0(\dot{u}(t))\,\mathrm{d}t$.*

Concerning the last convergence of the dissipation, we emphasize that the convergence only occurs after integration. The integrand $2\Psi_\varepsilon(\dot{u}_\varepsilon(t))$ develops many sharp peaks at distance proportional to ε that have a width proportional to $\varepsilon^{\alpha+1}$ and a height of order $\varepsilon^{-\alpha}$, each of which corresponds to a fast viscous jump from one wiggle to the next.

3.4 E-Convergence Using Evolutionary Variational Estimates

Here we derive evolutionary Γ-convergence results based on the *integrated evolutionary variational estimate (IEVE)*, which was initiated in [2, Chap. 4.0] and is further developed [15, 70]. There, the name *evolutionary variational inequality* and the abbreviation EVI is used. However, to distinguish this notion from the evolutionary (quasi-) variational inequalities (see (3.5.4) and e.g. [12, 33]) we stick with our name IEVE.

The main advantage of reformulating gradient flows $\dot{u} = \mathrm{D}_\xi \mathscr{R}^*(u, -\mathrm{D}\mathscr{E}(u))$ in terms of an IEVE is that the latter is derivative free (i.e. no occurrence of $\dot{u}$, $\mathrm{D}\mathscr{E}$, nor $\partial_\xi \mathscr{R}^*$), so we can use Γ-convergence for the functionals $\mathscr{E}_\varepsilon$ and $\mathscr{R}_\varepsilon$ more directly. However, the theory using IEVE is restricted to the case of quadratic dissipation potentials, which can be replaced by the associated dissipation distance $\mathscr{D}$, and an additional strong convexity condition, called geodesic λ-convexity, is needed as well, see Sect. 3.4.3.

We first motivate the equivalence of the IEVE and the EDP by looking at an abstract linear gradient flow. This will allow us to treat the one-dimensional homogenization problem with oscillatory coefficient $c_\varepsilon(x) = c(x/\varepsilon)$ for $\dot{u}$, see Sect. 3.4.2. Next, we discuss the general metric approach in Sect. 3.4.3, where the dissipation distance is a general geodesic distance. In Theorem 3.4.3 we then provide a general result showing E-convergence.

3.4.1 *The Simplest Case: Quadratic Convex Functionals*

We consider the simplest case that the functionals $\mathscr{E}$ and Ψ are quadratic, i.e. a gradient systems $(\boldsymbol{H}, \mathscr{E}, \Psi)$ where $\boldsymbol{H}$ is a Hilbert space, $\mathscr{E}(u) = \frac{1}{2}\langle \mathbb{L}u, u\rangle \geq 0$, and $\Psi(v) = \frac{1}{2}\langle \mathbb{G}v, v\rangle \geq 0$ with $\mathbb{G}^* = \mathbb{G}$ and $\mathbb{L}^* = \mathbb{L}$. The associated gradient flow is the linear equation $\mathbb{G}\dot{u} = -\mathbb{L}u$.

The following formulation in terms of the IEVE is a special case of the integral solutions developed in [6], which was developed for a general Banach space $\boldsymbol{X}$ and equations $\dot{u} + \boldsymbol{A}(u) \ni f(t)$, where $\boldsymbol{A}$ is a possibly multi-valued accretive operator (see [68, Sect. 9.6] for more details on integral solutions).

Proposition 3.4.1 IEVE in the convex and quadratic case) *For the above gradient system* $(\boldsymbol{H}, \mathscr{E}, \Psi)$ *the following formulations are equivalent:*

$$(FB) \iff (RE) \iff (EDP) \iff (IEVE)$$

where the Integrated Evolutionary Variational Estimate *formulation is given by*

$$\text{IEVE:} \qquad \begin{cases} \forall\, 0 \le s < t \le T\ \forall w \in \boldsymbol{H}: \\ \Psi(u(t) - w) - \Psi(u(s) - w) \le (t-s)\big(\mathscr{E}(w) - \mathscr{E}(u(t))\big). \end{cases} \tag{3.4.1}$$

The addition "Integrated" distinguishes this formulation from the differential evolutionary variational estimate (EVE) given by $\frac{\mathrm{d}}{\mathrm{d}t}\Psi(u(t) - w) \le \mathscr{E}(w) - \mathscr{E}(u(t))$.

Proof The equivalence between (FB), (RE), and (EDP) is already established in Sect. 3.3.1, so it remains to show (i) $\Leftrightarrow$ (IEVE).

"$\Rightarrow$" The quadratic form of Ψ and $\mathscr{E}$ and the linear form $\mathbb{G}\dot{u} = -\mathbb{L}u$ of (i) yield

$$\begin{aligned} \tfrac{\mathrm{d}}{\mathrm{d}t}\Psi(u - w) &= \langle \mathbb{G}\dot{u}, u - w\rangle \overset{\text{(i)}}{=} -\langle \mathbb{L}u, u - w\rangle \\ &= \tfrac{1}{2}\langle \mathbb{L}w, w\rangle - \tfrac{1}{2}\langle \mathbb{L}u, u\rangle - \tfrac{1}{2}\langle \mathbb{L}(u-w), u - w\rangle \le \mathscr{E}(w) - \mathscr{E}(u) - 0. \end{aligned}$$

Integration over time and the monotonicity of $\mathscr{E}$ along solutions give

$$\Psi(u(t) - w) - \Psi(u(s) - w) = \int_s^t \mathscr{E}(w) - \mathscr{E}(u(\tau))\mathrm{d}\tau \le (t-s)\big(\mathscr{E}(w) - \mathscr{E}(u(t))\big),$$

which is the desired IEVE.

"$\Leftarrow$" Rearrangement of quadratic expressions gives

$$\begin{aligned} \text{IEVE} \iff \quad \forall w \in \boldsymbol{H}: &\ \tfrac{1}{2}\langle \mathbb{G}(u(t) - u(s)), u(t) + u(s) - 2w\rangle \\ &\le \tfrac{t-s}{2}\langle \mathbb{L}(u(t) + w), w - u(t)\rangle. \end{aligned}$$

Choosing $s = t - h$, dividing by h, and letting $h \to 0_+$, we find $\langle \mathbb{G}\dot{u}(t), u(t) - w\rangle \le \frac{1}{2}\langle \mathbb{L}(u(t) + w), w - u(t)\rangle$ for all $w \in \boldsymbol{H}$. Setting $w = u(t) - \delta\widehat{v}$, dividing by δ and letting $\delta \to 0_+$ gives $\langle \mathbb{G}\dot{u}(t), \widehat{v}\rangle \le -\langle \mathbb{L}u(t), \widehat{v}\rangle$ for all $\widehat{v} \in \boldsymbol{H}$. Replacing $\widehat{v}$ by $-\widehat{v}$, we even have equality and conclude that (i) holds. □

This formulation leads to the following simple but new E-convergence result. For this, we introduce the domains $\mathrm{dom}(\mathscr{E}_\varepsilon) := \{\, u \in \boldsymbol{H} \mid \mathscr{E}_\varepsilon(u) < \infty \,\}$ and note that the solutions do not need $u_\varepsilon(0) \in \mathrm{dom}(\mathscr{E}_\varepsilon)$, instead $u_\varepsilon(0) \in \overline{\mathrm{dom}(\mathscr{E}_\varepsilon)}$ is sufficient.

Theorem 3.4.1 (E-convergence for quadratic case) *Assume that* $\mathscr{E}_\varepsilon$ *and* Ψ_ε *are quadratic functionals of a Hilbert space* $\boldsymbol{H}$. *Further assume that there is a Hilbert space* $\boldsymbol{Z} \Subset \boldsymbol{H}$ *(compact embedding) such that we have the coercivities*

$$\exists c > 0\ \forall \varepsilon \in [0,1],\ u \in \boldsymbol{H}: \ \ \mathscr{E}_\varepsilon(u) \ge c\|u\|_{\boldsymbol{Z}}^2,\ c\|u\|_{\boldsymbol{H}}^2 \le \Psi_\varepsilon(u) \le \frac{1}{c}\|u\|_{\boldsymbol{H}}^2. \tag{3.4.2}$$

If $\mathcal{E}_\varepsilon \xrightarrow{\mathsf{M}} \mathcal{E}_0$ *in* $\boldsymbol{H}$ *and* $\Psi_\varepsilon \xrightarrow{\mathsf{C}} \Psi_0$ *in* $\boldsymbol{H}$, *then* $(\boldsymbol{H}, \mathcal{E}_\varepsilon, \Psi_\varepsilon) \xrightarrow{\mathsf{E}} (\boldsymbol{H}, \mathcal{E}_0, \Psi_0)$ *(i.e. without well-prepared initial conditions), namely if* $u_\varepsilon(0) \in \overline{dom(\mathcal{E}_\varepsilon)}$ *for all* $\varepsilon \in [0, 1]$, *then*

$$u_\varepsilon(0) \to u_0(0) \text{ in } \boldsymbol{H} \quad \Longrightarrow \quad \forall t > 0 : u_\varepsilon(t) \rightharpoonup u_0(t) \text{ in } \boldsymbol{Z} \text{ and } \mathcal{E}_\varepsilon(u_\varepsilon(t)) \to \mathcal{E}_0(u_0(t)),$$

where $u_\varepsilon : [0, \infty[\to \boldsymbol{H}$ *are solutions for* $(\boldsymbol{H}, \mathcal{E}_\varepsilon, \Psi_\varepsilon)$.

The following proof is a simplified version of the more general proof for Theorem 3.4.3 in the metric setting for uniformly geodesically λ-convex gradient systems, which is given in [15, 70]. Comparing this result with Theorem 3.3.2, we see that the Mosco convergence of the energies $\mathcal{E}_\varepsilon$ is the same, but there the Mosco convergence $\Psi_\varepsilon \xrightarrow{\mathsf{M}} \Psi_0$ is much stronger than the continuous convergence $\Psi_\varepsilon \xrightarrow{\mathsf{C}} \Psi_0$ needed here.

Proof The solutions u_ε of $\mathbb{G}_\varepsilon \dot{u} = -\mathbb{L}_\varepsilon u$ with $u_\varepsilon(0) \in \boldsymbol{H}$ satisfy

$$\text{IEVE}_\varepsilon : \quad \Psi_\varepsilon(u_\varepsilon(t) - w) - \Psi_\varepsilon(u_\varepsilon(s) - w) \leq (t - s)\big(\mathcal{E}_\varepsilon(w) - \mathcal{E}_\varepsilon(u_\varepsilon(t))\big)$$

for $0 \leq s < t \leq T$ and $w \in \boldsymbol{H}$. Hence we find the following a priori estimates:

$$\Psi_\varepsilon(u_\varepsilon(t)) + t\mathcal{E}_\varepsilon(u_\varepsilon(t)) \leq \Psi_\varepsilon(u_\varepsilon(0)) \leq C_0, \tag{3.4.3a}$$

$$\mathcal{E}_\varepsilon(u_\varepsilon(t)) + \int_s^t \Psi_\varepsilon(u_\varepsilon) + \Psi_\varepsilon^*(\mathrm{D}\mathcal{E}_\varepsilon(u_\varepsilon))\mathrm{d}\tau = \mathcal{E}_\varepsilon(u_\varepsilon(s)) \leq C_0/s. \tag{3.4.3b}$$

$$\Psi_\varepsilon^*(\mathrm{D}\mathcal{E}_\varepsilon(u_\varepsilon(t))) \geq \Psi_\varepsilon^*(\mathrm{D}\mathcal{E}_\varepsilon(u_\varepsilon(s))) \text{ for } 0 < t < s \leq T. \tag{3.4.3c}$$

Here (3.4.3a) follows from the IEVE by setting $s = 0$ and $w = 0$ using $\mathcal{E}_\varepsilon(0) = 0$, while (3.4.3b) is a consequence of (EDP) and (3.4.3a). The monotonicity (3.4.3c) follows from $\frac{\mathrm{d}}{\mathrm{d}t}\Psi_\varepsilon^*(\mathrm{D}\mathcal{E}_\varepsilon(u_\varepsilon)) = -\langle \mathbb{L}_\varepsilon \dot{u}_\varepsilon, \dot{u}_\varepsilon \rangle \leq 0$. Choosing $s = t/2$ in (3.4.3b) and using (3.4.3c) we obtain $\Psi_\varepsilon^*(\mathrm{D}\mathcal{E}_\varepsilon(u_\varepsilon(t)) \leq 4C_0/t^2$.

Hence, for each $t_0 > 0$ we have a uniform bound for all u_ε in the spaces $\mathrm{H}^1([t_0, T]; \boldsymbol{H}) \subset \mathrm{C}^{1/2}([t_0, T]; \boldsymbol{H})$ and $\mathrm{L}^\infty([t_0, T]; \boldsymbol{Z})$. By Arzela-Ascoli's theorem we find a subsequence (not relabeled) such that $u_\varepsilon(t) \rightharpoonup U(t)$ in $\boldsymbol{Z}$ for all $t > 0$. We also set $U(0) = u_0(0)$ and prove now that $U : [0, T] \to \boldsymbol{H}$ is a solution of $\text{IEVE}_{\varepsilon=0}$ as follows.

For a given $\widehat{w}$ in $\boldsymbol{H}$ we choose a recovery sequence $\widehat{w}_\varepsilon \to \widehat{w}$ in $\boldsymbol{H}$ with $\mathcal{E}_\varepsilon(\widehat{w}_\varepsilon) \to \mathcal{E}_0(\widehat{w})$. Inserting $\widehat{w}_\varepsilon$ into IEVE_ε we can pass to the limit using $\mathcal{E}_\varepsilon \xrightarrow{\Gamma} \mathcal{E}_0$ and $\Psi_\varepsilon \xrightarrow{\mathsf{C}} \Psi_0$ and find

$$\text{IEVE}_0 : \quad \Psi_0(U(t) - \widehat{w}) - \Psi_0(U(s) - \widehat{w}) \leq (t - s)\big(\mathcal{E}_0(\widehat{w}) - \mathcal{E}_0(U(t))\big).$$

Thus, U solves IEVE_0. To conclude that it is indeed the unique solution u_0 of $\mathbb{G}_0 \dot{u} = -\mathbb{L}_0 u$ with $u(0) = u_0(0)$ we still need to show that it is continuous at $t = 0$, i.e. $U(t) \to U(0) = u_0(0)$. Taking $s = 0$ in IEVE_0 and $\widehat{w} \in \mathrm{dom}(\mathcal{E}_0)$ and using $\mathcal{E}_0(U(t)) \geq 0$ we find

$$
\begin{aligned}
c \limsup_{t \searrow 0} \|U(t) - \widehat{w}\|_{\boldsymbol{H}}^2 \leq \limsup_{t \searrow 0} \Psi_0(U(t) - \widehat{w}) & \\
&\overset{\text{IEVE}_0}{\leq} \limsup_{t \searrow 0} \Big(\Psi_0(U(0) - \widehat{w}) + t\big[\mathscr{E}_0(\widehat{w}) - \mathscr{E}_0(U(t))\big]\Big) \\
&\leq \Psi_0(U(0) - \widehat{w}) \ \leq \ \tfrac{1}{c}\|U(0) - \widehat{w}\|_{\boldsymbol{H}}^2.
\end{aligned}
$$

Choosing $\widehat{w} = w_k \to U(0)$ in $\boldsymbol{H}$, where we used $U(0) = u_0(0) \in \overline{\mathrm{dom}(\mathscr{E}_0)}$, we conclude $U(t) \to U(0)$ as desired. Hence, $U(t) = u_0(t)$ because of the uniqueness of solutions.

Finally, we show energy convergence along the solutions. From (3.41) we have derived the ε-independent bound $\|\mathrm{D}\mathscr{E}_\varepsilon(u_\varepsilon(t))\|_{\boldsymbol{H}^*} \leq C_*/t^2$, which implies

$$
\mathscr{E}_\varepsilon(u) \geq \mathscr{E}_\varepsilon(u_\varepsilon(t)) - \tfrac{C_*}{t^2}\|u - u_\varepsilon(t)\|_{\boldsymbol{H}}^2 \quad \text{for all } u \in \boldsymbol{H}.
$$

Fixing $t > 0$ and inserting a recovery sequence $u = \widehat{u}_\varepsilon \to u_0(t)$ in $\boldsymbol{H}$ with $\mathscr{E}_\varepsilon(\widehat{u}_\varepsilon) \to \mathscr{E}_0(u_0(t))$ into the above estimate yield $\mathscr{E}_0(u_0(t)) \geq \limsup_{\varepsilon \to 0} \mathscr{E}_\varepsilon(u_\varepsilon(t))$, because of $\|u_\varepsilon(t) - \widehat{u}_\varepsilon\|_H \to 0$. Since the opposite estimate holds by $\mathscr{E}_\varepsilon \xrightarrow{\mathsf{M}} \mathscr{E}_0$, we obtain $\mathscr{E}_\varepsilon(u_\varepsilon(t)) \to \mathscr{E}_0(u_0(t))$, and the proof is complete. □

3.4.2 Linear Parabolic Homogenization via E-Convergence

Here we show that the linear *parabolic homogenization* problem can be handled using the IEVE. We recall the linear parabolic equation

$$
c_\varepsilon \dot{u} = \mathrm{div}(a_\varepsilon \nabla u) - b_\varepsilon u \quad \text{in } \Omega, \qquad a_\varepsilon \nabla u \cdot \nu = 0 \text{ on } \partial\Omega, \tag{3.4.4}
$$

where $\Omega \subset \mathbb{R}^d$ is a bounded Lipschitz domain. The coefficients are $b_\varepsilon(x) = b(x/\varepsilon)$ and $c_\varepsilon(x) = c(x/\varepsilon) > 0$ and $a_\varepsilon(x) = a(x/\varepsilon) \in \mathbb{R}^{d\times d}_{\mathrm{spd}}$, where a, b, and c are 1-periodic.

The Allen–Cahn-type gradient structure is given by $\boldsymbol{H} = \mathrm{L}^2(\Omega)$ and the functionals

$$
\Psi_\varepsilon(v) = \int_\Omega \frac{c_\varepsilon(x)}{2} v(x)^2 \mathrm{d}x \ \text{ and } \ \mathscr{E}_\varepsilon(u) = \int_\Omega \frac{1}{2}\nabla u(x) \cdot a_\varepsilon(x)\nabla u(x) + \frac{b_\varepsilon(x)}{2} u(x)^2 \mathrm{d}x.
$$

Now the assumptions of Theorem 3.4.1 are satisfied with $\boldsymbol{Z} = \mathrm{H}^1(\Omega)$. In particular, we have $\Psi_\varepsilon \xrightarrow{\mathsf{C}} \Psi_0$ and $\mathscr{E}_\varepsilon \xrightarrow{\mathsf{M}} \mathscr{E}_0$ in $\boldsymbol{H}$ with

$$
\mathscr{E}_0(u) = \int_\Omega \frac{1}{2}\nabla u \cdot a_{\mathrm{eff}} \nabla u + \frac{b_{\mathrm{arith}}}{2} u^2 \mathrm{d}x \ \text{ and } \ \Psi_0(v) = \int_\Omega \frac{c_{\mathrm{arith}}}{2} v^2 \mathrm{d}x,
$$

where in the 1D case we have $a_{\rm ell} = a_{\rm harm}$, see Lemma 3.2.2. For homogenization with $d \geq 2$ we refer to [9, 14], where $\mathscr{E}_\varepsilon \xrightarrow{\Gamma} \mathscr{E}_0$ is shown in $\mathrm{H}^1(\Omega)$, which gives $\mathscr{E}_\varepsilon \xrightarrow{\mathrm{M}} \mathscr{E}_0$ in $\mathrm{L}^2(\Omega)$ via Proposition 3.2.1.

Theorem 3.4.1 guarantees that solutions u_ε of (3.4.4) converge to solutions u_0 of the effective equation

$$c_{\rm arith}\dot{u} = \mathrm{div}\,\big(a_{\rm eff}\nabla u\big) - b_{\rm arith}u \text{ in } \Omega, \quad a_{\rm eff}\nabla u\cdot\nu = 0 \text{ on } \partial\Omega, \tag{3.4.5}$$

in the following way:

$$u_\varepsilon(0) \xrightarrow{\mathrm{L}^2} u_0(0) \quad\Longrightarrow\quad \forall t>0:\ u_\varepsilon(t) \xrightarrow{\mathrm{H}^1} u_0(t) \text{ and } \mathscr{E}_\varepsilon(u_\varepsilon(t)) \to \mathscr{E}_0(u_0(t)). \tag{3.4.6}$$

Again, we emphasize that neither well-preparedness nor finite energy are needed. It can be shown that the *E-convergence* for $t>0$ implies $a_\varepsilon\nabla u_\varepsilon(t) \to a_{\rm eff}\nabla u(t)$ in $\mathrm{L}^2(\Omega;\mathbb{R}^d)$.

A second way of formulating (3.4.4) as a gradient system relates to the gradient structure for reaction-diffusion systems developed in [44] as a generalization of the Wasserstein gradient structure for diffusion equations of [32]. For this, we introduce the density $\rho = c_\varepsilon u$ and obtain an equation in the Onsager form (cf. Sect. 3.3.1)

$$\dot\rho = \mathrm{div}\,\Big(a_\varepsilon\nabla\big(\frac{\rho}{c_\varepsilon}\big)\Big) - \frac{b_\varepsilon}{c_\varepsilon}\rho = -\mathbb{K}_\varepsilon \mathrm{D}\mathscr{F}_\varepsilon(\rho),$$

where we introduced the Onsager operator $\mathbb{K}_\varepsilon$ and the free energy $\mathscr{F}_\varepsilon$ via

$$\mathbb{K}_\varepsilon\xi := -\,\mathrm{div}\,\big(a_\varepsilon\nabla\xi\big) + b_\varepsilon\xi \text{ and } \mathscr{F}_\varepsilon(\rho) = \int_\Omega \frac{1}{2c_\varepsilon}\rho^2\,\mathrm{d}x.$$

Here $\mathbb{K}_\varepsilon$ can be seen as the operator generating the dual dissipation potential $\mathscr{R}^*_\varepsilon$, viz.

$$\mathscr{R}^*_\varepsilon(\xi) = \frac12\langle\xi,\mathbb{K}_\varepsilon\xi\rangle := \int_\Omega \frac12\nabla\xi\cdot a_\varepsilon\nabla\xi + \frac{b_\varepsilon}{2}\xi^2\,\mathrm{d}x.$$

Because of this form, we see that the relevant Hilbert space $\widetilde{\boldsymbol{H}}$ is defined such that $\widetilde{\boldsymbol{H}}^* = \mathrm{H}^1(\Omega)$, i.e. $\widetilde{\boldsymbol{H}}$ is a closed subspace of $\mathrm{H}^{-1}(\Omega)$. Again, Theorem 3.4.1 is applicable with $\widetilde{\boldsymbol{Z}} = \mathrm{L}^2(\Omega)$, which is the space for which the functionals $\mathscr{F}_\varepsilon$ are equi-coercive. Moreover, we have the convergences $\mathscr{F}_\varepsilon \xrightarrow{\Gamma} \mathscr{F}_0$ in $\mathrm{L}^2(\Omega)$ and $\mathscr{R}^*_\varepsilon \xrightarrow{\Gamma} \mathscr{R}^*_0$ in $\mathrm{H}^1(\Omega)$, where

$$\mathscr{F}_0(\rho) = \int_\Omega \frac{1}{2c_{\rm arith}}\rho^2\,\mathrm{d}x \text{ and } \mathscr{R}^*_0(\xi) = \int_\Omega \frac12\nabla\xi\cdot A_{\rm eff}\nabla\xi + \frac{b_{\rm arith}}{2}\xi^2\,\mathrm{d}x.$$

By Proposition 3.2.1 we conclude $\mathscr{F}_\varepsilon \overset{\mathsf{M}}{\to} \mathscr{F}_0$ in $\widetilde{\boldsymbol{H}}$, and by Theorem 3.2.2 and Lemma 3.2.1(c) we have $\mathscr{R}_\varepsilon \overset{\mathsf{C}}{\to} \mathscr{R}_0$ in $\widetilde{\boldsymbol{H}}$.

Again, Theorem 3.4.1 is applicable and we obtain $(\widetilde{\boldsymbol{H}}, \mathscr{F}_\varepsilon, \mathscr{R}_\varepsilon) \overset{\mathsf{E}}{\to} (\widetilde{\boldsymbol{H}}, \mathscr{F}_0, \mathscr{R}_0)$. In particular, we conclude that the solutions ρ_ε converge to ρ_0 in the following sense:

$$\rho_\varepsilon(0) \overset{\mathrm{H}^{-1}}{\to} \rho_0(0) \implies \forall t > 0: \ \rho_\varepsilon(t) \overset{\mathrm{L}^2}{\rightharpoonup} \rho_0(t) \text{ and } \mathscr{F}_\varepsilon(\rho_\varepsilon(t)) \to \mathscr{F}_\varepsilon(\rho_0(t)). \tag{3.4.7}$$

The latter convergences imply $u_\varepsilon(t) = \rho_\varepsilon(t)/c_\varepsilon \to u_0(t) = \rho_0(t)/c_{\mathrm{arith}}$, where we used the original definition of ρ_ε.

Hence, we see that the second gradient structure yields a stronger convergence result, because the necessary assumptions on the initial conditions are much weaker. In the original variables u_ε the convergence in (3.4.7) means that we need $u_\varepsilon(0)/c_\varepsilon \to u_0(0)/c_{\mathrm{arith}}$ in $\widetilde{\boldsymbol{H}} \approx \mathrm{H}^{-1}(\Omega)$ to obtain the weak convergence $u_\varepsilon(t)/c_\varepsilon \to u_0(t)/c_{\mathrm{arith}}$ as well as $u_\varepsilon(t) \to u_0(t)$ in $\mathrm{L}^2(\Omega)$. Hence, we can apply the convergence (3.4.6) afterwards.

3.4.3 Metric Gradient Systems and IEVE$_\lambda$

We present here a few basic facts on the general theory of evolutionary variational estimates on metric spaces. We refer to [2, 15, 70] for the general theory. In the next section we then show that there is a natural approach to establish E-convergence via the IEVE.

A *metric gradient system* is given by a triple $(\mathscr{Q}, \mathscr{E}, \mathscr{D})$ where $(\mathscr{Q}, \mathscr{D})$ is a complete, geodesic space, that is a complete metric space admitting constant-speed geodesic curves $\gamma : [0, 1] \to \mathscr{Q}$ between every two points $u_0, u_1 \in \mathscr{Q}$, i.e.

$$\gamma(0) = u_0, \ \gamma(1) = u_1, \text{ and } \mathscr{D}(\gamma(s), \gamma(t)) = |s - t|\mathscr{D}(u_0, u_1) \text{ for } s, t \in [0, 1].$$

The notion of geodesic spaces is motivated by classical gradient systems $(\boldsymbol{X}, \mathscr{E}, \mathscr{R})$ with a quadratic dissipation potential $\mathscr{R}(u, v) = \frac{1}{2}\langle \mathbb{G}(u)v, v\rangle$. Under suitable conditions we can define the associated *dissipation distance* via

$$\mathscr{D}_{\mathscr{R}}(u_0, u_1)^2 = \inf\left\{ \int_0^1 2\mathscr{R}(\gamma(s), \dot\gamma(s))\,\mathrm{d}s \;\middle|\; \gamma(0) = u_0, \ \gamma(1) = u_1, \ \gamma \in \mathrm{W}^{1,1}([0, 1]; \boldsymbol{X}) \right\},$$

and the minimizers are constant-speed geodesic curves.

For general metric gradient systems the notion of gradient flows was introduced in [17], see [2] for historical remarks, using the *metric velocity* $|u'|_{\mathscr{D}}(t)$ and the *metric slope* $|\partial\mathscr{E}|_{\mathscr{D}}(u)$. These notions can be seen as generalizations of the terms $\big(2\mathscr{R}(u,\dot u)\big)^{1/2}$ and $\big(2\mathscr{R}^*(u,-\mathrm{D}\mathscr{E}(u))\big)^{1/2}$, respectively, and are defined via

$$|u'|_{\mathscr{D}}(t) := \lim_{h\to 0}\frac{\mathscr{D}(u(t+h),u(t))}{h} \quad\text{and}\quad |\partial\mathscr{E}|_{\mathscr{D}}(u) := \limsup_{w\to u}\frac{\max\{\mathscr{E}(w)-\mathscr{E}(u),0\}}{\mathscr{D}(w,u)}.$$

Both definitions do not use any linear structure on the metric space $(\mathscr{Q},\mathscr{D})$. We only take difference quotients in the real numbers. In [17] metric gradient flows were introduced by generalizing the energy-dissipation principle (EDP) (3.1.3) (cf. Sect. 3.3.1):

$$\mathscr{E}(u(T)) + \int_0^T \frac{1}{2}|u'|_{\mathscr{D}}(t)^2 + \frac{1}{2}|\partial\mathscr{E}|_{\mathscr{D}}(u(t))^2\,\mathrm{d}t = \mathscr{E}(u(0)). \tag{3.4.8}$$

As in Theorem 3.3.1 the EDP can be replaced by an Upper Energy-Dissipation Estimate (UEDE) if we have a suitable lower chain-rule estimate, namely

$$\begin{aligned} &|u'|_{\mathscr{D}}(\cdot),\ |\partial\mathscr{E}|_{\mathscr{D}}(u(\cdot)) \in \mathrm{L}^2([0,T]) \\ &\qquad\Longrightarrow \begin{cases} \mathscr{E}(u(\cdot)) \in \mathrm{W}^{1,1}([0,T]) \text{ and} \\ \frac{\mathrm{d}}{\mathrm{d}t}\mathscr{E}(u(t)) \geq -|u'|_{\mathscr{D}}(t)\,|\partial\mathscr{E}|_{\mathscr{D}}(u(t)) \text{ a.e.} \end{cases} \end{aligned} \tag{3.4.9}$$

Solutions of the UEDE are called *curves of maximal slope*.

The general class of metric gradient systems is especially adapted to construct solutions by taking limits $\tau \to 0$ for the time-discretized (backward-Euler) minimization problems

$$u_{k+1}^{\tau} \in \operatorname{Arg\,min}\{\, \mathscr{E}(u) + \tfrac{1}{2\tau}\mathscr{D}(u_k^{\tau},u)^2 \mid u \in \mathscr{Q}\} \text{ for } k = 1,\ldots,N, \tag{3.4.10}$$

where $\tau = T/N > 0$ is the time step. We refer to [2, 15, 51] for general surveys.

Remark 3.4.1 (Generalized metric gradient systems) The above metric gradient flow corresponds to classical gradient flows in the sense that the dissipation potential $\frac{1}{2}|u'|_{\mathscr{D}}(t)^2$ is quadratic in the velocity. Considering a general convex, lsc, and monotone function $\psi : [0,\infty[\to [0,\infty]$ we obtain a generalized metric gradient system $(\mathscr{Q},\mathscr{E},\mathscr{D},\psi)$. The associated generalized metric gradient flow is defined by the convex dual $\psi^*(\beta) = \sup\{\, \nu\beta - \psi(\nu) \mid \nu \geq 0\}$ and the EDP

$$\mathscr{E}(u(T)) + \int_0^T \psi\big(|u'|_{\mathscr{D}}(t)\big) + \psi^*\big(|\partial\mathscr{E}|_{\mathscr{D}}(u(t))\big)\,\mathrm{d}t = \mathscr{E}(u(0)).$$

The p-gradient systems are given by $\psi(\nu) = \nu^p/p$ and are studied in [2]. The case of rate-independent systems is given by $\psi(\nu) = c\,\nu$ and will be studied in more detail in Sect. 3.5. Rate-independent systems with small viscosity are given by the function $\psi(v) = \sigma_{\text{yield}} v + \delta v^2/2$. The limit $\delta \searrow 0$ is treated in [50, 52].

The important point is that convexity of functionals can be generalized to geodesic convexity on geodesic spaces $(\mathscr{Q}, \mathscr{D})$ as follows. A functional $\mathscr{E} : \mathscr{Q} \to \mathbb{R}_\infty$ is called *geodesically* λ *-convex* on $(\mathscr{Q}, \mathscr{D})$ if for all $u_0, u_1 \in \mathrm{dom}(\mathscr{E}) := \{ u \in \mathscr{Q} \mid \mathscr{E}(u) < \infty \}$, there exists a constant-speed geodesic $\gamma : [0,1] \to \mathscr{Q}$ with $\gamma(0) = u_0$ and $\gamma(1) = u_1$ such that the function

$$s \mapsto \mathscr{E}(\gamma(s)) + \frac{\lambda s(1-s)}{2} \mathscr{D}(\gamma(0), \gamma(1))^2 \text{ is convex on } [0,1]. \tag{3.4.11}$$

The case $\lambda = 0$ is simply called geodesic convexity, $\lambda > 0$ improves the convexity, and $\lambda < 0$ relaxes the convexity. A trivial but useful and important special case is given by the Hilbert space setting with $\mathscr{Q} = \boldsymbol{H}$ and $\Psi(v) = \frac{1}{2}\langle \mathbb{G}v, v\rangle$. Then, we have $\mathscr{D}(u_0, u_1) = (2\Psi(u_1 - u_0))^{1/2}$ and the constant-speed geodesics are $\gamma(s) = (1-s)u_0 + s\,u_1$. Moreover, for smooth energies $\mathscr{E}$, geodesic λ-convexity is equivalent to $\mathrm{D}^2\mathscr{E} \geq \lambda\mathbb{G}$.

For general geodesically λ-convex gradient systems $(\mathscr{Q}, \mathscr{E}, \mathscr{D})$ there is a rather complete existence and uniqueness theory, see [2, 15], which we summarize here: For each $u_0 \in \overline{\mathrm{dom}(\mathscr{E})}$ there exists a unique solution $u : [0, \infty[\to \mathscr{Q}$ of (3.4.8) with $u(0) = u_0$. Moreover the solutions depend Lipschitz continuously on the initial data, i.e. any two solutions u_1 and u_2 satisfy

$$\mathscr{D}(u_1(t), u_2(t)) \leq \mathrm{e}^{-\lambda(t-s)} \mathscr{D}(u_1(s), u_2(s)) \text{ for } 0 \leq s < t.$$

Moreover, under slightly stronger assumptions it can be shown that the time-incremental solutions of (3.4.10) converge strongly, namely

$$\mathscr{D}(u(k\tau), u_k^\tau) \leq C(u_0)\sqrt{\tau}\, \mathrm{e}^{-\lambda_\tau k\tau} \quad \text{for } k = 1, ..., N = T/\tau,$$

where $\lambda_\tau = \frac{1}{\tau}\log(1 - \lambda\tau)$. See [48] for an application in one-dimensional viscoelasticity with a true metric (not translation invariant).

The main tool for the analysis of geodesically λ-convex metric gradient systems is the reformulation in terms of the integrated evolutionary variational estimate IEVE_λ, where now Proposition 3.4.1 is generalized by the appearance of λ, which was chosen 0 in Sect. 3.4.1, and by replacing $\Psi(u - w)$ with $\frac{1}{2}\mathscr{D}(u, w)^2$. We have the following equivalence:

Theorem 3.4.2 ([15, Theorem 2.11]) *Assume that $(\mathscr{Q}, \mathscr{E}, \mathscr{D})$ is a geodesically λ-convex gradient system as introduced above. Then, every solution $u : [0, \infty[\to \mathscr{Q}$ of the metric EDP (3.4.8) is a solution of the Integrated Evolutionary Variational Estimate $(\mathrm{IEVE})_\lambda$:*

$$\forall w \in \mathscr{Q}\ \forall s \geq 0,\ \forall t > s:$$
$$\frac{\mathrm{e}^{\lambda(t-s)}}{2}\mathscr{D}(u(t),w)^2 - \frac{1}{2}\mathscr{D}(u(s),w)^2 \leq \mathsf{m}_\lambda(t-s)\big(\mathscr{E}(w) - \mathscr{E}(u(t))\big), \tag{3.4.12}$$

where $\mathsf{m}_\lambda(r) = \int_0^r \mathrm{e}^{\lambda\rho}\mathrm{d}\rho$. *Moreover, every solution* u *of (IEVE)*$_\lambda$ *with* $\mathscr{E}(u(0)) < \infty$ *is a solution of (3.4.8).*

Again, we see that the IEVE is a formulation without any derivative, i.e. we do neither need the metric velocity $|u'|_{\mathscr{D}}$ nor the slope $|\partial\mathscr{E}|_{\mathscr{D}}$. Thus, it is natural to study evolutionary Γ-limits $\varepsilon \to 0$ for families $(\mathscr{Q}, \mathscr{E}_\varepsilon, \mathscr{D}_\varepsilon)$ of metric gradient systems. The importance here is that we need to be able to choose $\lambda \in \mathbb{R}$ independent of ε, i.e. we need uniform geodesic convexity.

The crucial a priori estimate, which generalizes (3.41), is

$$\begin{aligned}&\frac{\mathrm{e}^{\lambda t}}{2}\mathscr{D}(u(t),w)^2 + \mathsf{m}_\lambda(t)\mathscr{E}(u(t)) + \frac{\mathsf{m}_\lambda(t)^2}{2}|\partial\mathscr{E}(u(t))|^2_{\mathscr{D}}\\ &\leq \frac{1}{2}\mathscr{D}(u(0),w)^2 + \mathsf{m}_\lambda(t)\mathscr{E}(w) \text{ for all } t > 0 \text{ and } w \in \mathscr{Q},\end{aligned} \tag{3.4.13}$$

see [15, Theorem 2.6].

3.4.4 E-Convergence for Metric Gradient Systems

The following result on evolutionary Γ-convergence is a slight variant of [15, Theorem 2.17] or [70, Theorem 6.2], since we allow $\mathscr{D}_\varepsilon$ to depend on ε, while it was assumed to be constant there. We refer to [70, Theorem 7.4, Corollary 8.6] for more general results including abstract Gromov-Hausdorff convergence of metric spaces.

We first list the precise assumptions on the gradient system $(\mathscr{Q}, \mathscr{E}_\varepsilon, \mathscr{D}_\varepsilon)_{\varepsilon\in[0,1]}$. Note that we assume that all the metrics $\mathscr{D}_\varepsilon$ on the space $\mathscr{Q}$ are equivalent, cf. (3.4.14a). Hence, we can write $u_n \xrightarrow{\mathscr{D}} u$ for $\mathscr{D}_{\varepsilon_n}(u_n, u) \to 0$ for $n \to \infty$, where $\varepsilon_n \in [0, 1]$ can be arbitrary. However, for the geodesic convexity of $\mathscr{E}_\varepsilon$ it is crucial to consider the dissipation distance $\mathscr{D}_\varepsilon$ with the same ε, see (3.4.14e):

$$\begin{aligned}&\exists c > 0\ \forall \varepsilon \in [0,1],\ u_1, u_2 \in \mathscr{Q}:\\ &\qquad c\mathscr{D}_0(u_1,u_2) \leq \mathscr{D}_\varepsilon(u_1,u_2) \leq \tfrac{1}{c}\mathscr{D}_0(u_1,u_2);\end{aligned} \tag{3.4.14a}$$

$$\mathscr{D}_\varepsilon \xrightarrow{\mathsf{C}} \mathscr{D}_0, \quad \text{i.e. } u_\varepsilon^{(j)} \xrightarrow{\mathscr{D}} u^{(j)} \Rightarrow \mathscr{D}_\varepsilon(u_\varepsilon^{(1)}, u_\varepsilon^{(2)}) \to \mathscr{D}_0(u^{(1)}, u^{(2)}); \tag{3.4.14b}$$

$$\begin{aligned}&\forall E > 0\ \exists \mathscr{C}_E \text{ compact in } (\mathscr{Q}, \mathscr{D}_0)\ \forall \varepsilon \in [0,1]:\\ &\qquad \{\, u \in \mathscr{Q} \mid \mathscr{E}_\varepsilon(u) \leq E \,\} \subset \mathscr{C}_E;\end{aligned} \tag{3.4.14c}$$

$$\mathscr{E}_\varepsilon \xrightarrow{\Gamma} \mathscr{E}_0 \text{ in } (\mathscr{Q}, \mathscr{D}_0); \tag{3.4.14d}$$

$$\exists \lambda \in \mathbb{R}\ \forall \varepsilon \in [0,1]:\ \mathscr{E}_\varepsilon \text{is geodesically } \lambda\text{-convex in } (\mathscr{Q}, \mathscr{D}_\varepsilon). \tag{3.4.14e}$$

Thus, the crucial assumptions are the uniform compactness (3.4.14c), the uniform geodesic λ-convexity (3.4.14e) as well as the convergences $\mathscr{E}_\varepsilon \xrightarrow{\Gamma} \mathscr{E}_0$ and $\mathscr{D}_\varepsilon \xrightarrow{\mathsf{C}} \mathscr{D}_0$ in $(\mathscr{Q}, \mathscr{D}_0)$.

Theorem 3.4.3 (E-convergence for IEVE) *If the assumptions (3.4.14) hold, then* $(\mathscr{Q}, \mathscr{E}_\varepsilon, \mathscr{D}_\varepsilon) \xrightarrow{\mathsf{E}} (\mathscr{Q}, \mathscr{E}_0, \mathscr{D}_0)$ *(no well-preparedness of init. cond.). More precisely,*

$$\begin{aligned} &\text{if } u_\varepsilon(0) \in \overline{\mathrm{dom}(\mathscr{E}_\varepsilon)} \text{ for } \varepsilon \in [0,1] \text{ and } u_\varepsilon(0) \xrightarrow{\mathscr{D}} u_0(0), \\ &\text{then } \forall t > 0: \ u_\varepsilon(t) \to u_0(t) \text{ and } \mathscr{E}_\varepsilon(u_\varepsilon(t)) \to \mathscr{E}_0(u_0(t)). \end{aligned}$$

We highlight that the whole sequence u_ε converges to u_0, because we have uniqueness of the solutions. Even though well-preparedness of the initial conditions is not needed, we obtain energy convergence for all $t > 0$.

Sketch of proof: For a full proof we refer to the above references. Here we simply highlight the main steps, which are exactly the same as for Theorem 3.4.1.

Based on the a priori estimate (3.4.13) evaluated for $t \in [t_0, T]$ with arbitrary $0 < t_0 < T$ and the uniform compactness (3.4.14c) we can extract a subsequence (not relabeled) with $u_\varepsilon(t) \xrightarrow{\mathscr{D}} U(t)$. Choosing a recovery sequence $\widehat{w}_\varepsilon \xrightarrow{\mathscr{D}} \widehat{w}$ with $\mathscr{E}_\varepsilon(\widehat{w}_\varepsilon) \to \mathscr{E}_0(\widehat{w})$ we can pass to the limit $\varepsilon \to 0$ in $(\mathrm{IEVE})^\varepsilon_\lambda$ and find that U is a solution of the limit equation $(\mathrm{IEVE})^0_\lambda$.

Continuity at $t = 0$ and energy convergence follow as for Theorem 3.4.1. □

For an application of this theory we refer to [26], where the heat equation on the torus $\mathbb{T}_d = \mathbb{R}^d/_{\mathbb{Z}^d}$ is considered as generated by the the Kantorovich-Wasserstein gradient system $(\mathrm{Prob}(\mathbb{T}_d), \mathscr{E}_0, \mathscr{D}_0)$ with the entropy functional $\mathscr{E}_0(\rho) = \int_{\mathbb{T}_d} \rho \log \rho \, \mathrm{d}x$ and $\mathscr{D}_0 = \mathscr{D}_{\mathrm{Wass}}$. It is obtained as evolutionary Γ-limit of discrete Markov chains in the periodic d-dimensional lattice $(\mathbb{Z}/_{N\mathbb{Z}})^d$ for $N \to \infty$.

An application to the justification of amplitude equations is given in [47]. In fact, this application is a Hilbert-space case, but nevertheless using the approach via the IEVE improves the results considerably. The original equation is the ε-dependent, fourth-order parabolic Swift-Hohenberg equation

$$\dot{u} = -\frac{1}{\varepsilon^2}(1 + \varepsilon^2 \partial_x^2)^2 u + \mu u - u^3 \ \text{ on } \mathbb{S} := \mathbb{R}/_{2\pi\mathbb{Z}}, \tag{3.4.15}$$

which is the L^2-gradient flow of the functional $\mathscr{F}_\varepsilon^{\mathrm{SH}}(u) = \int_{\mathbb{S}} \frac{1}{2\varepsilon^2}(u + \varepsilon^2 u_{xx})^2 - \frac{\mu}{2}u^2 + \frac{1}{4}u^4 \mathrm{d}x$. Because of the linear operator the typical solutions (e.g. well-prepared solutions) will spatially oscillate on the scale ε and are approximately of the form $u(t,x) \approx \mathrm{Re}\big(A(t,x)\mathrm{e}^{\mathrm{i}x/\varepsilon}\big)$. Using Theorem 3.4.3 it is possible to show that the Ginzburg-Landau equation

$$\dot{A} = 4A_{xx} + \mu A - \tfrac{3}{4}|A|^2 A$$

is the evolutionary Γ-limit. More precisely, using a bijection $\mathbb{M}_\varepsilon$, which satisfies $u = \mathrm{Re}\left((\mathbb{M}_\varepsilon u)\mathrm{e}^{\mathrm{i}x/\varepsilon}\right)$, it is shown in [47, Theorem 2.3] that the solutions u_ε of the Swift-Hohenberg equation converge to solutions A of Ginzburg-Landau equation in the following sense:

$$\mathbb{M}_\varepsilon u_\varepsilon(0) \overset{\mathrm{L}^2(\mathbb{S})}{\to} A(0) \quad \Longrightarrow \quad \forall t > 0 : \mathbb{M}_\varepsilon u_\varepsilon(t) \overset{\mathrm{H}^1(\mathbb{S})}{\rightharpoonup} A(t) \text{ and}$$
$$\mathscr{F}_\varepsilon^{\mathrm{SH}}(u_\varepsilon(t)) \to \mathscr{F}^{\mathrm{GL}}(A(t)).$$

The last application of the IEVE shows that even in the case of a Hilbert space with a quadratic and translation-invariant dissipation potential the metric concepts are extremely helpful and give a new look to semilinear parabolic equations, in particular concerning question of evolutionary Γ-convergence.

3.5 Rate-Independent Systems (RIS)

We again consider generalized gradient systems $(\boldsymbol{X}, \mathscr{E}, \mathscr{R})$ on a Banach space $\boldsymbol{X}$, but now we focus to the special case that the dissipation potential $\mathscr{R}$ is positively homogeneous of degree 1, i.e.

$$\forall u, v \in \boldsymbol{X}\ \forall \gamma > 0 : \quad \mathscr{R}(u, \gamma v) = \gamma^1 \mathscr{R}(u, v). \tag{3.5.1}$$

In particular, $\mathscr{R}(u, \cdot)$ cannot be smooth, but using convexity we still have a subdifferential $\partial_v \mathscr{R}(u, v)$ that is set-valued now. In particular, the relation between the rate $v = \dot{u}$ and the dissipative forces $\xi \in \partial \mathscr{R}(u, \dot{u})$ is positively homogeneous of degree 0, i.e. $\partial_v \mathscr{R}(u, \gamma v) = \gamma^0 \partial_v \mathscr{R}(u, v)$, which explains the name "rate independence" of the dissipative constitutive law. *Rate-independent systems* are ideal to describe hysteretic effects and occur in the case of Coulomb friction, in plasticity, magnetism, or phase transformations.

3.5.1 Subdifferential Formulation of RIS

The evolution equation in Biot's form (cf. (3.2.1)) is the subdifferential inclusion

$$0 \in \partial_{\dot{u}} \mathscr{R}(u(t), \dot{u}(t)) + \mathrm{D}_u \mathscr{E}(t, u(t)) \in \boldsymbol{X}^*, \tag{3.5.2}$$

where we now included a time-dependence into the energy functional $\mathscr{E}$, as in Sect. 3.2.1. This is essential for rate-independent systems because they do not have any internal time scale any more. So they do not move without an external time-dependent stimulus.

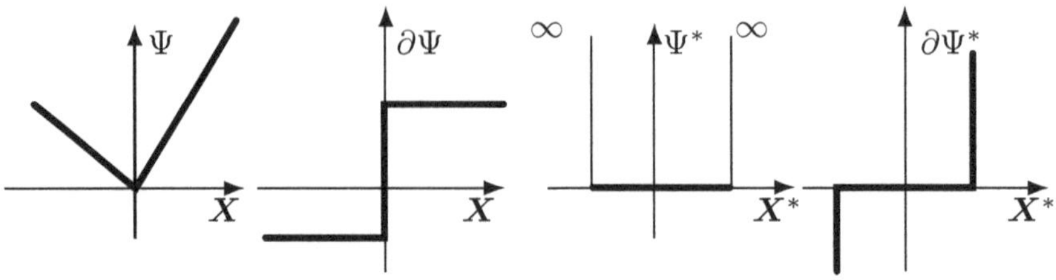

Fig. 3.4 Primal and dual dissipation potential for RIS

As before, Biot's equation (3.5.2) is only one of the three equivalent formulations of the problem. However, in the rate-independent case the 1-homogeneity of the dissipation induces a very particular structure, see also Fig. 3.4.

Lemma 3.5.1 *Assume that* $\Psi : \boldsymbol{X} \to [0, \infty]$ *is convex, lsc, and 1-homogeneous, and set* $K := \partial\Psi(0)$. *Then,*

$$\xi \in \partial\Psi(v) \iff \begin{cases} \xi \in K \\ \langle \xi, v \rangle = \Psi(v) \end{cases} \quad \text{and} \quad \Psi^*(\xi) = \begin{cases} 0 & \text{if } \xi \in K, \\ \infty & \text{else.} \end{cases}$$

To understand the rate equation $\dot{u} \in \partial_\eta \mathscr{R}^*(u, -\mathrm{D}\mathscr{E}(t, u))$, which is now often called "flow rule", we introduce the closed and convex set $K(u) := \partial_{\dot{u}} \mathscr{R}(u, 0) \subset \boldsymbol{X}^*$ and its outer normal cone

$$\mathrm{N}_{K(u)}(\eta) := \{ v \in \boldsymbol{X} \mid \forall \xi \in K(u) : \langle \xi - \eta, v \rangle \geq 0 \},$$

then we have $\partial_\eta \mathscr{R}^*(u, -\eta) = -\mathrm{N}_{K(u)}(\eta)$ (cf. [66]) and find the rate equation

$$\dot{u}(t) \in \partial_\eta \mathscr{R}^*\big(u(t), -\mathrm{D}\mathscr{E}(t, u(t))\big), \quad \text{or} \quad -\dot{u}(t) \in \mathrm{N}_{K(u(t))}(\mathrm{D}\mathscr{E}(t, u(t))). \quad (3.5.3)$$

Using the characterization of $\partial_{\dot{u}}\mathscr{R}$ in Lemma 3.5.1 and the definition of $\mathrm{N}_{K(u)}$ we can rewrite the two equivalent equations (3.5.2) and (3.5.3) as so-called an evolutionary quasi-variational inequalities (cf. [12, 33, 41]):

$$(3.55) \iff \forall v \in \boldsymbol{X} : \langle \mathrm{D}_u \mathscr{E}(t, u), v - \dot{u} \rangle + \mathscr{R}(u, v) - \mathscr{R}(u, \dot{u}) \geq 0, \quad (3.5.4a)$$

$$(3.56) \iff \forall \xi \in K(u) \subset \boldsymbol{X}^* : \langle \mathrm{D}_u \mathscr{E}(t, u) - \xi, \dot{u} \rangle \geq 0. \quad (3.5.4b)$$

For treating more general systems the energy-dissipation principle (3.2.5) is more relevant. However, for RIS the dual dissipation potential $\mathscr{R}^*$ does not give a contribution to the balance, because it only attains the values 0 or $+\infty$. The latter value leads to a constraint. Thus, the EDP takes the form

$$\forall t \in [0, T] : -\mathrm{D}\mathscr{E}(t, u(t)) \in K(u(t)) = \partial_{\dot{u}} \mathscr{R}(u(t), 0), \quad (3.5.5a)$$

$$\mathscr{E}(T, u(T)) + \int_0^T \mathscr{R}(u, \dot{u})\mathrm{d}t = \mathscr{E}(0, u(0)) + \int_0^T \partial_t \mathscr{E}(t, u(t))\mathrm{d}t. \tag{3.5.5b}$$

Here (3.5.5a) is a totally static *local stability* condition saying that the potential restoring force is balanced by one of the possible dissipative friction forces, while (3.5.5b) is a reduced energy balance, where the last integral is the work of the external forces.

3.5.2 Energetic Solutions of RIS

Since the dissipation $\int_0^T \mathscr{R}(u(t), \dot{u}(t))\mathrm{d}t$ controls the BV-norm of u with respect to the time only, in general cases the solutions u will not be absolutely continuous such that $\dot{u}$ is not properly defined because of jumps with $u(t^-) \neq u(t^+)$. The following definition of *energetic solutions* is adapted to this situation, because it does not need any continuity or differentiability with respect to time. We use a general state space $\mathscr{Q}$, which can be a general topological space without a differential structure, and a general dissipation distance $\mathscr{D} : \mathscr{Q} \times \mathscr{Q} \to [0, \infty]$ which is assumed to satisfy the triangle inequality $\mathscr{D}(u_1, u_3) \leq \mathscr{D}(u_1, u_2) + \mathscr{D}(u_2, u_3)$ (but not necessarily the symmetry and positivity of usual distances, see [41, Chap. 5] or [43]). We call a triple $(\mathscr{Q}, \mathscr{E}, \mathscr{D})$ an *energetic RIS*. Starting from a rate-independent dissipation potential $\mathscr{R}$, it is possible to find the associated dissipation distance $\mathscr{D}(u_0, u_1)$ by minimizing $\int_0^1 \mathscr{R}(u(s), \dot{u}(s))\mathrm{d}s$ along all curves connecting u_0 and u_1, which is similar but not equal to the definition of $\mathscr{D}$ in Sect. 3.4.3.

Definition 3.5.1 (*Energetic solutions for RIS*) A function $u : [0, T] \to \mathscr{Q}$ is called an *energetic solution* for the RIS $(\mathscr{Q}, \mathscr{E}, \mathscr{D})$ if **stability (S)** and **energy balance (E)** hold:

$$\textbf{(S)} \quad \mathscr{E}(t, u(t)) \leq \mathscr{E}(t, w) + \mathscr{D}(u(t), w) \text{ for all } t \in [0, T] \text{ and } w \in \mathscr{Q},$$

$$\textbf{(E)} \quad \mathscr{E}(T, u(T)) + \mathrm{Diss}_{\mathscr{D}}(u, [0, T]) = \mathscr{E}(0, u(0)) + \int_0^T \partial_t \mathscr{E}(t, u(t))\mathrm{d}t. \tag{3.5.6}$$

In many cases this definition is equivalent to the notion of *quasistatic (irreversible) evolutions* studied in [13, 21, 60].

The dissipated energy can be expressed solely by the dissipation distance

$$\mathrm{Diss}_{\mathscr{D}}(u, [0, T]) := \sup \Big\{ \sum_{j=1}^N \mathscr{D}(u(t_{j-1}), u(t_j)) \;\Big|\; \text{all partitions of } [0, T] \Big\}$$

and coincides with $\int_0^T \mathscr{R}(u(t), \dot{u}(t))\mathrm{d}t$ in the smooth case.

The major advantage of the concept of energetic solutions is that it is a derivative-free formulation, which only features the values of the functionals $\mathscr{E}$ and $\mathscr{D}$ and the function values $u(t)$. Thus, it shares the same properties as the IEVE discussed in Sect. 3.4. In particular, existence results and evolutionary Γ-convergence can be attacked by tools from the calculus of variations. In fact, energetic solutions are the limits of the incremental minimization problems with time step $\tau = T/N \to 0$ (cf. [41, Theorem 5.2]):

$$u_k^\tau \in \operatorname{Arg\,Min}\{\, \mathscr{E}(k\tau, u) + \mathscr{D}(u_{k-1}^\tau, u) \mid u \in \mathscr{Q}\,\}. \tag{3.5.7}$$

In contrast to the backward-Euler algorithm (3.4.10) for the metric gradient flow, now there is no explicit dependence on the time step τ (reflecting rate independence) and the dissipation distance has the power 1.

In fact, the rate formulations (3.5.2)–(3.5.5) and the energetic solutions are equivalent in good cases, but are different in general, cf. [53, Chap. 3].

Proposition 3.5.1 (Equivalence of formulations for convex energies) *Consider a RIS $(\boldsymbol{X}, \mathscr{E}, \mathscr{D})$, where $\boldsymbol{X}$ is a Banach space and assume that the energies $\mathscr{E}(t,\cdot) : \boldsymbol{X} \to \mathbb{R}_\infty$ are convex for all $t \in [0,T]$. Further assume that the dissipation distance $\mathscr{D}$ has the form $\mathscr{D}(u_0,u_1) = \Psi(u_1 - u_0)$ for a lsc, convex and 1-homogeneous function $\Psi : \boldsymbol{X} \to [0,\infty]$. Then, $u \in \mathrm{W}^{1,1}([0,T]; \boldsymbol{X})$ is an energetic solution if and only if u solves (3.5.5).*

3.5.3 pE-Convergence for Energetic Solutions

We now consider a family $(\mathscr{Q}, \mathscr{E}_\varepsilon, \mathscr{D}_\varepsilon)$ of RIS, where for the simplicity of the presentation we restrict the discussion to the case that $\mathscr{Q}$ is a reflexive Banach space $\boldsymbol{X}$ and the energies $\mathscr{E}_\varepsilon$ have the form

$$\mathscr{E}_\varepsilon(t,u) = \mathscr{F}_\varepsilon(u) - \langle \ell_\varepsilon(t), u\rangle. \tag{3.5.8a}$$

We still keep general dissipation distances $\mathscr{D}_\varepsilon$ and do not assume convexity of $\mathscr{F}_\varepsilon$. Thus, all the results can be generalized to the general topological case, see [53, 55]. Our precise assumptions are as follows:

$$\exists\, c, C > 0\ \forall\, \varepsilon \geq 0,\ u \in \boldsymbol{X} :\ \mathscr{F}_\varepsilon(u) \geq c\|u\|^2 - C; \tag{3.5.8b}$$

$$\forall\, \varepsilon \geq 0 :\quad \mathscr{F}_\varepsilon : \boldsymbol{X} \to \mathbb{R}_\infty \text{ is weakly lsc;} \tag{3.5.8c}$$

$$\exists\, C > 0\ \forall\, \varepsilon \geq 0 :\ \|\ell_\varepsilon\|_{\mathrm{C}^1([0,T])} \leq C; \tag{3.5.8d}$$

$$\forall\, t \in [0,T] :\qquad \dot\ell_\varepsilon(t) \to \dot\ell_0(t) \text{ in } \boldsymbol{X}^*; \tag{3.5.8e}$$

$$\forall\, \varepsilon \geq 0\ \forall\, u_j \in \boldsymbol{X} :\ \begin{cases} \mathscr{D}_\varepsilon(u_1,u_3) \leq \mathscr{D}_\varepsilon(u_1,u_2) + \mathscr{D}_\varepsilon(u_2,u_3), \\ \qquad \mathscr{D}_\varepsilon(u_1,u_2) = 0 \implies u_1 = u_2. \end{cases} \tag{3.5.8f}$$

In general, these conditions are not strong enough to show existence of solutions. Indeed, even if we assume existence of solutions for $\varepsilon > 0$, we cannot expect to conclude existence of solutions for $\varepsilon = 0$ by a limit argument. We need additional properties, e.g. the weak continuity of $\mathscr{D}_0$.

However, we can already address another nice property of the energetic formulation, namely the general validity of an appropriate generalization of the chain rule (3.3.1), which allows us to turn an upper energy estimate into an energy-dissipation balance as in Theorem 3.3.1. If $(\boldsymbol{X}, \mathscr{E}_0, \mathscr{D}_0)$ satisfies (3.5.8) and $u : [0, T] \to \boldsymbol{X}$ satisfies the global stability (S) in (3.5.6), then we have the lower energy estimate (see [41, Proposition 5.7])

$$\mathscr{E}_0(t, u(t)) + \mathrm{Diss}_{\mathscr{D}_0}(u, [s, t]) \geq \mathscr{E}_0(s, u(s)) + \int_s^t \partial_r \mathscr{E}(r, u(r))\mathrm{d}r, \tag{3.5.9}$$

We present two results for evolutionary Γ-convergence. The first assumes that the dissipation distances $\mathscr{D}_\varepsilon$ weakly continuously converge to $\mathscr{D}_0$.

Theorem 3.5.1 (pE-convergence for RIS [55]) *Assume that the RIS* $(\boldsymbol{X}, \mathscr{E}_\varepsilon, \mathscr{D}_\varepsilon)$ *satisfies (3.5.8) and* $\mathscr{E}_\varepsilon \xrightharpoonup{\Gamma} \mathscr{E}_0$ *and* $\mathscr{D}_\varepsilon \xrightharpoonup{\mathsf{C}} \mathscr{D}_0$ *in* $\boldsymbol{X}$*, then* $(\boldsymbol{X}, \mathscr{E}_\varepsilon, \mathscr{D}_\varepsilon) \xrightharpoonup{\mathsf{pE}} (\boldsymbol{X}, \mathscr{E}_0, \mathscr{D}_0)$*, i.e. for energetic solutions* $u_\varepsilon : [0, T] \to \boldsymbol{X}$ *the following holds:*

$$\begin{aligned} &\text{if } u_\varepsilon(0) \rightharpoonup u_0(0) \text{ and } \mathscr{E}_\varepsilon(0, u_\varepsilon(0)) \to \mathscr{E}_0(0, u_0(0)) < \infty, \\ &\text{then } u_\varepsilon(t) \rightharpoonup u_0(t) \text{ and } \mathscr{E}_\varepsilon(t, u_\varepsilon(t)) \to \mathscr{E}_0(t, u_0(t)) \text{ for } 0 < t \leq T. \end{aligned} \tag{3.5.10}$$

Proof From (3.5.8b) and (3.5.8d) we find $C_1, \Lambda > 0$ such that the power control $|\partial \mathscr{E}_\varepsilon(t, u)| \leq \Lambda\big(\mathscr{E}_\varepsilon(t, u) + C_1\big)$ holds. Inserting this estimate into the energy balance (E) we obtain the a priori bound

$$\mathscr{E}_\varepsilon(t, u_\varepsilon(t)) + \mathrm{Diss}_{\mathscr{D}_\varepsilon}(u_\varepsilon, [0, t]) \leq \mathrm{e}^{\Lambda t}\big(\mathscr{E}_\varepsilon(0, u_\varepsilon(0)) + C_1\big) - C_1 \leq E_*,$$

where we used the well-preparedness $\mathscr{E}_\varepsilon(0, u_\varepsilon(0)) \to \mathscr{E}_0(0, u_0(0)) < \infty$. Using (3.5.8b) once again we find $\|u_\varepsilon(t)\| \leq C_2$ for all t and $\varepsilon > 0$, and Helly's selection principle (cf. [41, Theorem 5.1]) allows us to extract a (not relabeled) subsequence such that $u_\varepsilon(t) \rightharpoonup u_0(t)$ for all t.

Next, we show that all $u_0(t)$ satisfy the stability condition (S). We know that (S) is true for $\varepsilon > 0$, i.e.

$$\mathscr{E}_\varepsilon(t, u_\varepsilon(t)) \leq \mathscr{E}_\varepsilon(t, w_\varepsilon) + \mathscr{D}_\varepsilon(u_\varepsilon(t), w_\varepsilon) \text{ for all } t \in [0, T] \text{ and } w_\varepsilon \in \boldsymbol{X}. \tag{3.5.11}$$

For a given test state w we choose a recovery sequence $w_\varepsilon \rightharpoonup w$ with $\mathcal{E}_\varepsilon(t, w_\varepsilon) \to \mathcal{E}_0(t, w)$ and pass to the limit, viz.

$$\mathcal{E}_0(t, u_0(t)) \overset{\Gamma}{\leq} \liminf_{\varepsilon\to 0} \mathcal{E}_\varepsilon(t, u_\varepsilon(t)) \leq \mathcal{E}_0(t, w) + \mathcal{D}_0(u_0(t), w),$$

where we used $\mathcal{E}_\varepsilon \overset{\Gamma}{\rightharpoonup} \mathcal{E}_0$ and $\mathcal{D}_\varepsilon \overset{\mathrm{C}}{\rightharpoonup} \mathcal{D}_0$. Thus, stability of $u_0(t)$ is established.

In the energy balance $(\mathrm{E})_\varepsilon$ we can pass to the limit using (3.5.8d) and (3.5.8e) to obtain

$$\int_0^T \partial_t \mathcal{E}_\varepsilon(t, u_\varepsilon(t))\,\mathrm{d}t = -\int_0^T \langle \dot{\ell}_\varepsilon, u_\varepsilon\rangle\,\mathrm{d}t \to -\int_0^T \langle \dot{\ell}_0, u_0\rangle\,\mathrm{d}t = \int_0^T \partial_t \mathcal{E}_0(t, u_0(t))\,\mathrm{d}t.$$

Since the initial energies converge, the right-hand sides in $(\mathrm{E})_\varepsilon$ converge to that of $(\mathrm{E})_0$.

Using $\mathcal{D}_\varepsilon \overset{\mathrm{C}}{\rightharpoonup} \mathcal{D}_0$ we estimate the total dissipation on the interval $[0, t]$ via

$$\lim_{\varepsilon\to 0} \sum_{j=1}^N \mathcal{D}_\varepsilon(u_\varepsilon(t_{j-1}), u_\varepsilon(t_j)) = \sum_{j=1}^N \mathcal{D}_0(u_0(t_{j-1}), u_0(t_j)) \geq \mathrm{Diss}_{\mathcal{D}_0}(u_0, [0, T]) - \rho,$$

where $\rho > 0$ can be made arbitrary small by choosing a suitable partition $0 = t_0 < t_1 < \cdots < t_{N-1} < t_N{=}t$. Hence, $\liminf_{\varepsilon\to 0} \mathrm{Diss}_{\mathcal{D}_\varepsilon}(u_\varepsilon, [0, t]) =: D_0(t) \geq d_0(t) := \mathrm{Diss}_{\mathcal{D}_0}(u_0, [0, t])$. Because of $\mathcal{E}_\varepsilon \overset{\Gamma}{\rightharpoonup} \mathcal{E}_0$ for all $t \in [0, T]$ we have

$$e_0(t) := \mathcal{E}_0(t, u_0(t)) \leq E_0(t) := \liminf_{\varepsilon\to 0} \mathcal{E}_\varepsilon(t, u_\varepsilon(t)) \leq E_1(t) := \limsup_{\varepsilon\to 0} \mathcal{E}_\varepsilon(t, u_\varepsilon(t)).$$

Using the upper energy estimate on $[0, t]$ yields

$$e_0(t) + d_0(t) \leq E_1(t) + D_0(t) \leq e_0(0) + \int_0^t \partial_s \mathcal{E}_0(s, u_0(s))\,\mathrm{d}s \leq e_0(t) + d_0(t),$$

where the last estimate follows from the lower energy estimate (3.5.9). Hence, we conclude $e_0(t) = E_0(t) = E_1(t)$ which is the desired energy convergence in (3.5.10). □

We apply this result to a rate-independent homogenization problem with non-quadratic energies. We first show that the result fails, even if $\mathcal{D}_\varepsilon = \mathcal{D}_0$, e.g. if $\mathcal{D}_0$ is not weakly continuous.

Example 3.5.1 (Counterexample) We consider $\boldsymbol{X} = \mathrm{L}^2(\Omega)$,

$$\mathcal{E}_\varepsilon(t,u) = \int_\Omega \frac{1}{2} a(\frac{1}{\varepsilon}x)u(x)^2 - \ell(t)u(x)\,\mathrm{d}x, \quad \text{and } \Psi_\varepsilon(v) = \int_\Omega c(\frac{1}{\varepsilon}x)|v(x)|\,\mathrm{d}x.$$

Clearly, $\mathcal{E}_\varepsilon \xrightarrow{\Gamma} \mathcal{E}_0 : u \mapsto \int_\Omega \frac{a_{\mathrm{harm}}}{2}u^2 - \ell u\,\mathrm{d}x$ while $\Psi_\varepsilon \xrightarrow{\Gamma} \Psi_{\mathrm{arith}} := c_{\mathrm{arith}}\|\cdot\|_{\mathrm{L}^1}$ and $\Psi_\varepsilon \xrightharpoonup{\Gamma} \Psi_{\min} = c_{\min}\|\cdot\|_{\mathrm{L}^1}$, if the set $\{\, y \in [0,1]^d \mid c(y) = c_{\min} \,\}$ has positive $\mathcal{L}^d$ measure.

For the loading we assume $\ell(0) = 0$ and $\dot{\ell}(t) > 0$ for $t > 0$. Starting from the well-prepared initial condition $u_\varepsilon(0,x) = 0$ we find the unique solution

$$u_\varepsilon(t,x) = \max\{0, \ell(t) - c(\frac{1}{\varepsilon}x)\}/a(\frac{1}{\varepsilon}x).$$

Hence, for $c \equiv \text{const}$ we obtain weak evolutionary Γ-convergence to $(\mathrm{L}^2(\Omega), \mathcal{E}_0, \Psi_0)$ with $\Psi_0 = \Psi_{\mathrm{arith}} = \Psi_{\min}$, while for $c(\cdot)$ nonconstant the weak limit $U(t)$ of $u_\varepsilon(t)$ cannot be described by a RIS system $(\mathrm{L}^2(\Omega), \widehat{\mathcal{E}}_0, \widehat{\Psi}_0)$ for any $\widehat{\mathcal{E}}_0$ and $\widehat{\Psi}_0$.

Example 3.5.2 (Homogenization of RIS) We consider $\boldsymbol{X} = \mathrm{H}^1_0(\Omega)$ for a bounded Lipschitz domain $\Omega \subset \mathbb{R}^d$, and energy functional $\mathcal{F}_\varepsilon(u) = \int_\Omega \frac{1}{2}\nabla u \cdot A(\frac{1}{\varepsilon}x)\nabla u + F(\frac{1}{\varepsilon}x, u)\,\mathrm{d}x$, and a dissipation distance $\mathcal{D}_\varepsilon(u_1,u_2) = \Psi_\varepsilon(u_2 - u_1)$ with $\Psi_\varepsilon(v) = \int_\Omega c(\frac{1}{\varepsilon}x)|v(x)|\mathrm{d}x$, where $F(y,\cdot) \geq 0$ is continuous and convex and A, c, and $F(\cdot,u)$ are 1-periodic in $y = \frac{1}{\varepsilon}x$.

By Proposition 3.5.1, the energetic solutions solve the differential inclusion

$$0 \in c(\frac{1}{\varepsilon}x)\mathrm{Sign}(\dot{u}) - \mathrm{div}\,\big(A(\frac{1}{\varepsilon}x)\nabla u\big) + \partial_u F(\frac{1}{\varepsilon}x, u) - \ell_\varepsilon(t).$$

Weak convergence in $\mathrm{H}^1(\Omega)$ implies strong convergence in $\mathrm{L}^2(\Omega)$, so in $\boldsymbol{X}$ we obtain the convergences

$$\Psi_\varepsilon \xrightharpoonup{\mathsf{C}} \Psi_0 : v \mapsto \int_\Omega c_{\mathrm{arith}}|v|\mathrm{d}x \quad \text{and} \quad \mathcal{F}_\varepsilon \xrightharpoonup{\Gamma} \mathcal{F}_0 : u \mapsto \int_\Omega \frac{1}{2}\nabla u \cdot A_{\mathrm{eff}}\nabla u + F_{\mathrm{arith}}(u)\,\mathrm{d}x.$$

Theorem 3.5.1 is applicable giving $(\boldsymbol{X}, \mathcal{E}_\varepsilon, \Psi_\varepsilon) \xrightharpoonup{\mathsf{pE}} (\boldsymbol{X}, \mathcal{E}_0, \Psi_0)$, where the limit equation reads $0 \in c_{\mathrm{arith}}\,\mathrm{Sign}(\dot{u}) - \mathrm{div}\,\big(A_{\mathrm{eff}}\nabla u\big) + \partial_u F_{\mathrm{arith}}(u) - \ell_0(t)$.

Our second result on evolutionary Γ-convergence does not need the weak continuity of $\mathcal{D}_0$, but requires the Hilbert-space setting with

$$\begin{aligned} &\mathcal{F}_\varepsilon(u) = \frac{1}{2}\langle A_\varepsilon u, u\rangle \geq c\|u\|_H^2 \quad \text{and} \\ &\mathcal{D}_\varepsilon(u_1,u_2) = \Psi_\varepsilon(u_2 - u_1) \text{ with } \Psi_\varepsilon(\gamma v) = \gamma^1 \Psi_\varepsilon(v). \end{aligned} \tag{3.5.12}$$

This case is important in classical models of elastoplasticity, and for fixed ε it can be reduced to the analysis of monotone operators, see [1]. For evolutionary Γ-convergence with Ψ_ε truely dependent on ε the notion of energetic solutions is more flexible. The following convergence result was developed for two-scale homogenization and dimension reduction in elastoplastic models in [58] and [37], respectively.

The homogenization of RIS without compactness does not work directly (see Example 3.5.1), however the method of periodic unfolding developed for two-scale homogenization turns weak Γ-convergence of the functionals into Mosco-convergence in the two-scale setting. Hence, the homogenization results for elastoplasticity in [24, 29, 58] can be derived using the following abstract evolutionary Γ-convergence result.

Theorem 3.5.2 (pE-convergence for quadratic RIS) *Assume that* $\boldsymbol{H}$ *is a Hilbert space and that the RIS* $(\boldsymbol{H}, \mathscr{E}_\varepsilon, \mathscr{D}_\varepsilon)$, $\varepsilon \in [0, 1]$, *satisfy (3.5.8) and* (3.5.12). *If additionally*

$$\mathscr{E}_\varepsilon \overset{\mathsf{M}}{\to} \mathscr{E}_0, \qquad \Psi_\varepsilon \overset{\mathsf{C}}{\to} \Psi_0, \quad \textit{and} \quad \Psi_\varepsilon \overset{\Gamma}{\rightharpoonup} \Psi_0 \quad \textit{in } \boldsymbol{H},$$

then $(\boldsymbol{H}, \mathscr{E}_\varepsilon, \mathscr{D}_\varepsilon) \overset{\mathsf{pE}}{\rightharpoonup} (\boldsymbol{H}, \mathscr{E}_0, \mathscr{D}_0)$ *as in (3.5.10).*

Proof We can proceed as in the proof of Theorem 3.5.1 for all parts except for the stability of the weak limits $u_0(t)$, since now weak convergence is not enough to pass to the limit in $\Psi_\varepsilon(w_\varepsilon - u_\varepsilon(t))$, but for the upper energy estimate we need $\Psi_\varepsilon \overset{\Gamma}{\rightharpoonup} \Psi_0$.

We prove the desired stability of $u_0(t)$ under the additional assumption $\mathscr{F}_\varepsilon(u) \leq C_2\|u\|^2$. For the general case we refer to [37] and [53, Chap. 3.5.4]. Then, for a test state v we can choose the recovery sequence $v_\varepsilon = A_\varepsilon^{-1}A_0 v$. Indeed, since v_ε minimizes the functional $u \mapsto \mathscr{F}_\varepsilon(u) - \langle A_0 w, u\rangle$ and $\mathscr{F}_\varepsilon \overset{\mathsf{M}}{\to} \mathscr{F}_0$ we conclude $v_\varepsilon \to v$ and $\mathscr{F}_\varepsilon(v_\varepsilon) \to \mathscr{F}_0(v)$ by a variant of Theorem 3.2.1.

Using the stability of $u_\varepsilon(t)$ we have, for all w_ε,

$$0 \leq \mathscr{E}_\varepsilon(t, w_\varepsilon) - \mathscr{E}_\varepsilon(t, u_\varepsilon) + \Psi_\varepsilon(w_\varepsilon - u_\varepsilon(t)). \tag{3.5.13}$$

For a given limit test state w we choose $w_\varepsilon = u_\varepsilon + A_\varepsilon^{-1}A_0(w - u_0)$, which guarantees

$$w_\varepsilon \rightharpoonup w \quad \text{and} \quad w_\varepsilon - u_\varepsilon \to w - u_0, \text{ giving } \Psi_\varepsilon(w_\varepsilon - u_\varepsilon) \to \Psi_0(w - u_0), \tag{3.5.14}$$

because of $\Psi_\varepsilon \overset{\mathsf{C}}{\to} \Psi_0$. Moreover, using the quadratic structure of $\mathscr{E}_\varepsilon$ we have

$$\begin{aligned} \mathscr{E}_\varepsilon(t, w_\varepsilon) - \mathscr{E}_\varepsilon(t, u_\varepsilon) &= \frac{1}{2}\langle A_\varepsilon(w_\varepsilon - u_\varepsilon), w_\varepsilon + u_\varepsilon\rangle - \langle \ell_\varepsilon, w_\varepsilon - u_\varepsilon\rangle \\ &= \frac{1}{2}\langle A_0(w - u_0), w_\varepsilon + u_\varepsilon\rangle - \langle \ell_\varepsilon, w_\varepsilon - u_\varepsilon\rangle \end{aligned}$$

$$\to \frac{1}{2}\langle A_0(w-u_0), w_0+u_0\rangle - \langle \ell_\varepsilon, w_0-u_0\rangle \;=\; \mathcal{E}_0(t,w)-\mathcal{E}_0(t,u_0).$$

Using this convergence and (3.5.14) in (3.5.13) yields the desired stability of $u_0(t)$. □

The following simple example from [55] shows that Mosco convergence of $\mathcal{E}_\varepsilon$ and Ψ_ε is not sufficient for evolutionary Γ-convergence even in finite dimensions.

Example 3.5.3 For $\varepsilon > 0$ consider the RIS $(\boldsymbol{H}, \mathcal{E}_\varepsilon, \Psi_\varepsilon)$ with

$$\boldsymbol{H} = \mathbb{R}^2, \quad \mathcal{E}_\varepsilon(t,q) = \frac{1}{2}q_1^2 + \frac{1}{2}(q_1 - \frac{q_2}{\varepsilon})^2 - tq_1, \quad \mathcal{R}_\varepsilon(v) = |v_1| + \tfrac{1}{\varepsilon^2}|v_2|.$$

In $\mathbb{R}^2$ all Γ-limits are Mosco limits, i.e. we have $\mathcal{E}_\varepsilon \overset{\mathrm{M}}{\to} \mathcal{E}_0$ and $\mathcal{R}_\varepsilon \overset{\mathrm{M}}{\to} \mathcal{R}_0$ with

$$\mathcal{E}_0(t,q) = \begin{cases} \frac{1}{2}q_1^2 - tq_1 & \text{for } q_2 = 0, \\ \infty & \text{for } q_2 \neq 0; \end{cases} \quad \text{and } \mathcal{R}_0(v) = \begin{cases} |v_1| & \text{for } v_2 = 0, \\ \infty & \text{for } v_2 \neq 0. \end{cases}$$

For the unique solutions q_ε with $q_\varepsilon(0) = 0$ we find

$$q_0(t) = \begin{pmatrix} \max\{t-1, 0\} \\ 0 \end{pmatrix} \neq \lim_{\varepsilon\to 0} q_\varepsilon(t) = \begin{pmatrix} \max\{0, t/2 - 1\} \\ 0 \end{pmatrix},$$

which contradicts evolutionary Γ-convergence.

Remark 3.5.1 (Mutual recovery sequences) In both convergence results the important step in the proof is the stability of the limits $u_0(t)$. The same problem already appears in the existence theory via the time-incremental minimization (3.5.7). Thus, in [53, 55] the notion of *mutual recovery sequences* $(w_\varepsilon)_{\varepsilon>0}$ is introduced for a given stable sequence u_ε and a test state w by asking for the condition

$$\limsup_{\varepsilon\to 0} \big(\mathcal{E}_\varepsilon(t,w_\varepsilon) - \mathcal{E}_\varepsilon(t,u_\varepsilon) + \mathcal{D}_\varepsilon(w_\varepsilon,u_\varepsilon)\big) \;\le\; \mathcal{E}_0(t,w) - \mathcal{E}_0(t,u_0) + \mathcal{D}_0(w,u_0),$$

see the occurrence of w_ε in (3.5.11) and (3.5.13) in the proofs of Theorems 3.5.1 and 3.5.2, respectively. An important nontrivial construction in that direction was the so-called *jump transfer lemma* for rate-independent models for brittle fracture, see [13, 21].

The above abstract results have a variety of applications. In [56] it is shown that linearized elastoplasticity can be obtained by pE-convergence from finite-strain elastoplasticity in the limit of small loadings and small yield stress. In [22] brittle fracture is obtained as evolutionary Γ-limit of the Ambrosio-Tortorelli approximation, while [23] discusses homogenization for fracture evolution.

Remark 3.5.2 (Numerical approximation of RIS)

Both of the above convergence results can be used to establish convergence of numerical schemes for a given RIS $(\boldsymbol{X}, \mathcal{E}, \mathcal{D})$, see [42] for various versions. The main idea is to consider the incremental minimization problem (3.5.7) restricted to finite-dimensional subspaces $\boldsymbol{X}_m$ of the underlying Banach space $\boldsymbol{X}$, to keep the distance $\mathcal{D}$ or Ψ independent of m, while defining $\mathcal{E}_m(t,u) = \mathcal{E}(t,u)$ for $u \in \boldsymbol{X}_m$ and $\mathcal{E}_m(t,u) = \infty$ for $u \in \boldsymbol{X} \setminus \boldsymbol{X}_m$.

We say that the subspaces $\boldsymbol{X}_m$ approximate $\boldsymbol{X}$, if for each $u \in \boldsymbol{X}$ there exists $u_m \in \boldsymbol{X}_m$ such that $u_m \to u$ for $m \to \infty$. Under mild conditions on $\mathcal{E}(t,\cdot)$ one can show that $\mathcal{E}_m \overset{\mathsf{M}}{\to} \mathcal{E}$ in $\boldsymbol{X}$, and then the above theorems are applicable.

Piecewise constant interpolant $\overline{u}^\tau$

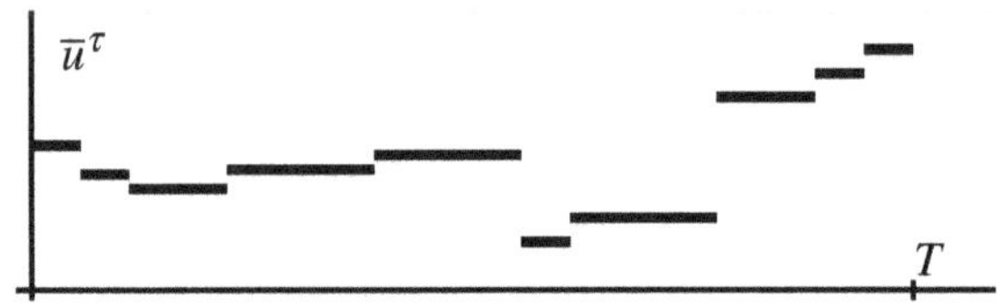

In fact, one can strengthen the result for the piecewise constant approximants $\overline{u}^{\tau,m}$ by proving that the joint limit $\tau \to 0$ and $m \to \infty$ always has a subsequence converging to a limit u_0 which is an energetic solution for $(\boldsymbol{X}, \mathcal{E}, \mathcal{D})$, cf. [25, 42, 55].

In [10] the general interplay between incremental minimization with time step $\tau > 0$ and Γ-convergence of the functionals $\mathcal{E}_\varepsilon$ and Ψ_ε for $\varepsilon \to 0$ is studied.

A completely different approach to pE-convergence for RIS is developed in [60] which involves the notion of parametrized solutions for RIS, which is a variant of the *balanced-viscosity solutions* defined in [50, 52]. The convergence result in [60] is based on an adaptation of the theory in Sect. 3.3.3 using the energy-dissipation principle.

Acknowledgments The DFG within the SFB 910 (subproject A5) and by the ERC under AdG 267802 AnaMultiScale. These notes are based on lectures given at Twente University in June 2012 and at the Centro De Giorgi (Pisa) in November 2013. The author gives special thanks to Karoline Disser, Matthias Liero, Michiel Renger, Sina Reichelt, and Marita Thomas for reading and considerably improving earlier versions of this work. Moreover, he is grateful for stimulating discussions with Dorothee Knees, Mark Peletier, Riccarda Rossi, Tomáš Roubíček, Giuseppe Savaré, Ulisse Stefanelli, Lev Truskinovsky, and Augusto Visintin.

References

1. Alber, H.-D.: Materials with memory. Lecture Notes in Mathematics, vol. 1682. Springer-Verlag, Berlin (1998)
2. Ambrosio, L., Gigli, N., Savaré, G.: Gradient flows in metric spaces and in the space of probability measures. Lectures in Mathematics ETH Zürich. Birkhäuser Verlag, Basel (2005)
3. Arnrich, S., Mielke, A., Peletier, M.A., Savaré, G., Veneroni, M.: Passing to the limit in a Wasserstein gradient flow: from diffusion to reaction. Calc. Var. Part. Diff. Eqns. **44**, 419–454 (2012)

4. Attouch, H.: Variational Convergence of Functions and Operators. Pitman Advanced Publishing Program, Pitman (1984)
5. Bellettini, G., Bertini, L., Mariani, M., Novaga, M.: Convergence of the one-dimensional Cahn-Hilliard equation. SIAM J. Math. Anal. **44**(5), 3458–3480 (2012)
6. Bénilan, P.: Solutions intégrales d'équations d'évolution dans un espace de Banach, C. R. Acad. Sci. Paris Sér. A-B **274**, A47–A50 (1972)
7. Biot, M.A.: Variational principles in irreversible thermodynamics with applications to viscoelasticity. Phys. Rev. **97**(6), 1463–1469 (1955)
8. Braides, A.: G-Convergence for Beginners. Oxford University Press (2002)
9. Braides, A.: A handbook of g-convergence. In: Handbook of Differential Equations. Stationary Partial Differential Equations, vol. 3 (2006)
10. Braides, A.: Local minimization, variational evolution and gamma-convergence. Lecture Notes in Mathematics, vol. 2094. Springer (2013)
11. Brézis, H.: Opérateurs Maximaux Monotones et Semi-groupes de Contractions dans les espaces de Hilbert. North-Holland Publishing Co., Amsterdam (1973)
12. Brokate, M., Krejč'i, P., Schnabel, H.: On uniqueness in evolution quasivariational inequalities. J. Convex Anal. **11**, 111–130 (2004)
13. Dal Maso, G., Francfort, G.A., Toader, R.: Quasistatic crack growth in nonlinear elasticity. Arch. Rational Mech. Anal. **176**, 165–225 (2005)
14. Dal Maso, G.: An Introduction to G-convergence. Birkhäuser Boston Inc., Boston (1993)
15. Daneri, S., Savaré, G.: Lecture notes on gradient flows and optimal transport (2010)
16. De Giorgi, E., Franzoni, T.: Su un tipo di convergenza variazionale. Atti Accad. Naz. Lincei Rend. Cl. Sci. Fis. Mat. Natur. (8) **58**(6), 842–850 (1975)
17. De Giorgi, E., Marino, A., Tosques, M.: Problems of evolution in metric spaces and maximal decreasing curve. Atti Accad. Naz. Lincei Rend. Cl. Sci. Fis. Mat. Natur. (8) **68**(3), 180–187 (1980)
18. Ekeland, I., Temam, R.: Convex Analysis and Variational Problems. North Holland (1976)
19. Feireisl, E., Novotný, A. :Singular Limits in Thermodynamics of Viscous Fluids. Birkhäuser (2009)
20. Fenchel, W.: On conjugate convex functions. Canadian J. Math. **1**, 73–77 (1949)
21. Francfort, G.A., Larsen, C.J.: Existence and convergence for quasi-static evolution of brittle fracture. Commun. Pure Appl. Math. **56**, 1495–1500 (2003)
22. Giacomini, A.: Ambrosio-Tortorelli approximation of quasistatic evolution of brittle fractures. Calc. Var. Part. Diff. Eqns. **22**, 129–172 (2005)
23. Giacomini, A., Ponsiglione, M.: AG-convergence approach to stability of unilateral minimality properties in fracture mechanics and applications. Arch. Ration. Mech. Anal. **180**(3), 399–447 (2006)
24. Giacomini, A., Musesti, A.: Two-scale homogenization for a model in strain gradient plasticity. ESAIM Control Optim. Calc. Var. **17**(4), 1035–1065 (2011). Published online. doi:10.1051/cocv/2010036
25. Giacomini, A., Ponsiglione, M.: Discontinuous finite element approximation of quasistatic crack growth in nonlinear elasticity. Math. Models Meth. Appl. Sci. (M3AS) **16**, 77–118 (2006)
26. Gigli, N., Maas, J.: Gromov-Hausdorff convergence of discrete transportation metrics. SIAM J. Math. Anal. **45**(2), 879–899 (2013)
27. Glitzky, A.: An electronic model for solar cells including active interfaces and energy resolved defect densities. SIAM J. Math. Anal. **44**, 3874–3900 (2012)
28. Glitzky, A., Mielke, A.: A gradient structure for systems coupling reaction-diffusion effects in bulk and interfaces. Z. Angew. Math. Phys. (ZAMP) **64**, 29–52 (2013)
29. Hanke, H.: Homogenization in gradient plasticity. Math. Models Meth. Appl. Sci. **21**(8), 1651–1684 (2011)
30. Ioffe, A.D.: On lower semicontinuity of integral functionals. I. SIAM J. Control Optim. **15**(4), 521–538 (1977)
31. James, R.D.: Hysteresis in Phase Transformations, Iciam 95 (hamburg, 1995), pp. 135–154 1996

32. Jordan, R., Kinderlehrer, D., Otto, F.: The variational formulation of the Fokker-Planck equation. SIAM J. Math. Anal. **29**(1), 1–17 (1998)
33. Krejč'i, P.: Evolution variational inequalities and multidimensional hysteresis operators. In: Nonlinear Differential Equations, pp. 47–110 (1999)
34. Le, N.Q.: A gamma-convergence approach to the Cahn-Hilliard equation. Calc. Var. Part. Diff. Eqns. **32**(4), 499–522 (2008)
35. Liero, M.: Variational methods for evolution. Ph.D. Thesis (2012)
36. Liero, M.: Passing from bulk to bulk-surface evolution in the Allen-Cahn equation. Nonl. Diff. Eqns. Appl. (NoDEA) **20**(3), 919–942 (2013)
37. Liero, M., Mielke, A.: An evolutionary elastoplastic plate model derived via G-convergence. Math. Models Meth. Appl. Sci. **21**(9), 1961–1986 (2011)
38. Liero, M., Mielke, A., Peletier, M.A., Renger, D.R.M.: On the microscopic origin of generalized gradient structures. In preparation (2015)
39. Liero, M., Mielke, A., Savaré, G. :Optimal transport in competition with reaction-the Hellinger-Kantorovich distance. In preparation (2014)
40. Maas, J.: Gradient flows of the entropy for finite Markov chains. J. Funct. Anal. **261**, 2250–2292 (2011)
41. Mielke, A.: Evolution in rate-independent systems (Ch. 6). In: Handbook of Differential Equations, Evolutionary Equations, vol. 2, pp. 461–559 (2005)
42. Mielke, A., Roubíček, T.: Numerical approaches to rate-independent processes and applications in inelasticity. Math. Model. Numer. Anal. **43**, 399–428 (2009)
43. Mielke, A.: Differential, energetic, and metric formulations for rate-independent processes. Nonlinear pde's and applications, pp. 87–170 (2011). (C.I.M.E. Summer School, Cetraro, Italy 2008, Lect. Notes Math. Vol. 2028)
44. Mielke, A.: A gradient structure for reaction-diffusion systems and for energy-drift-diffusion systems. Nonlinearity **24**, 1329–1346 (2011)
45. Mielke, A.: Emergence of rate-independent dissipation from viscous systems with wiggly energies. Contin. Mech. Thermodyn. **24**(4), 591–606 (2012)
46. Mielke, A.: Thermomechanical modeling of energy-reaction-diffusion systems, including bulk-interface interactions. Discr. Cont. Dynam. Systems Ser. S **6**(2), 479–499 (2013)
47. Mielke, A.: Deriving amplitude equations via evolutionary G-convergence. Discr. Cont. Dynam. Systems Ser. A **35**(6) (2015)
48. Mielke, A., Ortner, C., Şengül, Y.: An approach to nonlinear viscoelasticity via metric gradient flows. SIAM J. Math. Analysis **46**(2), 1317–1347 (2014)
49. Mielke, A., Peletier, M., Renger, M.: On the relation between gradient flows and the largedeviation principle, with applications to Markov chains and diffusion. Potential Anal. **41**(4), 1293–1327 (2014)
50. Mielke, A., Rossi, R., Savaré, G.: Modeling solutions with jumps for rate-independent systems on metric spaces. Discr. Cont. Dynam. Systems Ser. A **25**(2), 585–615 (2009)
51. Mielke, A., Rossi, R., Savaré, G.: Nonsmooth analysis of doubly nonlinear evolution equations. Calc. Var. Part. Diff. Eqns. **46**(1–2), 253–310 (2013)
52. Mielke, A., Rossi, R., Savaré, G.: Balanced-viscosity (BV) solutions to infinite-dimensional rate-independent systems. J. Europ. Math. Soc. (2014). To appear. (WIAS preprint 1845). http://arxiv.org/abs/1309.6291
53. Mielke, A., Roubíček, T.: Rate-Independent Systems: Theory and Application. Springer (2015)
54. Mielke, A., Roubíček, T., Thomas, M.: From damage to delamination in nonlinearly elastic materials at small strains. J. Elasticity **109**, 235–273 (2012)
55. Mielke, A., Roubíček, T., Stefanelli, U.: G-limits and relaxations for rate-independent evolutionary problems. Calc. Var. Part. Diff. Eqns. **31**, 387–416 (2008)
56. Mielke, A., Stefanelli, U.: Linearized plasticity is the evolutionary G-limit of finite plasticity. J. Eur. Math. Soc. **15**(3), 923–948 (2013)
57. Mielke, A., Thomas, M.: GENERIC-a powerful tool for thermomechanical modeling. In preparation (2014)

58. Mielke, A., Timofte, A.M.: Two-scale homogenization for evolutionary variational inequalities via the energetic formulation. SIAM J. Math. Anal. **39**(2), 642–668 (2007)
59. Mielke, A., Truskinovsky, L.: From discrete visco-elasticity to continuum rate-independent plasticity: rigorous results. Arch. Rational Mech. Anal. **203**(2), 577–619 (2012)
60. Negri, M.: Quasi-static rate-independent evolutions: characterization, existence, approximation and application to fracture mechanics. ESAIM Control Optim. Calc. Var. **20**(4), 983–1008 (2014)
61. Onsager, L., Machlup, S.: Fluctuations and irreversible processes. Phys. Rev. **91**(6), 1505–1512 (1953)
62. Onsager, L.: Reciprocal relations in irreversible processes, I+II. Phys. Rev. **37**, 405–426. (part II, 38:2265–2279) (1931)
63. Otto, F.: The geometry of dissipative evolution equations: the porous medium equation. Commun. Partial Different. Eqns. **26**, 101–174 (2001)
64. Prandtl, L.: Gedankenmodel zur kinetischen Theorie der festen Körper. Z. angew. Math. Mech. (ZAMM) **8**, 85–106 (1928)
65. Puglisi, G., Truskinovsky, L.: Thermodynamics of rate-independent plasticity. J. Mech. Phys. Solids **53**, 655–679 (2005)
66. Rockafellar, R.T.: Convex Analysis. Princeton University Press (1970)
67. Rossi, R., Savaré, G.: Gradient flows of non convex functionals in Hilbert spaces and applications. ESAIM Control Optim. Calc. Var. **12**, 564–614 (2006)
68. Roubíček, T.: Nonlinear partial differential equations with applications. Birkhäuser Verlag, Basel (2005)
69. Sandier, E., Serfaty, S.: Gamma-convergence of gradient flows with applications to Ginzburg-Landau. Comm. Pure Appl. Math. **LVII**, 1627–1672 (2004)
70. Savaré, G.: Gradient flows and diffusion semigroups in metric spaces under lower curvature bounds (2011)
71. Serfaty, S.: Gamma-convergence of gradient flows on Hilbert spaces and metric spaces and applications. Discr. Cont. Dynam. Syst. Ser. A **31**(4), 1427–1451 (2011)
72. Stefanelli, U.: The Brezis-Ekeland principle for doubly nonlinear equations. SIAM J. Control Optim. **47**(3), 1615–1642 (2008)
73. Tartar, L.: Nonlocal effects induced by homogenization. Partial Different. Eqns. Calcul. Variat. **2**, 925–938 (1989)
74. Tartar, L.: Memory effects and homogenization. Arch. Rational Mech. Anal. **111**(2), 121–133 (1990)

Chapter 4
Homoclinic Points of Principal Algebraic Actions

Martin Göll and Evgeny Verbitskiy

Abstract The 1999 paper by D. Lind and K. Schmidt on homoclinic points of a special class of dynamical systems—the so called algebraic $\mathbb{Z}^d$-actions—attracted a lot of interest to the study of homoclinic points. In the present paper we review the developments over the past 15 years. Major progress has been made in questions of existence of homoclinic points for $\mathbb{Z}^d$-actions. More recently, first results were obtained for actions of non-abelian discrete groups. Summable homoclinic points were successfully used in the computation of entropy and in the study of probabilistic properties of dynamical systems, e.g., Central Limit Theorems. Moreover, homoclinic points allow one to construct coding maps which link certain particle systems to algebraic dynamical systems.

4.1 Introduction

4.1.1 Historic Perspective

The notion of homoclinic points goes back to Poincaré's work on the stability analysis of the three-body problem at the end of the nineteenth century, see [1, 45] for a historical account. Two points x, y of a metric phase space (X, d) are called *homoclinic* under the transformation $T : X \longrightarrow X$ if

---

MG gratefully acknowledges support by a Huygens Fellowship from Leiden University.
MG and EV are grateful to M. Gordin, D. Lind and K. Schmidt for illuminating discussions.

M. Göll (✉) · E. Verbitskiy
Mathematical Institute, University of Leiden, Box 9512, 2300 RA
Leiden, The Netherlands
e-mail: gollm@math.leidenuniv.nl

E. Verbitskiy
Johann Bernoulli Institute for Mathematics and Computer Science,
University of Groningen, Box 407, 9700 AK Groningen, The Netherlands
e-mail: e.a.verbitskiy@rug.nl

A. Muntean et al. (eds.), *Macroscopic and Large Scale Phenomena: Coarse Graining, Mean Field Limits and Ergodicity*, Lecture Notes in Applied Mathematics and Mechanics 3, DOI 10.1007/978-3-319-26883-3_4

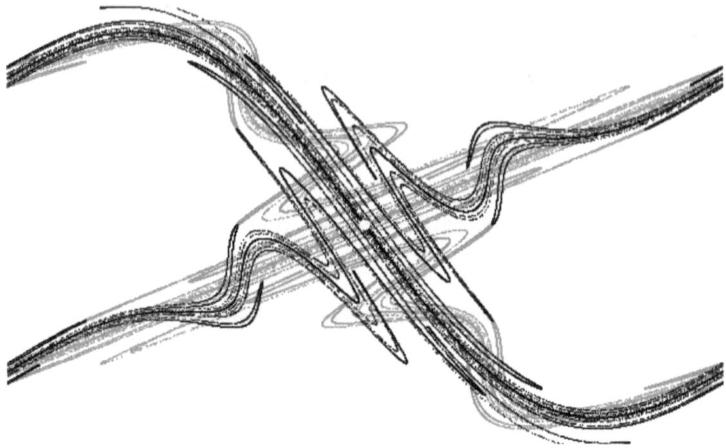

Fig. 4.1 Homoclinic tangle for the hyperbolic standard map on the two-dimensional torus $r_{n+1} = r_n - \frac{K}{2\pi}\sin(2\pi\theta_n) \mod 1, \theta_{n+1} = \theta_n + r_n \mod 1$. The white dot indicates the location of a hyperbolic fixed point. The image was created using the Mathematica package accompanying [49]

$$d(T^t(x), T^t(y)) \to 0 \text{ as } t \to \pm\infty. \tag{4.1.1}$$

Often, one of the points (say y) is fixed by the dynamics, and one is interested in all points x which are doubly asymptotic to y in the sense of (4.1.1). Poincaré observed that if for a two-dimensional diffeomorphism, stable and unstable manifolds of a hyperbolic fixed point intersect transversally, then the intersection point is obviously homoclinic to the fixed point, but more importantly, then there is also a countable set of points homoclinic to the fixed point. This is known as a homoclinic tangle which is formed by an intricate web of intersections of stable and unstable manifolds, see Fig. 4.1. Homoclinic tangles give rise to a Smale horseshoe, and hence, to chaotic behaviour. Homoclinic points are studied extensively in the theory of smooth dynamical systems. However, the notion has proved to be useful in symbolic dynamics as well: e.g., the equilibrium Gibbs measures on symbolic spaces can be defined naturally using homoclinic points [9, 40, 44].

4.1.2 Spatial Dynamics and Group Actions

In the field of Dynamics as well as in Physics one calls models in which states evolve over time dynamical systems. Choose a set X—the *state space*, *phase space* or *configuration space*—and an invertible transformation $T : X \longrightarrow X$. The additive group $\mathbb{Z}$ acts in a canonical way on the system, i.e., for each $n \in \mathbb{Z}$, $n \mapsto T^n$. Obviously, the state space X is *invariant* under this action of $\mathbb{Z}$.

Let us now demonstrate spatial dynamical systems by means of examples. We start with quite trivial models.

Example 4.1.1 In Coding and Information Theory one usually represents information by (infinite) strings of symbols. Let Σ_N be the set of all sequences $(x_n)_{n\in\mathbb{Z}}$ with $x_n \in \{0, 1, \ldots, N\}$, for every $n \in \mathbb{Z}$. Hence each site of $\mathbb{Z}$ is in one of $N + 1$ possible states represented by the set $\{0, 1, \ldots, N\}$. Sequences of zeroes and ones (i.e., Σ_1) can be used to model the information saved on a hard disc. The information stored in DNA can be represented by Σ_3. In both examples the information is stored at ('spatial') sites of a stripe, which we identify with $\mathbb{Z}$.

Fix Σ_N with $N \in \mathbb{N}$ and $m \in \mathbb{Z}$. One can consider a translate $\sigma^m x$ of a sequence $x = (x_n)_{n\in\mathbb{Z}} \in \Sigma_N$ which is given by $(\sigma^m x)_n = x_{n+m}$ for each $n \in \mathbb{Z}$. The result of this shift is again an allowed configuration in Σ_N which usually represents different information.

We give now examples of spatially extended symbolic systems.

Example 4.1.2 Let each site of the lattice $\mathbb{Z}^2$ be in one of the two states $\{0, 1\}$. Denote the set of all configurations by X_2. Models which can be described by such configurations are spin models in Statistical Mechanics or a two-dimensional data storage medium, which is actually more natural than the representation via stripes. The group $\mathbb{Z}^2$ acts on X_2 via translations σ, i.e., for each $\mathbf{m} \in \mathbb{Z}^2$, $\mathbf{m} \mapsto \sigma^{\mathbf{m}}$; where $(\sigma^{\mathbf{m}} x)_{\mathbf{n}} = x_{\mathbf{m}+\mathbf{n}}$, for each $x \in X_2$.

The next specific example is of great interest in Statistical Mechanics and the field of Combinatorics (cf. [16]).

Example 4.1.3 Tile the lattice $\mathbb{Z}^2$ with dominoes in such a way that each vertex of $\mathbb{Z}^2$ is occupied with exactly one half of a domino (cf. Fig. 4.2).

Consider the set $\mathscr{D}$ of all possible domino tilings of the lattice $\mathbb{Z}^2$. Let η be a typical element in the state space $\mathscr{D}$. It is clear that any translation of η by a vector in $\mathbb{Z}^2$ will lead to a configuration in $\mathscr{D}$. Hence, $\mathscr{D}$ is invariant under translations and $(\mathscr{D}, \sigma_{\mathscr{D}})$ forms a dynamical system, where $\sigma_{\mathscr{D}}$ denotes the shift action on $\mathscr{D}$.

The Ising model and other models in Statistical Mechanics as well as certain systems in Probability Theory, like percolation models, are typical examples of spatially extended lattice systems. In the passage to multi-parameter systems new phenomena, like phase transitions which are completely absent for 1-dimensional systems with

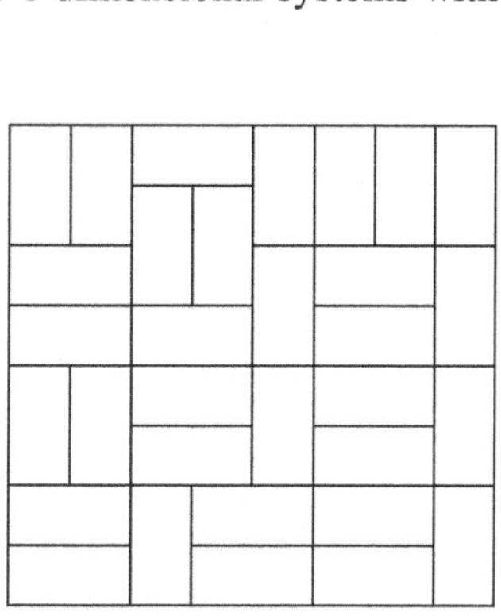

Fig. 4.2 A possible domino tiling of the 8×8 lattice

short range interactions, arise. Hence, the study of equilibrium measures (which are called equilibrium states in Statistical Mechanics), those translation invariant measures which maximize entropy, becomes complicated for multi-dimensional systems. The presence of correlations over long distances in particle systems with short range interaction is another phenomenon of interest which only occurs in higher dimensions. Moreover, the transition to higher-dimensional lattices leads to certain undecidability problems (cf. [2]) which are mainly of mathematical interest.

Beside purely temporal and purely spatial dynamical systems there are temporal spatial dynamical systems as well. This class of dynamical systems contains certain cellular automata (cf. Example 4.7.3) and coupled map lattices (cf. [7]).

Other temporal spatial dynamical systems are generated from systems of ordinary or partial differential equations. In the study of reaction-diffusion systems invariant subspaces arise which display (temporal spatial) chaotic behaviour (cf. [54]). We would like to point out that in [38, Sect. 2] spatial homoclinic points are the subject of investigation. In particular, the appearance of chaos is linked to the existence of (transversal) homoclinic points.

In [37] the Ginzburg-Landau formalism is mentioned several times in the context of Γ-convergence. For the Ginzburg-Landau equation homoclinic orbits were investigated in [17] in order to understand solutions that are aperiodic in time and space.

Interesting examples of dynamical systems are not restricted to actions of the groups $\mathbb{Z}$ and $\mathbb{R}$, but also include multivariate actions ($\mathbb{Z}^d$, $\mathbb{R}^d$), as well as actions of non-commutative groups (e.g., SO(3) or the discrete Heisenberg group, which will be discussed in greater detail below).

Definition 4.1.1 Let X be a set and denote by $\mathrm{Bij}(X)$ the group of all bijections of X. Further, let Γ be a group. A Γ-action (X, α) is a homomorphism $\alpha : \Gamma \longrightarrow \mathrm{Bij}(X)$, i.e.,

$$\alpha(\gamma\gamma') = \alpha(\gamma) \circ \alpha(\gamma') \tag{4.1.2}$$

for every $\gamma, \gamma' \in \Gamma$.

Example 4.1.4 Fix $N \in \mathbb{N}$. The shift σ^n is clearly a bijection on Σ_N, for every $n \in \mathbb{Z}$. The shift action is given by $\sigma : n \mapsto \sigma(n) = \sigma^n$. Hence, (Σ_N, σ) is a $\mathbb{Z}$-action. Condition (4.1.2) means that for every $x \in \Sigma_N$ and every $k, l, m \in \mathbb{Z}$

$$(\sigma^k \sigma^l x)_m = (\sigma^{k+l} x)_m.$$

In other words, the result of the composition of two translations σ^k and σ^l is the same as the translation by σ^{k+l}.

Dynamics and in particular Modern Dynamics deals with the asymptotic behaviour of Γ-actions on spaces X via bijective transformations $T : X \longrightarrow X$, which leave some structure of X invariant.

Topological Dynamics deals with actions of groups on (compact) metric spaces (X, d) via homeomorphisms, hence, those transformations which leave the topology invariant.

In Ergodic Theory one is interested in actions of a group Γ on a measure space (X, μ) via invertible measure-preserving transformations. Ergodic Theory arose from certain problems in Statistical Mechanics. One of the main ideas in Ergodic Theory and Dynamics is to study and give information about the average behaviour of a system. Fluctuations of the dynamics are usually averaged out by a law of large number. This principle was already pointed out and applied in [26]. Several techniques and tools presented in other chapters of the book follow the same approach. Assume that one wants to describe an inhomogeneous material. One way to model this situation is to consider PDEs with rapidly oscillating coefficients. The homogenization procedure (cf. [37]) consists of cancelling out fluctuations to obtain averaged equations. In other words, one replaces the oscillating parameters by uniform ones.

In order to complete the picture we mention Γ-actions, where Γ is neither $\mathbb{Z}^d$ nor $\mathbb{R}^d$, which are of great interest in Modern Dynamics due to their applications in other fields. Certain actions of $\mathrm{SL}(2, \mathbb{R})$ on Teichmüller spaces are used to describe billiards. Moreover, actions of certain subgroups of semisimple Lie groups or of linear algebraic groups over local fields were successfully applied in Number Theory to (partially) solve various problems involving

- Diophantine approximation;
- quantum chaos, quantum unique ergodicity on arithmetic surfaces;
- counting problems.

4.1.3 Algebraic Dynamics

We introduce in this subsection Γ-actions which have a state space with an algebraic structure. The study of homoclinic points is particularly fruitful in case of *algebraic dynamical systems*. For such systems, the underlying phase space X is a compact abelian additive group equipped with a translation invariant metric $d : X \times X \longrightarrow \mathbb{R}_+$, i.e., for each $x, y, z \in X$ one has $d(x + z, y + z) = d(x, y)$. We denote by λ_X the *Haar measure* on X, i.e., the unique non-trivial measure (up to a multiplicative positive constant) which is invariant under translations: for all $x \in X$ and all measurable (Borel) sets A, $\lambda_X(A) = \lambda_X(x + A)$.

The group $\mathrm{Aut}(X)$ consists of all continuous bijective homomorphisms of the group X. The dynamics is given by an action of a countable discrete group Γ by continuous group automorphisms of X, i.e., a group homomorphism

$$\alpha : \Gamma \longrightarrow \mathrm{Aut}(X) : \gamma \mapsto \alpha^{\gamma}.$$

Here, $\alpha^{\gamma} \in \mathrm{Aut}(X)$ denotes the image of $\gamma \in \Gamma$.

We will refer to the pair (X, α) as an *algebraic dynamical system* or *algebraic action* of Γ. In this review we will consider only finitely generated countable groups Γ.

The advantages of algebraic actions are based on the fact that they are connected to other branches of mathematics like Harmonic Analysis and Algebra. These interesting links allow one to use tools and methods which are not available for general Γ-actions. In the case of algebraic $\mathbb{Z}^d$-actions methods from Commutative Algebra and Fourier Analysis were successfully used to understand most of their dynamical properties.

The text book examples of algebraic dynamical systems are commuting toral automorphisms and in particular actions of $\mathbb{Z}$ via a single toral automorphism.

Denote the torus by $\mathbb{T} = \mathbb{R}/\mathbb{Z} \simeq [0, 1)$. A (linear) toral automorphisms $T_A : \mathbb{T}^d \longrightarrow \mathbb{T}^d$ is given by

$$T_A(a_1, a_2, \dots, a_d) = A\mathbf{a} \mod 1,$$

for every $\mathbf{a} = (a_1, a_2, \dots, a_d)^\top \in \mathbb{T}^d$, where A is a $d \times d$ matrix with integer entries and $\det A = \pm 1$ (hence, A is invertible and volume preserving). Let us verify that these systems describe an algebraic $\mathbb{Z}$-action. Here, the phase space X is given by the compact abelian group $\mathbb{T}^d$ and the Haar measure λ_X is the Lebesgue-measure on $\mathbb{T}^d$. The group $\Gamma = \mathbb{Z}$ acts via iterations of A on $\mathbb{T}^d$. In other words, the action is given by a homomorphism α_{T_A} which maps $n \in \mathbb{Z}$ to $\alpha_{T_A}^n := T_A^n$. This is clearly a homomorphism because, for any $n, m \in \mathbb{Z}$,

$$\alpha_{T_A}^n \circ \alpha_{T_A}^m = T_A^n \circ T_A^m = T_A^{n+m} = \alpha_{T_A}^{n+m}.$$

Toral automorphisms naturally appear in Classical Mechanics, e.g., in the study of phase-diagrams. Consider

$$\begin{pmatrix} p_1 \\ x_1 \end{pmatrix} = A \begin{pmatrix} p_0 \\ x_0 \end{pmatrix} = \begin{pmatrix} a & b \\ c & d \end{pmatrix} \begin{pmatrix} p_0 \\ x_0 \end{pmatrix} \mod 1,$$

where $ad - bc = \pm 1$. The dynamics are given by iterations, hence after n time steps

$$\begin{pmatrix} p_n \\ x_n \end{pmatrix} = A^n \begin{pmatrix} p_0 \\ x_0 \end{pmatrix} \mod 1.$$

The invertibility of A allows us to consider negative n as well. Due to periodicity the phase-diagram will be on the two-torus $\mathbb{T}^2 = \mathbb{R}^2/\mathbb{Z}^2$.

Set $a = c = d = 1$ and $b = 0$. The corresponding dynamical system $n \mapsto A^n$ is Hamiltonian and describes the motion of a free particle since

$$p_n = p_0 \quad \text{, i.e., the momentum is constant and} \quad x_n = x_0 + np_0,$$

for every $n \in \mathbb{Z}$.

Consider next hyperbolic dynamical systems, i.e., the presence of expanding and contracting directions, which play a special role in Dynamics and Mechanics because they are typical examples of dynamical systems with chaotic behaviour. A toral

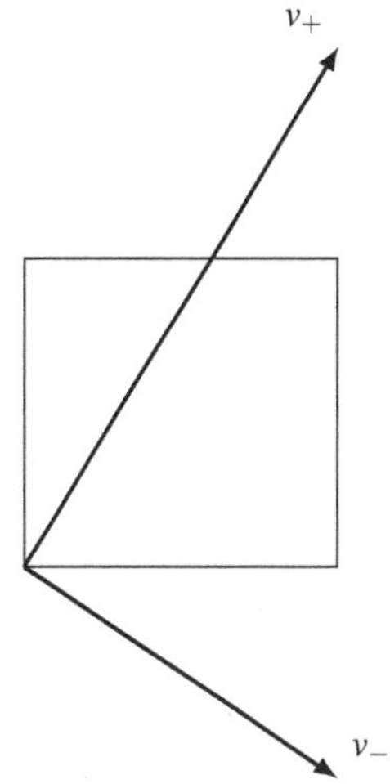

Fig. 4.3 The two eigenvectors of A

automorphism T_A is hyperbolic if and only if A has no eigenvalues of absolute value 1.

Example 4.1.5 Let $T_A : \mathbb{T}^2 \longrightarrow \mathbb{T}^2$ be the toral automorphism given by

$$T_A(a_1, a_2) = A \begin{pmatrix} a_1 \\ a_2 \end{pmatrix} = \begin{pmatrix} 0 & 1 \\ 1 & 1 \end{pmatrix} \begin{pmatrix} a_1 \\ a_2 \end{pmatrix} = \begin{pmatrix} a_2 \\ a_1 + a_2 \end{pmatrix} \mod 1.$$

Obviously, $\det A = -1$. The characteristic polynomial of A is given by

$$f_A(t) = t^2 - t - 1.$$

The eigenvalues of A are $\lambda_+ = \frac{1+\sqrt{5}}{2} > 1$ and $\lambda_- = \frac{1-\sqrt{5}}{2} < 1$ and the corresponding eigenvectors are given by $v_+ = (1, \lambda_+)^\top$ and $v_- = (1, \lambda_-)^\top$, respectively. Hence, $T_A : \mathbb{T}^2 \longrightarrow \mathbb{T}^2$ is a hyperbolic linear toral automorphism. The contracting (or stable) eigenspace E_- of $A : \mathbb{R}^2 \longrightarrow \mathbb{R}^2$ is given by $\{\lambda v_- : \lambda \in \mathbb{R}\}$ and the expanding (or unstable) eigenspace E_+ is given by $\{\lambda v_+ : \lambda \in \mathbb{R}\}$ (Fig. 4.3).

Note that f_A is also the characteristic polynomial of the linear recurrence relation given by

$$a_n = a_{n-1} + a_{n-2} \mod 1$$

and moreover

$$A \begin{pmatrix} a_n \\ a_{n+1} \end{pmatrix} = \begin{pmatrix} a_{n+1} \\ a_n + a_{n+1} \end{pmatrix} = \begin{pmatrix} a_{n+1} \\ a_{n+2} \end{pmatrix} \mod 1.$$

Hence, the characteristic polynomial f_A relates two different dynamical systems:

1. the $\mathbb{Z}$-action via the toral automorphism $T_A : \mathbb{T}^2 \longrightarrow \mathbb{T}^2$;
2. the algebraic dynamical system (X_{f_A}, α_{f_A}) defined as follows

$$X_{f_A} = \{(a_n)_{n\in\mathbb{Z}} \,:\, a_n \in \mathbb{T} \text{ and } a_{n+2} = a_{n+1} + a_n \mod 1, \text{ for every } n \in \mathbb{Z}\}.$$

One can easily check that for $a, b \in X_{f_A}$,

$$(a+b)_n = a_n + b_n \mod 1,$$

i.e., coordinate-wise addition, defines an element $a + b \in X_{f_A}$. Hence, X_{f_A} is an additive group. The shift action $\alpha_{f_A} : \mathbb{Z} \longrightarrow \mathrm{Aut}(X_{f_A})$ is given by $(\alpha_{f_A}^n a)_m = a_{n+m}$ for each $a \in X_{f_A}$ and $m \in \mathbb{Z}$.

The relation between these two systems can be written down explicitly

$$\begin{pmatrix} (\sigma_A^n a)_m \\ (\sigma_A^n a)_{m+1} \end{pmatrix} = \begin{pmatrix} a_{m+n} \\ a_{m+n+1} \end{pmatrix} = A^n \begin{pmatrix} a_m \\ a_{m+1} \end{pmatrix},$$

for every $a \in X_{f_A}$.

To make this more precise define the projection map $\pi_n : X_{f_A} \longrightarrow \mathbb{T}^2$ given by

$$\pi_n(a) = (a_n, a_{n+1})^\top \quad \text{for every } a \in X_{f_A}.$$

Since, for an arbitrary $n \in \mathbb{Z}$, $a \in X_{f_A}$ is fully determined by (a_n, a_{n+1}), π_n defines an isomorphism which respects the dynamics, i.e., the following diagram commutes

$$\begin{array}{ccc} X_{f_A} & \xrightarrow{\alpha_{f_A}} & X_{f_A} \\ {\scriptstyle \pi_n}\downarrow & & \downarrow{\scriptstyle \pi_n} \\ \mathbb{T}^2 & \xrightarrow{A} & \mathbb{T}^2 \end{array}$$

for every $n \in \mathbb{Z}$.

Remark 4.1.1 Example 4.1.5 can be generalized. Consider polynomials $f = c_0 + c_1 u + c_2 u^2 + \cdots + c_d u^d$ with integer coefficients $c_n, n \in \{1, 2, \ldots, d\}$, which fulfill $|c_0| = c_d = 1$. One can consider the toral automorphism $T_{C_f} : \mathbb{T}^d \longrightarrow \mathbb{T}^d$ defined by the companion matrix of f, i.e.,

$$C_f = \begin{pmatrix} 0 & 1 & 0 & 0 & \cdots & 0 \\ 0 & 0 & 1 & 0 & \cdots & \vdots \\ \vdots & \vdots & 0 & 1 & \ddots & \vdots \\ \vdots & \vdots & \vdots & \ddots & \ddots & 0 \\ 0 & 0 & 0 & 0 & \cdots & 1 \\ -c_0 & -c_1 & -c_2 & -c_3 & \cdots & -c_{d-1} \end{pmatrix},$$

as well as the $\mathbb{Z}$-action (X_f, α_f) as in Example 4.1.5. Moreover, the same construction as in Example 4.1.5 shows that these two dynamical systems associated to f are equivalent.

The algebraic dynamical system (X_f, α_f) can be defined for an *arbitrary* polynomial f with integer coefficients and will be called *principal algebraic* $\mathbb{Z}$-action.

The focus of this review will be on the class of principal algebraic Γ-actions. For a general discrete countable group Γ, and any element f in the integer group ring of Γ (cf. Sect. 4.2), one can associate an algebraic dynamical system (X_f, α_f) in a canonical way.

As in Example 4.1.5 the element f in the group ring will define a constraint on a full shift. These restrictions on the degrees of freedom provide the simplest sort of constrained systems. The dynamics will be given by translations of configurations. Hence, one expects these systems to be deterministic. But under very mild extra conditions (i.e., completely positive entropy) these systems are 'as random as a coin flip' (they are Bernoulli systems) as it was shown in [43].

The main advantage (convenience) of studying principal actions lies in the fact that interesting properties of the dynamics can be expressed in terms of algebraic properties of f (cf. [46]) like in Example 4.1.5.

We will see in the following sections that basically all dynamical properties and quantities of a principal Γ-action (X_f, α_f), such as

1. expansiveness (cf. Sect. 4.3);
2. the growth rate of the number of periodic points (cf. Sect. 4.7);
3. entropy (cf. Sect. 4.7);
4. mixing properties,

are encoded in an algebraic way in f.

Concepts and tools which were developed in order to understand hyperbolic toral automorphism play an important role in Classical Dynamics. Many ideas (e.g. symbolic covers) were picked up for the analysis in a more general setting. Our hope is that algebraic Γ-actions and in particular principal actions provide a similar groundwork for multi-parameter—and hence more complicated—dynamics.

We have already listed many spatially extended dynamical systems. Although these systems are all more or less of the same nature there is no common theory to describe and study these systems. Hence, it is very benificial to have classes of spatial dynamical systems (like principal algebraic dynamical systems) which are fairly well understood. Principal algebraic Γ-actions have the potential to serve as major examples of dynamical systems to investigate phenomena which only occure in higher dimensions. As we will see in the final Section of this contribution one can find explicit links of principal algebraic actions to *non*-algebraic systems which are driven by different dynamics than translations. These links are strong enough to deduce certain properties of the non-algebraic system from the study of the corresponding principal algebraic action.

4.1.4 Homoclinic Points in Algebraic Dynamical Systems

It is easy to see that for algebraic actions two points $x, \tilde{x} \in X$ are homoclinic if and only if their difference $x - \tilde{x}$ and 0_X are homoclinic. This allows to study points which are homoclinic to 0_X. Moreover, such points form a subgroup of X, the group of homoclinic points

$$\Delta_\alpha(X) = \{x \in X \,:\, d(\alpha^\gamma x, 0_X) \to 0 \text{ as } \gamma \to \infty\}.$$

If x is a homoclinic point, then so is every point in $\mathcal{O}_\Gamma(x) = \{\alpha^\gamma x \mid \gamma \in \Gamma\}$—the Γ-orbit of x. The point $x \in \Delta_\alpha(X)$ is called a *fundamental homoclinic point* if the homoclinic group $\Delta_\alpha(X)$ is generated by its orbit $\mathcal{O}_\Gamma(x)$, i.e., every homoclinic point is a finite linear combination of translates of x.

Often, one is interested in homoclinic points with sufficiently rapid decay of $d(\alpha^\gamma x, 0_X)$ as $|\gamma| \to \infty$. Namely, the group

$$\Delta^1_\alpha(X) = \{x \in X : \sum_{\gamma\in\Gamma} d(\alpha^\gamma x, 0_X) < \infty\}.$$

of *summable homoclinic points*, which plays an important role in constructing symbolic covers and other applications (see Sect. 4.7).

Similarly, one can define the group of p-summable homoclinic points for any $p \geq 1$

$$\Delta^p_\alpha(X) = \{x \in X : \sum_{\gamma\in\Gamma} \Big(d(\alpha^\gamma x, 0_X)\Big)^p < \infty\}.$$

In the case of a $\mathbb{Z}$-action via a single toral automorphism one easily obtains homoclinic points: every point in the intersection of the stable and unstable eigenspaces (after mapping them to $\mathbb{T}^2$) leads to a homoclinic point.

Example 4.1.6 Consider the matrix A which we studied in Example 4.1.5. Let

$$\tilde{E}_+ = \{\mathbf{a} \in \mathbb{T}^2 : \mathbf{a} = \mathbf{v} \mod 1, \text{ for some } \mathbf{v} \in E_+\},$$
$$\tilde{E}_- = \{\mathbf{a} \in \mathbb{T}^2 : \mathbf{a} = \mathbf{v} \mod 1, \text{ for some } \mathbf{v} \in E_-\}.$$

Then every point in $\tilde{E}_+ \cap \tilde{E}_-$ is homoclinic.

The principal question is whether a given algebraic dynamical system (X, α) admits relatively rich homoclinic structures. For example, under which conditions are the groups $\Delta_\alpha(X)$ and $\Delta^1_\alpha(X)$ non-empty, dense, etc.?

The importance of homoclinic groups is easily demonstrated by the following results, which imply strong mixing properties of the underlying algebraic actions:

- for $\Gamma = \mathbb{Z}$ and if $\Delta_\alpha(X)$ is dense in X, then due to [19] the action α on X is *mixing*, i.e., for all λ_X-measurable sets A, B the following holds

$$\lim_{n\to\infty} \lambda_X(\alpha^n(A) \cap B) = \lambda_X(A)\lambda_X(B).$$

  Even more, for general Γ the action is mixing of all orders (w.r.t. the Haar measure on X) due to [5, Proposition 4.6];
- if $\Delta^1_\alpha(X) \neq \varnothing$, then the action α has positive entropy. This result was communicated to us in private conversations by Lind and Schmidt. A similar result was established in [10] (see Sect. 4.7.4);
- if $\Gamma = \mathbb{Z}$ and $\Delta^1_\alpha(X) \neq \varnothing$, then any real-valued function ϕ on X of the form

$$\phi(z) = \Phi(z + x) - \Phi(z),$$

  where Φ is sufficiently smooth and $x \in \Delta^1_\alpha(X)$, satisfies the Central Limit Theorem:

$$\frac{1}{\sqrt{n}} \sum_{k=0}^{n-1} \phi(\alpha^k(z)) \to \mathcal{N}(0, \sigma^2).$$

We will discuss these and other properties which follow from the existence of large homoclinic groups extensively in the subsequent sections.

4.1.5 Outline

The outline of the present review is as follows. In Sect. 4.2 we define a large class of algebraic actions—the so-called principal actions. In Sect. 4.3 we discuss expansive algebraic actions. As we shall see, expansiveness implies existence of large homoclinic groups. Thus deciding on whether a (principal) action is expansive or not is an important question. For $\mathbb{Z}^d$-actions a complete characterization of expansive actions is available. Much less is known for non-abelian groups; we discuss some recent progress in case of the discrete Heisenberg group. In Sect. 4.4 we discuss results on the existence of summable homoclinic points for non-expansive actions. The structure of the homoclinic group of principal actions will be the subject of Sect. 4.6. Finally, in Sect. 4.7 we present a number of results and methods relying on existence of homoclinic points, e.g., construction of symbolic covers, counting periodic points, topological mixing (specification property), and Gordin's homoclinic approach to Central Limit Theorems.

4.2 Principal Algebraic Actions

In this section we introduce the class of principal algebraic actions. After giving the rather abstract definition of a principal algebraic dynamical system we will study some examples.

4.2.1 Integer Group Rings

For any set $\mathscr{A}$ and a group Γ we denote by $\mathscr{A}^\Gamma$ the set of functions $\Gamma \longrightarrow \mathscr{A}$.

The *integer group ring* $\mathbb{Z}[\Gamma] \subset \mathbb{Z}^\Gamma$ is the set of integer valued functions on Γ with finite support. Every $g \in \mathbb{Z}[\Gamma]$ is written as a formal sum

$$g = \sum_\gamma g_\gamma \cdot \gamma,$$

where g_γ denotes the value of the function g at γ, for each $\gamma \in \Gamma$.

Example 4.2.1 Let $\Gamma = \mathbb{Z}^d$. We will first introduce a notation which allows us to interpret the abelian additive group $\mathbb{Z}^d$ as a multiplicative group. The unit standard vectors in $\mathbb{Z}^d$ will be denoted by u_j with $j \in \{1, \ldots, d\}$. An arbitrary element $\mathbf{n} = (n_1, n_2, \ldots, n_d) \in \mathbb{Z}^d$ is written as $u_1^{n_1} u_2^{n_2} \cdots u_d^{n_d}$. The sum of two elements $\mathbf{n}, \mathbf{m} \in \mathbb{Z}^d$ is given by the product

$$u_1^{n_1} u_2^{n_2} \cdots u_d^{n_d} u_1^{m_1} u_2^{m_2} \cdots u_d^{m_d} = u_1^{n_1+m_1} u_2^{n_2+m_2} \cdots u_d^{n_d+m_d}.$$

The inverse of $u_1^{n_1} u_2^{n_2} \cdots u_d^{n_d}$ in $\mathbb{Z}^d$ is $u_1^{-n_1} u_2^{-n_2} \cdots u_d^{-n_d}$ since

$$u_1^{n_1} u_2^{n_2} \cdots u_d^{n_d} u_1^{-n_1} u_2^{-n_2} \cdots u_d^{-n_d} = u_1^0 u_2^0 \cdots u_d^0$$

which corresponds to the identity element in $\mathbb{Z}^d$.

The formal series expansion of an element f in $\mathbb{Z}[\mathbb{Z}^d]$ is then given by

$$\sum_{\mathbf{n} \in \mathbb{Z}^d} f_{\mathbf{n}} u_1^{n_1} \cdots u_d^{n_d},$$

this is a polynomial with integer coefficients.

Put $f = 2 - u_1 - u_2^{-1}$. This is a function on $\mathbb{Z}^2$ which has the values 2 at $(0, 0)$, -1 at $(1, 0)$, -1 at $(0, -1)$ and 0 otherwise.

Example 4.2.2 The discrete Heisenberg group $\Gamma = \mathbb{H}$ is the multiplicative group of upper-triangular matrices of the form

$$\mathbb{H} = \left\{ \begin{pmatrix} 1 & a & b \\ 0 & 1 & c \\ 0 & 0 & 1 \end{pmatrix} : a, b, c \in \mathbb{Z} \right\} \tag{4.2.1}$$

A canonical generating set of $\mathbb{H}$ is $S = \{x, x^{-1}, y, y^{-1}\}$, where

$$x = \begin{pmatrix} 1 & 1 & 0 \\ 0 & 1 & 0 \\ 0 & 0 & 1 \end{pmatrix}, \quad y = \begin{pmatrix} 1 & 0 & 0 \\ 0 & 1 & 1 \\ 0 & 0 & 1 \end{pmatrix}.$$

The center of $\mathbb{H}$ is generated by

$$z = xyx^{-1}y^{-1} = \begin{pmatrix} 1 & 0 & 1 \\ 0 & 1 & 0 \\ 0 & 0 & 1 \end{pmatrix}.$$

Equivalently, the discrete Heisenberg group can be defined as the group generated by x, y, z subject to the following commutation relations

$$xz = zx, \quad yz = zy, \quad xy = yxz. \tag{4.2.2}$$

Every group element $\gamma \in \mathbb{H}$ can be written as a product of the generating set, i.e., $\gamma = x^k y^l z^m$ for some $(k, l, m) \in \mathbb{Z}^3$.

The element $g = 3x^3 + xy - 3z \in \mathbb{Z}[\mathbb{H}]$ is the formal series expansion of the function g which fulfills: $g(x^3) = 3$, $g(xy) = 1$, $g(z) = -3$ and $g(\gamma) = 0$ for every $\gamma \in \mathbb{H} \smallsetminus \{x^3, xy, z\}$.

The formal series expansion will make the following calculations more transparent. The product fg of two elements $f = \sum_\gamma f_\gamma \cdot \gamma$ and $g = \sum_\gamma g_\gamma \cdot \gamma$ in $\mathbb{Z}[\Gamma]$ is given by *convolution*:

$$fg = \sum_{\gamma,\gamma' \in \Gamma} f_\gamma g_{\gamma'} \cdot \gamma\gamma' = \sum_{\gamma \in \Gamma} \Big[\sum_{\delta \in \Gamma} f_{\gamma\delta^{-1}} g_\delta \Big] \cdot \gamma = \sum_{\delta \in \Gamma} \Big[\sum_{\gamma \in \Gamma} f_\gamma g_{\gamma^{-1}\delta} \Big] \cdot \delta.$$

Example 4.2.3 Consider $f = 1 - u_1^{-3} u_2^2$ and $g = u_1 + u_1^3 u_2^{-2}$ in $\mathbb{Z}[\mathbb{Z}^2]$, then

$$\begin{aligned} fg &= u_1 - u_1^{-3} u_2^2 u_1 + u_1^3 u_2^{-2} - u_1^{-3} u_2^2 u_1^3 u_2^{-2} \\ &= u_1 - u_1^{-3+1} u_2^2 + u_1^3 u_2^{-2} - u_1^{-3+3} u_2^{2-2} \\ &= u_1 - u_1^{-2} u_2^2 + u_1^3 u_2^{-2} - 1. \end{aligned}$$

This is just multiplication of the two polynomials f and g.

Example 4.2.4 Set $f = 2 - x - y \in \mathbb{Z}[\mathbb{H}]$ and $g = x + y + y^{-1} - 3z \in \mathbb{Z}[\mathbb{H}]$. Then,

$$fg = 2x - x^2 - yx + 2y - xy - y^2 + 2y^{-1} - xy^{-1} - yy^{-1} - 6z + 3xz + 3yz \tag{4.2.3}$$

$$= 2x - x^2 + 2y - xy(1 + z^{-1}) - y^2 + 2y^{-1} - xy^{-1} - 1 - 6z + 3xz + 3yz. \tag{4.2.4}$$

Hence, convolution of f and g is a 'product of polynomials'. In (4.2.4) we used the commutation relations defined in (4.2.2).

4.2.2 The Definition of a Principal Algebraic Action

A large class of examples of algebraic Γ-actions arises from left principal ideals of $\mathbb{Z}[\Gamma]$, i.e., ideals of $\mathbb{Z}[\Gamma]$ which are of the form

$$\mathbb{Z}[\Gamma]f = \{gf : g \in \mathbb{Z}[\Gamma]\},$$

for some $f \in \mathbb{Z}[\Gamma]$. Let us describe these actions more explicitly. Set $\mathbb{T} = \mathbb{R}/\mathbb{Z}$ and define the *left* and *right shift-actions* λ and ρ of Γ on $\mathbb{T}^\Gamma$ (the functions from $\Gamma \longrightarrow \mathbb{T}$) by setting

$$(\lambda^\gamma x)_{\gamma'} = x_{\gamma^{-1}\gamma'}, \qquad (\rho^\gamma x)_{\gamma'} = x_{\gamma'\gamma}, \tag{4.2.5}$$

for every $\gamma \in \Gamma$ and $x = (x_{\gamma'})_{\gamma' \in \Gamma} \in \mathbb{T}^\Gamma$. The Γ-actions λ and ρ extend to actions of $\mathbb{Z}[\Gamma]$ on $\mathbb{T}^\Gamma$: for $f \in \mathbb{Z}[\Gamma]$, $f = \sum_{\gamma \in \Gamma} f_\gamma \cdot \gamma$, put

$$\lambda^f = \textstyle\sum_{\gamma \in \Gamma} f_\gamma \lambda^\gamma, \qquad \rho^f = \textstyle\sum_{\gamma \in \Gamma} f_\gamma \rho^\gamma. \tag{4.2.6}$$

Finally, define the abelian additive group X_f as the kernel of ρ^f on $\mathbb{T}^\Gamma$:

$$X_f := \ker \rho^f = \Big\{x \in \mathbb{T}^\Gamma : \sum_{\gamma' \in \Gamma} f_{\gamma'} x_{\gamma\gamma'} = 0 \text{ for every } \gamma \in \Gamma\Big\}. \tag{4.2.7}$$

Since the Γ-actions λ and ρ on $\mathbb{T}^\Gamma$ commute, the group $X_f = \ker \rho^f \subset \mathbb{T}^\Gamma$ is invariant under λ, and we denote by α_f the restriction of λ to X_f. The pair (X_f, α_f) forms an algebraic dynamical system which we call the *principal algebraic Γ-action* defined by $f \in \mathbb{Z}[\Gamma]$—the reason for this name will become clear after the discussion at the beginning of Sect. 4.6 where the group X_f will be identified with the dual of the quotient of $\mathbb{Z}[\Gamma]/\mathbb{Z}[\Gamma]f$.

4.2.3 Examples

We will now give some examples of principal algebraic actions which will serve as guiding examples for this article.

Example 4.2.5 Consider first $f_A = u_1^2 - u_1 - 1 \in \mathbb{Z}[\mathbb{Z}]$. The associated principal algebraic action (X_{f_A}, α_{f_A}) was already defined in Example 4.1.5.

Example 4.2.6 Put $f = 2 - u_1 - u_2^{-1} \in \mathbb{Z}[\mathbb{Z}^2]$. Let us determine how the elements of X_f look like. For this purpose interpret a typical element η of $\mathbb{T}^{\mathbb{Z}^2}$ as a configuration on the lattice $\mathbb{Z}^2$, i.e., at each site (k, l) of the lattice we put the value of the function η at (k, l).

If the configuration η is an element of X_f, then, by definition (cf. (4.2.7)), it has to fulfill

$$2\eta_{(k,l)} = \eta_{(k+1,l)} + \eta_{(k,l-1)} \mod 1, \tag{4.2.8}$$

for each point $(k, l) \in \mathbb{Z}^2$. Hence, X_f is a subset of $\mathbb{T}^{\mathbb{Z}^2}$ and the restriction is defined by the linear recurrence relations (4.2.8).

This allows us to generate an easily interpreted graphic image. Choose some element $(k, l) \in \mathbb{Z}^2$. And fix $\eta_{(k+n,l)} = s_n \in \mathbb{T}$, for each $n \geq 1$. Then the configuration η is fully determined by the values s_n for each $n \in \mathbb{N}$ at those sites (m_1, m_2) of the lattice $\mathbb{Z}^2$ with $m_1 > k$ and $m_2 \leq l$.

$$\begin{array}{ccccccc}
 & \vdots & \vdots & \vdots & \vdots & \vdots & \\
\cdots & s_1 & s_2 & s_3 & s_4 & s_5 & \cdots \\
\cdots & 2s_1 - s_2 & 2s_2 - s_3 & 2s_3 - s_4 & 2s_4 - s_5 & 2s_5 - s_6 & \cdots \\
\cdots & 4s_1 + s_3 & 4s_2 + s_4 & 4s_3 + s_5 & 4s_4 + s_6 & 4s_5 + s_7 & \cdots \\
 & \vdots & \vdots & \vdots & \vdots & \vdots &
\end{array}$$

The action α_f acts via translations on X_f, i.e., for any $\mathbf{n} \in \mathbb{Z}^2$,

$$(\alpha_f^{\mathbf{n}} \eta)_{\mathbf{m}} = \eta_{\mathbf{m}-\mathbf{n}},$$

for every $\mathbf{m} \in \mathbb{Z}^2$ and $\eta \in X_f$.

Example 4.2.7 Consider the polynomial

$$l_{\mathbb{Z}^2} = 4 - u_1 - u_2 - u_1^{-1} - u_2^{-1}$$

the discrete Laplacian on $\mathbb{Z}^2$. The principal algebraic action $(X_{l_{\mathbb{Z}^2}}, \alpha_{l_{\mathbb{Z}^2}})$ associated to $l_{\mathbb{Z}^2}$ is the so called *harmonic model* (cf. [47]). Any configuration $\eta \in X_{l_{\mathbb{Z}^2}}$ fulfills:

$$4\eta_{(k,l)} = \eta_{(k+1,l)} + \eta_{(k,l+1)} + \eta_{(k-1,l)} + \eta_{(k,l-1)} \quad \mod 1,$$

for every $(k, l) \in \mathbb{Z}^2$. At each site (k, l) of the lattice the value of $\eta_{(k,l)}$ times 4 coincides with the sum of its 4 nearest neighbours modulo 1:

$$\begin{array}{ccc} & a & \\ b & \frac{a+b+c+d}{4} & d \\ & c & \end{array}$$

Hence, $X_{l_{\mathbb{Z}^2}}$ consists of the harmonic functions of $\mathbb{T}^{\mathbb{Z}^2}$, in particular, all constant configurations of $\mathbb{T}^{\mathbb{Z}^2}$ are elements of $X_{l_{\mathbb{Z}^2}}$.

The dynamical systems defined in Examples 4.2.6 and 4.2.7 can both be seen as multidimensional versions of the systems (X_{f_A}, α_{f_A}) defined in Example 4.1.5 and Remark 4.1.1.

4.3 Expansive Algebraic Actions

An algebraic action α of a countable group Γ on a compact abelian group X with identity element 0_X and a translation invariant metric d is called *expansive* if there exists an $\varepsilon > 0$ such that

$$\sup_{\gamma\in\Gamma} d(\alpha^\gamma x, \alpha^\gamma y) = \sup_{\gamma\in\Gamma} d\big(\alpha^\gamma (x - y), 0_X\big) \geq \varepsilon \tag{4.3.1}$$

for all distinct $x, y \in X$. Equivalently, an algebraic Γ-action α on a compact abelian group X is *expansive* if there exists an open neighbourhood $\mathscr{U}$ of the identity 0_X in X with

$$\bigcap_{\gamma\in\Gamma} \alpha^\gamma(\mathscr{U}) = \{0_X\}.$$

For expansive systems, any 'error' in the initial conditions x and y will be amplified to at least ε under the dynamics of the action. In this sense, expansiveness of the dynamics can be seen as the weakest form of sensitive dependence on initial conditions. However, in defining chaos in dynamical systems, one often requires that the divergence is exponential in time.

Example 4.3.1 Let T_A be a d—dimensional toral automorphism defined by a matrix A. The system $(\mathbb{T}^d, T_A)$ is expansive if and only if A has no eigenvalues which are elements of $\mathbb{S}$, where $\mathbb{S} := \{e^{2\pi i t} : t \in \mathbb{T}\}$. In other words, the characteristic polynomial $f_A(t)$ of A does not vanish on $\mathbb{S}$ (cf. [53, p. 143]).

Consider the case

$$A = \begin{pmatrix} 0 & 1 \\ 1 & 1 \end{pmatrix}.$$

As we have seen in Example 4.1.5 the two eigenvalues λ_+ and λ_- of A fulfill $|\lambda_+| \neq 1$ and $|\lambda_-| \neq 1$. Hence, $(\mathbb{T}^2, T_A)$ is an expansive dynamical system.

As already mentioned above, expansiveness of a principal action has important consequences for the study of homoclinic groups.

Let us first recall that a group is called *residually finite* if the intersection of all its finite-index subgroups is trivial. The class of residually finite groups is very large, and contains $\mathbb{Z}^d$ and $\mathbb{H}$.

Theorem 4.3.1 (Deninger and Schmidt [12]) *Let Γ be a countable residually finite group and $f \in \mathbb{Z}[\Gamma]$ is such that (X_f, α_f) is expansive. Then the homoclinic group $\Delta_{\alpha_f}(X_f)$ is countable and dense in X_f. Moreover,*

$$\Delta_{\alpha_f}(X_f) = \Delta^1_{\alpha_f}(X_f).$$

The second part of the Theorem 4.3.1 was generalized in [10, Theorem 5.6] to all countable discrete groups Γ.

Example 4.3.2 It easily follows from the discussion in Examples 4.2.6 and 4.2.7 that X_g and $X_{l_{\mathbb{Z}^2}}$ corresponding to the polynomials $g = 2 - u_1 - u_2^{-1}$ and $l_{\mathbb{Z}^2}$, respectively, contain the constant configurations of $\mathbb{T}^{\mathbb{Z}^2}$.

If X_f contains the constant configurations, then (X_f, α_f) is non-expansive. Indeed, for all $\varepsilon > 0$ one can find $c_\varepsilon \in \mathbb{T}$ such that

$$d(\eta^{(c_\varepsilon)}, 0_{X_f}) < \varepsilon, \tag{4.3.2}$$

where $(\eta^{(c_\varepsilon)})_{(k,l)} = c_\varepsilon$ for all $(k, l) \in \mathbb{Z}^2$. Clearly, for all $\mathbf{n} \in \mathbb{Z}^2$

$$d(\alpha_f^{\mathbf{n}}\eta^{(c_\varepsilon)}, \alpha_f^{\mathbf{n}}0_{X_f}) = d(\eta^{(c_\varepsilon)}, 0_{X_f}) < \varepsilon, \tag{4.3.3}$$

because any translate of a constant configuration coincides with the configuration.

In Example 4.3.2 both dynamical systems contained all constant configurations in $\mathbb{T}^{\mathbb{Z}^d}$ and by a simple minded argument it was easy to show the non-expansiveness of the two systems. In general, a non-expansive principal algebraic action does not necessarily contain the constant functions. The principal algebraic action associated to the polynomial $3 - u_1 - u_2 - u_1^{-1} - u_2^{-1} \in \mathbb{Z}[\mathbb{Z}^2]$ is non-expansive as well (cf. Example 4.3.4), but does not contain any constant configuration of $\mathbb{T}^{\mathbb{Z}^2}$.

In view of Theorem 4.3.1, it is important to be able to decide whether a particular principal action (X_f, α_f) is expansive or not.

Let us discuss alternative characterizations of expansiveness of principal algebraic actions. The main goal is to formulate algebraic characterizations of expansiveness.

The first result of this kind was established by Schmidt for $\Gamma = \mathbb{Z}^d$. The *variety* of $f \in \mathbb{Z}[\mathbb{Z}^d]$ is defined as

$$\mathsf{V}(f) := \{\mathbf{z} \in (\mathbb{C}^\times)^d : f(\mathbf{z}) = 0\},$$

i.e., the set of roots of the polynomial f. Further, define the *unitary variety* $\mathsf{U}(f)$ as the set $\mathsf{V}(f) \cap \mathbb{S}^d$.

Theorem 4.3.2 (Theorem 6.5, [46]) *For $f \in \mathbb{Z}[\mathbb{Z}^d]$, the corresponding $\mathbb{Z}^d$-action (X_f, α_f) is expansive if and only if $\mathsf{U}(f) = \varnothing$.*

Example 4.3.3 Let $f = 2 - u_1 - u_2^{-1} \in \mathbb{Z}[\mathbb{Z}^2]$ and $l_{\mathbb{Z}^2}$ the Graph-Laplacian of $\mathbb{Z}^2$. Since

$$\mathsf{U}(f) = \mathsf{U}(l_{\mathbb{Z}^2}) = \{(1, 1)\}$$

Theorem 4.3.2 leads to an alternative proof of the non-expansiveness of the systems (X_f, α_f) and $(X_{l_{\mathbb{Z}^2}}, \alpha_{l_{\mathbb{Z}^2}})$.

Example 4.3.4 One can easily check that $(1, e^{\frac{\pi}{4}})$ is a root of $g = 3 - u_1 - u_2 - u_1^{-1} - u_2^{-1} \in \mathbb{Z}[\mathbb{Z}^2]$ which also lies on $\mathbb{S}^2$. Hence, (X_g, α_g) is not expansive.

Schmidt's result can be reinterpreted in terms of Fourier analysis: for

$$f(u_1, \ldots, u_d) = \sum_{\mathbf{n} \in \mathbb{Z}^d} f_{\mathbf{n}} u_1^{n_1} \cdots u_d^{n_d} \in \mathbb{Z}[\mathbb{Z}^d]$$

the condition $\mathsf{U}(f) = \varnothing$ is equivalent to requiring that the Fourier transform $\hat{f} : \mathbb{T}^d \to \mathbb{C}$ of f given by

$$\hat{f}(\theta) = \sum_{\mathbf{n} \in \mathbb{Z}^d} f_{\mathbf{n}} e^{2\pi i <\mathbf{n}, \theta>} = f(e^{2\pi i \theta_1}, \ldots, e^{2\pi i \theta_d})$$

does not vanish on $\mathbb{T}^d$.

Let f be an element of the *group algebra* $\ell^1(\Gamma, \mathbb{R})$ of absolutely summable functions from $\Gamma \longrightarrow \mathbb{R}$. The norm of $f = \sum_{\gamma \in \Gamma} f_\gamma \gamma$ is given by $\|f\|_1 = \sum_{\gamma \in \Gamma} |f_\gamma|$. The adjoint f^* of f is given by $f^*_\gamma = f_{\gamma^{-1}}$ for all $\gamma \in \Gamma$. We say that f is invertible if there exists an element $g \in \ell^1(\Gamma, \mathbb{R})$ such that $fg = gf = e_\Gamma$, where e_Γ is the identity element of Γ. By a classical result of Wiener, the condition that $f \in \mathbb{Z}[\mathbb{Z}^d]$ has a non vanishing Fourier-transform is equivalent to the invertibility of f in $\ell^1(\mathbb{Z}^d, \mathbb{R})$.

Example 4.3.5 Set $h = 3 + u_1 + u_2 \in \mathbb{Z}[\mathbb{Z}^2]$, then the Fourier-transform of h is given by $\hat{h}(\theta_1, \theta_2) = 3 + e^{2\pi i \theta_1} + e^{2\pi i \theta_2}$. It is clear that, for all $(\theta_1, \theta_2) \in \mathbb{T}^2$, $|\hat{h}(\theta_1, \theta_2)| > 0$. Hence, (X_h, α_h) is an expansive dynamical system.

This connection to Fourier analysis, suggests that there are representation theoretical characterizations of expansiveness.

A unitary representation π of a countable group Γ maps $\gamma \in \Gamma$ to a unitary operator $\pi(\gamma)$ on some Hilbert space $\mathscr{H}_\pi$, such that $\pi(\gamma\gamma') = \pi(\gamma)\pi(\gamma')$ for all $\gamma, \gamma' \in \Gamma$.

A unitary representation π of Γ can be extended to a representation of $\mathbb{Z}[\Gamma]$. For an element $f \in \mathbb{Z}[\Gamma]$, $\pi(f)$ is a bounded operator on $\mathscr{H}_\pi$ given by

$$\pi(f) = \pi\left(\sum_{\gamma\in\Gamma} f_\gamma \gamma\right) = \sum_{\gamma\in\Gamma} f_\gamma \pi(\gamma).$$

Example 4.3.6 Let $\Gamma = \mathbb{Z}^d$. For every $\theta \in \mathbb{T}^d$ we can define a unitary representation of $\mathbb{Z}^d$ on the Hilbert space $\mathbb{C}$ which is given by $\pi_\theta(\mathbf{n})v = e^{2\pi i\langle \mathbf{n},\theta\rangle}v$, for every $\mathbf{n} \in \mathbb{Z}^d$ and $v \in \mathbb{C}$. Hence, for every $\mathbf{n} \in \mathbb{Z}^d$, $\pi_\theta(\mathbf{n})$ is a multiplication operator with a complex number.

Note that for $f \in \mathbb{Z}[\mathbb{Z}^d]$ the Fourier-transform at θ is given by:

$$\hat{f}(\theta) = \sum_{\mathbf{n}\in\mathbb{Z}^d} f_{\mathbf{n}}\pi_\theta(\mathbf{n}) = \pi_\theta(f).$$

Hence, the system (X_f, α_f) is non-expansive if and only there exists a $\theta \in \mathbb{T}^d$ such that $\pi_\theta(f) = 0$.

Miles [36] showed that for an abelian group Γ and $f \in \mathbb{Z}[\Gamma]$, the algebraic system (X_f, α_f) is expansive if and only if for all one-dimensional representations π (hence, those π with $\dim \mathcal{H}_\pi = 1$), one has $\pi(f)v \neq 0$ for every $v \in \mathcal{H}_\pi$.

The upper central series of a group Γ is given by a sequence

$$\{e_\Gamma\} = Z_0 \leq Z_1 \leq Z_2 \leq \cdots \leq Z_n \leq \cdots,$$

where Z_1 is the center of Γ and Z_{i+1} is the subgroup of Γ such that Z_{i+1}/Z_i is the center of Γ/Z_i. A group Γ is *nilpotent* if there is some n such that $Z_n = \Gamma$. For a nilpotent group Γ, expansiveness of principal actions can, in principle, be validated using methods of representation theory as well.

Theorem 4.3.3 (Einsiedler and Rindler [13, Theorem 8.2]) *Let Γ be a countably infinite discrete nilpotent group and $f \in \mathbb{Z}[\Gamma]$. Then the following statements are equivalent:*

1. *The principal algebraic action α_f is non-expansive;*
2. *There exists a (irreducible) unitary representation π of Γ on a Hilbert space $\mathcal{H}$ and a unit vector $v \in \mathcal{H}$ such that $\pi(f)v = 0$.*

However, one is forced to consider *all* (irreducible) unitary representations of Γ. Unfortunately, for many nilpotent groups determining all (irreducible) representations is prohibitively complex. Moreover, the representations may act on infinite-dimensional Hilbert spaces, which makes the analysis even more difficult.

The next line of attack is to consider generalizations suggested by the link between expansiveness and Wiener's Lemma mentioned above. It turns out that it is possible to relate expansiveness of a principal action (X_f, α_f) to invertibility of the polynomial f in the group algebra $\ell^1(\Gamma, \mathbb{R})$.

Theorem 4.3.4 (Deninger and Schmidt [12]) *Let Γ be a countably infinite discrete group and $f \in \mathbb{Z}[\Gamma]$. Then the following statements are equivalent:*

1. *The principal algebraic Γ-action (X_f, α_f) is expansive;*
2. *f is invertible in $\ell^1(\Gamma, \mathbb{R})$.*

Moreover, due to a Wiener Lemma which was established in [12] as well, the invertibilty of $f \in \mathbb{Z}[\Gamma]$ is equivalent to the triviality of the kernel of the right convolution operator $C_f : \ell^\infty(\Gamma, \mathbb{R}) \to \ell^\infty(\Gamma, \mathbb{R})$ defined by $C_f(v) = vf^*$ for every $v \in \ell^\infty(\Gamma, \mathbb{R})$. Furthermore, if there exists a formal inverse w_f of f, i.e. $w_f f = f w_f = e_\Gamma$, such that $w_f \in \ell^\infty(\Gamma, \mathbb{R}) \setminus \ell^1(\Gamma, \mathbb{R})$, then the principal action is not expansive.

Example 4.3.7 Let us study the convolution with $l_{\mathbb{Z}^2}$ on $\ell^\infty(\mathbb{Z}^2, \mathbb{R})$. It is clear that all constant functions are elements of $\ell^\infty(\mathbb{Z}^d, \mathbb{R})$ and of the kernel of $C_{l_{\mathbb{Z}^2}}$. Hence, $l_{\mathbb{Z}^2}$ is not invertible and therefore, as we already know, $(X_{l_{\mathbb{Z}^2}}, \alpha_{l_{\mathbb{Z}^2}})$ is non-expansive.

Even for one of the simplest non-abelian groups like the discrete Heisenberg group $\mathbb{H}$ explicit conditions for expansiveness are available only in some particular cases [13]. Recently we have achieved some progress in case of the discrete Heisenberg group [24, 25].

Define, for every $\theta, s, t \in \mathbb{T}$, a representation $\pi_\theta^{(s,t)}$ of $\mathbb{H}$ given by

$$(\pi_\theta^{(s,t)}(x)F)(u) = e^{2\pi i s} F(u+\theta),\ (\pi_\theta^{(s,t)}(y)F)(u) = e^{2\pi i(u+t)} F(u)$$
$$\text{and} \quad (\pi_\theta^{(s,t)}(z)F)(u) = e^{2\pi i\theta} F(u),$$

for every $F \in L^2(\mathbb{T}, \mathbb{C})$ and $u \in \mathbb{T}$.

Theorem 4.3.5 *The action α_f is expansive if and only if $\pi_\theta^{(s,t)}(f)$ is an invertible operator on $L^2(\mathbb{T}, \mathbb{C})$ for every $\theta, s, t \in \mathbb{T}$.*

Nevertheless, an important question of D. Lind and K. Schmidt, on whether there is an *'algorithmic test'* for expansiveness is still open.

In [25] we gave another representation theoretical characterization of expansiveness for principal Γ-actions of a nilpotent group Γ, which allows one to consider only 'nice' representations.

Theorem 4.3.6 *The principal Γ-action α_f is non-expansive if and only if there exists an irreducible representation π, which is induced from a one-dimensional representation of a subgroup of Γ, such that $\pi(f)$ has no bounded inverse.*

We refer to [35] for a description of the inducing procedure, which allows one to generate representations of Γ from representations of subgroups of Γ. It is a feasible task to find the subgroups of Γ and their one-dimensional representations.

4.4 Homoclinic Points of Non-expansive Actions

As we have seen in the previous section, under very mild conditions, i.e. expansiveness, algebraic dynamical systems, and in particular principal actions, have large non-trivial groups of (summable) homoclinic points.

The natural question is whether the same (albeit under some extra conditions) can be true for non-expansive systems as well?

In this section we will demonstrate that the rich homoclinic structures are present in a large class of non-expansive principal actions. For integer lattices $\Gamma = \mathbb{Z}^d$, we now have a *complete* picture. We will also discuss the first results for nilpotent groups.

4.4.1 Lattices $\Gamma = \mathbb{Z}^d$

Consider $f \in \mathbb{Z}[\Gamma]$. A function $w : \Gamma \longrightarrow \mathbb{R}$ is called the fundamental solution of f if it satisfies

$$w \cdot f^* = \delta_{\mathbf{0}}, \text{ where } \delta_{\mathbf{0}}(\mathbf{n}) = \begin{cases} 1, & \mathbf{n} = \mathbf{0}, \\ 0, & \mathbf{n} \neq \mathbf{0}. \end{cases}$$

If α_f is expansive, and hence f is invertible in $\ell^1(\mathbb{Z}^d, \mathbb{R})$, then the fundamental solution is unique and is necessarily summable. Therefore $x = (x_{\mathbf{n}})_{\mathbf{n} \in \mathbb{Z}^d}$ with $x_{\mathbf{n}} = w_{\mathbf{n}}$ mod 1 is a summable homoclinic point in X_f. Moreover, x is a fundamental homoclinic point, i.e., the orbit of x generates $\Delta_{\alpha_f}(X_f) = \Delta^1_{\alpha_f}(X_f)$. More specifically,

$$\Delta_{\alpha_f}(X_f) = \Delta^1_{\alpha_f}(X_f) = \{g \cdot x : g \in \mathbb{Z}[\mathbb{Z}^d]\}.$$

For non-expansive f, $f \neq 0$, fundamental solutions exist, but they are not unique and are not summable. Nevertheless, the natural question is whether for such f's one is still able to find polynomials $g \in \mathbb{Z}[\mathbb{Z}^d]$, with the property $g \cdot w \in \ell^1(\mathbb{Z}^d, \mathbb{R})$. Any such g will be called a multiplier for w. For any such g, $x^{(g)} = (x^{(g)}_{\mathbf{n}})$ with $x^{(g)}_{\mathbf{n} \in \mathbb{Z}^2} = (g \cdot w)_{\mathbf{n} \in \mathbb{Z}^2}$ mod 1 is a summable homoclinic point in X_f. Note that $x^{(g)}$ will be trivial, i.e. $x^{(g)} = 0$, if g is a multiple of f.

Let us start with some illustrative examples.

Example 4.4.1 ([30]) Consider $\Gamma = \mathbb{Z}^2$ and $f = 2 - u_1^{-1} - u_2^{-1} \in \mathbb{Z}[\mathbb{Z}^2]$. The corresponding algebraic dynamical system (X_f, α_f) is not expansive, as $\mathsf{U}(f) = \{(1, 1)\}$ is non-trivial (cf. Theorem 4.3.2). Alternatively, one easily checks that the fundamental solution $w = (w_{\mathbf{n}})_{\mathbf{n} \in \mathbb{Z}^2}$ of $w \cdot f^* = 1_{\mathbb{Z}[\mathbb{Z}^d]}$, given by

$$w_{\mathbf{n}} = w_{n_1, n_2} = \begin{cases} \dfrac{1}{2^{n_1+n_2+1}} \dbinom{n_1 + n_2}{n_1} & \text{if } n_1 \geq 0 \text{ and } n_2 \geq 0, \\ 0 & \text{otherwise,} \end{cases} \tag{4.4.1}$$

is bounded ($w \in \ell^{\infty}(\mathbb{Z}^2)$), but not summable. Note that w has non-negative values, and hence $v = (v_{\mathbf{n}})_{\mathbf{n}\in\mathbb{Z}^2}$ and $v' = (v'_{\mathbf{n}})_{\mathbf{n}\in\mathbb{Z}^2}$

$$v_{\mathbf{n}} = w_{\mathbf{n}} - w_{\mathbf{n}+e_1}, \quad v'_{\mathbf{n}} = w_{\mathbf{n}} - w_{\mathbf{n}+e_2}, \quad \mathbf{n} \in \mathbb{Z}^2,$$

will be "smaller". It turns out, that in order to achieve summability one has to consider third partial differences: for

$$v'' = (1-u_1)^3 \cdot w, \text{ i.e. } v''_{\mathbf{n}} = w_{\mathbf{n}} - 3w_{\mathbf{n}+e_1} + 3w_{\mathbf{n}+2e_1} - w_{\mathbf{n}+3e_1},$$

one has $v'' \in \ell^1(\mathbb{Z}^2)$. A similar conclusion holds for $(1-u_2)^3 \cdot w$, and $(1-u_1)^2(1-u_2) \cdot w$. Note that if for some polynomial $g \in \mathbb{Z}[\mathbb{Z}^2]$, $v = g \cdot w$ is absolutely summable ($v \in \ell^1(\mathbb{Z}^2)$), then $\eta = (\eta_{\mathbf{n}})_{\mathbf{n}\in\mathbb{Z}^2}$ with $\eta_{\mathbf{n}} = v_{\mathbf{n}} \mod 1$ for all $\mathbf{n} \in \mathbb{Z}^2$, is an absolutely summable homoclinic point in X_f.

Example 4.4.2 (Discrete Laplacians [47]) Consider the lattice $\mathbb{Z}^d$ and the corresponding discrete Laplacian

$$l_{\mathbb{Z}^d} = 2d - \sum_{j=1}^{d}(u_j + u_j^{-1}).$$

The fundamental solution (Green's function) for $l_{\mathbb{Z}^d}$ is given by

1. for $d = 2$,

$$w^{(2)}_{\mathbf{n}} = \int_{\mathbb{T}^2} \frac{e^{2\pi i\langle \mathbf{n},\theta\rangle} - 1}{4 - 2\cos(2\pi\theta_1) - 2\cos(2\pi\theta_2)}\, d\theta \text{ for every } \mathbf{n} \in \mathbb{Z}^2.$$

2. for $d \geq 3$,

$$w^{(d)}_{\mathbf{n}} = \int_{\mathbb{T}^d} \frac{e^{2\pi i\langle \mathbf{n},\theta\rangle}}{2d - 2\sum_{j=1}^{d}\cos(2\pi\theta_j)}\, d\theta \text{ for every } \mathbf{n} \in \mathbb{Z}^d.$$

The difference in these definitions for $d = 2$ and $d > 2$ is related to the fact that the simple random walk on $\mathbb{Z}^2$ is recurrent, while on higher dimensional lattices it is transient.

One can check that any polynomial g, which is not a multiple of $l_{\mathbb{Z}^d}$ and which is in one of the ideals $(1-u_i)(1-u_j)(1-u_k)\mathbb{Z}[\mathbb{Z}^d]$ where $i, j, k \in \{1, \cdots, d\}$, gives rise to a summable homoclinic point $x^{(g)} \in l_{\mathbb{Z}^d}$.

Note that in the examples above, the multiplier $g \in \mathbb{Z}[\mathbb{Z}^d]$ vanishes at points of the unitary variety $\mathsf{U}(f)$. In fact, the way to identify such multipliers in general is to require that the function

$$H(\theta) = \frac{\hat{g}(\theta)}{\hat{f}(\theta)} = \frac{\sum_{\mathbf{n}} g_{\mathbf{n}} e^{2\pi i\langle \mathbf{n},\theta\rangle}}{\sum_{\mathbf{n}} f_{\mathbf{n}} e^{2\pi i\langle \mathbf{n},\theta\rangle}}, \quad \theta \in \mathbb{T}^d,$$

is sufficiently smooth on $\mathbb{T}^d$. This automatically implies that $\mathsf{U}(f) \subseteq \mathsf{U}(g)$. Any polynomial g which is a multiple of f clearly satisfies $\mathsf{U}(f) \subseteq \mathsf{U}(g)$. However, a multiple of f will not produce a non-trivial homoclinic point. Hence we have to find multipliers g which are not multiples of f, i.e., do not belong to the principal ideal generated by f. This is indeed possible provided the unitary variety of f is not too big.

Theorem 4.4.1 *[31] Let $d \geq 2$, and let $f \in \mathbb{Z}[\mathbb{Z}^d]$ be a non-zero irreducible Laurent polynomial. The following conditions are equivalent.*

1. *f is atoral, i.e., $\mathsf{U}(f)$ is contained in an algebraic subset of $\mathbb{S}^d$ with dimension at most $d-2$;*
2. *the group of summable homoclinic points is non-trivial: $\Delta^1_{\alpha_f}(X_f) \neq \{0\}$.*

Moreover, if f is atoral, then $\Delta^1_{\alpha_f}(X_f)$ is also dense in X_f and there exists a point $x \in X_f$ and an ideal $\mathscr{I}_f \subset \mathbb{Z}[\mathbb{Z}^d]$ such that

$$\Delta^1_{\alpha_f}(X_f) = \{g \cdot x : g \in \mathscr{I}_f\}.$$

4.4.2 Non-abelian Groups

Very little is known about the existence of (summable) homoclinic points of non-expansive principal actions of general non-commutative groups Γ. The only exception are the actions corresponding to the so-called well-balanced polynomials introduced by Bowen and Li [5]. Well-balanced polynomials can be viewed as generalizations of discrete Laplacians discussed above.

Definition 4.4.1 A polynomial $f \in \mathbb{Z}[\Gamma]$ is called *well-balanced* if

1. $\sum_{\gamma \in \Gamma} f_\gamma = 0$,
2. $f_\gamma \leq 0$ for every $\gamma \in \Gamma \setminus \{e\}$,
3. $f = f^*$,
4. $S =$ supp(f)—the support of f, generates Γ.

Example 4.4.3 The element $f \in \mathbb{Z}[\mathbb{Z}^2]$ given by $4 - u_1 - u_1^{-1} - u_1 u_2^{-1} - u_1^{-1} u_2$ is well-balanced.

Example 4.4.4 The Graph-Laplacian of $\mathbb{H}$ given by $l_{\mathbb{H}} = 4 - x - x^{-1} - y - y^{-1} \in \mathbb{Z}[\mathbb{H}]$ is well-balanced.

For well-balanced polynomials, the homoclinic group turns out to be sufficiently rich.

Theorem 4.4.2 ([5]) *Let Γ be a countably infinite group such that Γ is not virtually $\mathbb{Z}$ or $\mathbb{Z}^2$ (i.e., does not have any finite-index normal subgroup isomorphic to $\mathbb{Z}$ or $\mathbb{Z}^2$). Let $f \in \mathbb{Z}[\Gamma]$ be well-balanced. Then the homoclinic group $\Delta_{\alpha_f}(X_f)$ is dense in X_f.*

An important ingredient of the proof of this result is the link to the theory of random walks on groups. Note that $\mu = -(f - f_e)/f_e$ defines a finite probability measure on $S \setminus \{e\}$. The condition that Γ is not virtually $\mathbb{Z}$ or $\mathbb{Z}^2$ ensures that the random walk generated by μ is transient. The existence of summable homoclinic points in this case remains an open problem. For the graph Laplacian on $\mathbb{Z}^2$ Theorem 4.4.2 holds as well.

4.4.3 Discrete Heisenberg Group

The natural question is whether a multiplier approach which is so successful for the integer lattices $\Gamma = \mathbb{Z}^d$ case, can be extended to non-abelian discrete groups as well. In particular, the discrete Heisenberg group—being one of the simplest non-abelian groups—is a reasonable starting point.

Let $q = z^{-1}$. By the well-known q-binomial theorem, for any $n \in \mathbb{N}$,

$$(x + y)^n = \sum_{k=0}^{n} \begin{bmatrix} n \\ k \end{bmatrix}_q x^k y^{n-k}$$

where $\begin{bmatrix} n \\ k \end{bmatrix}_q$ is the so-called q-binomial coefficient

$$\begin{bmatrix} n \\ k \end{bmatrix}_q = \prod_{i=0}^{k-1} \frac{1 - q^{n-i}}{1 - q^{i+1}} \quad \textit{for } 0 \le k \le n. \tag{4.4.2}$$

which turns out to be a polynomial in $q = z^{-1}$ with non-negative integer coefficients.

Consider $f = 2 - x^{-1} - y^{-1} \in \mathbb{Z}[\mathbb{H}]$. The corresponding action is not expansive since the constant configurations are annihilated by f^*, hence the kernel of f^* in $\ell^\infty(\mathbb{H}, \mathbb{R})$ is not empty. The q-binomial theorem allows us to obtain expression for w—the fundamental solution of f^*, similar to (4.4.1):

$$w = \sum_{n=0}^{\infty} \frac{1}{2^{n+1}} \sum_{k=0}^{n} x^k y^{n-k} \begin{bmatrix} n \\ k \end{bmatrix}_q . \tag{4.4.3}$$

Since q-binomial coefficients are polynomials in $q = z^{-1}$ with non-negative (integer) coefficients, multiplication with $(1 - z^{-1})$, equivalently with $(1 - z)$, should reduce the ℓ^1-norm of w. Let us demonstrate this by means of an example.

Example 4.4.5 One can easily compute that

$$\begin{bmatrix} 4 \\ 2 \end{bmatrix}_q = 1 + q + 2q^2 + q^3 + q^4,$$

which has ℓ^1-norm 6. Convolution with $(1-q)$ reduces the norm to 4, since

$$(1-q)(1+q+2q^2+q^3+q^4) = 1+q^2-q^3-q^5.$$

Moreover, there are some technical reasons for working with *central* multipliers: the homoclinic points obtained via central multipliers (if existing), are more convenient when one wants to construct symbolic covers (cf. Sect. 4.7.1).

Indeed, as the next result shows a sufficiently large power of $(1-z)$ is a desired multiplier.

Theorem 4.4.3 ([24]) *Let w be the formal inverse (fundamental solution) of $f^* = 2-x-y$ given by (4.4.3). Then $(1-z)^2 \cdot w \in \ell^1(\mathbb{H}, \mathbb{R})$. Hence the point $x^{((1-z)^2)} = (x_\gamma^{((1-z)^2)})_{\gamma\in\mathbb{H}} \in \mathbb{T}^{\mathbb{H}}$ with*

$$x_\gamma^{((1-z)^2)} = ((1-z)^2 \cdot w)_\gamma \mod 1 \quad \textit{for all } \gamma \in \mathbb{H},$$

is a summable homoclinic point in X_f.

Example 4.4.6 For the Laplacian $l_{\mathbb{H}} = 4-x-x^{-1}-y-y^{-1} \in \mathbb{Z}[\mathbb{H}]$, using rather abstract methods of Harmonic analysis, we are currently able to show that $(1-z)^9$ is indeed a multiplier leading to a summable homoclinic point. We conjecture that the third power (i.e., $(1-z)^3$) will suffice.

4.5 Topological Entropy of Algebraic Actions

In this section we introduce the notion of topological entropy without giving a precise definition. The *topological entropy* $h(X,\alpha)$ or $h(\alpha)$, of an algebraic Γ-action (X,α) is a non-negative number which describes the complexity of the system. Actions (X,α) with positive entropy $h(\alpha)$ can be characterized as chaotic.

The topological entropy of a Γ-action is defined for countable *amenable* groups Γ only. Recall that a group Γ is amenable if for every $\varepsilon > 0$ and every finite subset F of Γ there exists a non-empty subset $F' \subseteq \Gamma$ such that

$$\frac{|F' \smallsetminus \gamma F'|}{|F'|} < \varepsilon \quad \text{for all } \gamma \in F.$$

All abelian groups are amenable. Moreover, every solvable and hence nilpotent group is amenable. The typical example of a non-amenable group is the free group with two generators.

By a variational principle, the topological entropy $h(\alpha_f)$ of a principal algebraic action agrees with the measure-theoretical (Kolmogorov-Sinai) entropy with respect to the Haar measure of X_f.

The calculation of entropy for an arbitrary dynamical system is in general very difficult. In consideration of this fact the next result is rather striking. For principal algebraic $\mathbb{Z}^d$-actions (X_f, α_f) the entropy is fully determined by f. An expression for the topological entropy has been found by Lind et al. [32]: for every non-zero $f \in \mathbb{Z}[\mathbb{Z}^d]$:

$$h(\alpha_f) = \log M(f),$$

where

$$M(f) := \exp\left[\int_{\mathbb{S}^d} \log|f(\mathbf{s})|ds\right],$$

is the so-called *Mahler measure* of the polynomial $f \in \mathbb{Z}[\mathbb{Z}^d]$.

Example 4.5.1 The topological entropy of (X_f, α_f) with $f = 2 - x - y^{-1}$ is given by

$$\begin{aligned} h(\alpha_f) &= \log M(f) \\ &= \int_{\mathbb{T}}\int_{\mathbb{T}} \log|2 - e^{2\pi i s_1} - e^{-2\pi i s_2}|\, ds_1 ds_2 \\ &= \log 2. \end{aligned}$$

For more general groups, the entropy of an expansive principal Γ-action α_f is given by the logarithm of the Fuglede-Kadison determinant of a convolution operator associated to f.

Theorem 4.5.1 *[12] Suppose the group Γ is residually finite and amenable and $f \in \mathbb{Z}[\Gamma]$ is such that α_f is expansive. Denote by R^f the right-convolution operator acting on $\ell^2(\Gamma, \mathbb{C})$*

$$R^f g = gf \quad g \in \ell^2(\Gamma, \mathbb{C}).$$

Then

$$h(\alpha_f) = \log \det{}_{\mathcal{N}\Gamma}(R^f), \tag{4.5.1}$$

where the Fuglede-Kadison determinant $\det_{\mathcal{N}\Gamma}$ of an invertible operator $A \in \mathcal{N}\Gamma$—the group von Neumann algebra of left Γ-equivariant bounded operators on $\ell^2(\Gamma, \mathbb{C})$—is defined as

$$\det{}_{\mathcal{N}\Gamma}(A) = \exp\left(\frac{1}{2}\langle(\log AA^*)\mathbf{1}, \mathbf{1})\rangle\right),$$

here $\langle\cdot,\cdot\rangle$ is the standard inner product on $\ell^2(\Gamma, \mathbb{C})$.

Recently, Li and Thom [28] were able to strengthen the result by proving (4.5.1) for *all* (hence, not necessarily expansive) principal actions α_f of amenable groups Γ, for which f is not a zero-divisor.

The Fuglede-Kadison determinant of a given operator is typically hard to compute.

Finally, let us introduce the notion of *completely positive entropy*. In Ergodic Theory, a measure theoretical dynamical system (X, T, μ) has completely positive entropy if it does not possess any non-trivial subsystem with zero entropy—we refer to [53, Chap. 4] for an exact definition and rigorous treatment. The condition of completely positive is stronger than having just positive entropy. Systems with completely positive entropy are called Kolmogorov (or K-systems). Such systems have very strong mixing properties.

4.6 Structure and Associated Dynamics of the Homoclinic Group

In the preceding sections criteria for large groups of (summable) homoclinic points already have been given. We are now interested in results on the structure of homoclinic groups $\Delta_{\alpha_f}(X_f)$, $\Delta^1_{\alpha_f}(X_f)$. We will also discuss actions on the dual of homoclinic groups. We make use of basic results from Pontryagin dualty theory of locally compact abelian groups. The reader who is probably unfamiliar with this theory will still be able to follow since we will write down for all objects its corresponding dual explicity.

First, let us give an alternative definition of the group X_f. For every locally compact abelian group G denote by $\hat{G}$ its dual group; $\hat{G}$ is a locally compact abelian group as well. The dual group of $\hat{G}$ is isomorphic to G.

The pairing

$$\langle f, x\rangle = e^{2\pi i \sum_{\gamma\in\Gamma} f_\gamma x_\gamma},$$

for $f = \sum_{\gamma\in\Gamma} f_\gamma \cdot \gamma \in \mathbb{Z}[\Gamma]$ and $x = (x_\gamma) \in \mathbb{T}^\Gamma$, identifies $\mathbb{Z}[\Gamma]$ with the dual group $\widehat{\mathbb{T}^\Gamma}$ of $\mathbb{T}^\Gamma$. It easy to check that under this identification,

$$\begin{aligned} X_f &= \ker\rho^f = \left\{x \in \mathbb{T}^\Gamma : \sum\nolimits_{\gamma'\in\Gamma} f_{\gamma'} x_{\gamma\gamma'} = 0 \text{ for every } \gamma \in \Gamma\right\} \\ &= (\mathbb{Z}[\Gamma]f)^\perp = \widehat{\mathbb{Z}[\Gamma]/\mathbb{Z}[\Gamma]f} \subset \widehat{\mathbb{Z}[\Gamma]} = \mathbb{T}^\Gamma. \end{aligned} \tag{4.6.1}$$

For every $\gamma \in \Gamma$, denote by $\widehat{\alpha_f}^\gamma$ the automorphism of $\widehat{X_f}$ dual to α_f^γ: namely the automorphism defined by the following relation

$$\langle h, \alpha_f^\gamma x\rangle = \langle \widehat{\alpha_f}^\gamma h, x\rangle, \quad h \in \widehat{X_f} \text{ and } x \in X_f.$$

For principal $\mathbb{Z}^d$-actions $\widehat{\alpha_f}^{\mathbf{n}}$ corresponds to the multiplication with the units $u_1^{n_1}\cdots u_d^{n_d}$ on $\widehat{X_f} \simeq \mathbb{Z}[\mathbb{Z}^d]/\mathbb{Z}[\mathbb{Z}^d]f$.

4.6.1 Structure of the Homoclinic Group

Suppose that the principal $\mathbb{Z}^d$-action defined by $f \in \mathbb{Z}[\mathbb{Z}^d]$ is expansive. We know already that for expansive principal actions there exists a fundamental homoclinic point x which generates the homoclinic groups. Moreover, x is annihilated only by those polynomials $g \in \mathbb{Z}[\mathbb{Z}^d]$ which are elements of the principal ideal generated by f. Therefore,

$$\begin{aligned}\Delta_{\alpha_f}(X_f) &= \Delta^1_{\alpha_f}(X_f) = \{g \cdot x : g \in \mathbb{Z}[\mathbb{Z}^d]\} \\ &= \{g \cdot x : g \in \mathbb{Z}[\mathbb{Z}^d]/\mathbb{Z}[\mathbb{Z}^d]f\} \simeq \mathbb{Z}[\mathbb{Z}^d]/\mathbb{Z}[\mathbb{Z}^d]f \simeq \widehat{X_f}.\end{aligned}$$

Denote by β the restriction of α_f to $\Delta_{\alpha_f}(X_f)$. It is easy to see that the dynamical system $(\Delta_{\alpha_f}(X_f), \beta)$ is algebraically isomorphic to $(\widehat{X_f}, \hat{\alpha}_f)$, hence there exists a group isomorphism

$$K : \Delta^1_{\alpha_f}(X_f) \to \widehat{X_f}$$

such that

$$u^{\mathbf{n}} K(x) = K(\beta^{\mathbf{n}} x)$$

for all $x \in \Delta^1_{\alpha_f}(X_f)$ and $\mathbf{n} \in \mathbb{Z}^d$.

The proof of this fact, which can be found in [29, Lemma 4.5], relies only on the existence of a fundamental homoclinic point. We have already seen in previous sections (cf. Examples 4.4.1 and 4.4.2) that expansiveness is not necessary for the existence of a fundamental homoclinic point.

Moreover, the set of multipliers—polynomials g such that $g \cdot w_f$ is summable—forms an ideal of $\mathbb{Z}[\mathbb{Z}^d]$ (denoted by $\mathscr{I}_f$). Even stronger, each summable homoclinic point corresponds to such a multiplier. Hence,

$$\begin{aligned}\Delta^1_{\alpha_f}(X_f) &= \{g \cdot x : g \in \mathbb{Z}[\mathbb{Z}^d]\} \\ &= \{g \cdot x : g \in \mathbb{Z}[\mathbb{Z}^d] \setminus \mathbb{Z}[\mathbb{Z}^d]f\} \simeq \mathscr{I}_f \subsetneq \mathbb{Z}[\mathbb{Z}^d]/\mathbb{Z}[\mathbb{Z}^d]f \simeq \widehat{X_f}.\end{aligned}$$

Therefore, the structure of $\Delta^1_{\alpha_f}(X_f)$ can be linked to an ideal of $\widehat{X_f}$ and in some cases this ideal can be written down explicitly as we have seen in Sect. 4.4.

Much less is known about the structure of a homoclinic group $\Delta_{\alpha_f}(X_f)$ for non-expansive actions.

An example of a principal algebraic action, where the homoclinic group is not isomorphic to $\widehat{X_f}$ was given by Lind and Schmidt in [29, Example 7.3]; we will briefly discuss this in Remark 4.7.2. This principal algebraic action has uncountably many homoclinic points and hence cannot be isomorphic to the dual of $\widehat{X_f}$ which is countable.

In [11] Clark and Fokkink discuss another example of an algebraic action—originally suggested by H. Lenstra—for which the homoclinic group is not isomorphic to the dual of the phase space.

In any case, it is interesting to determine conditions for $f \in \mathbb{Z}[\mathbb{Z}^d]$ under which $\Delta_{\alpha_f}(X_f)$ is isomorphic to $\widehat{X_f}$. For example, this the case for Laplacians on $\mathbb{Z}^d$ for $d \geq 3$, and we suspect, is not the case for $d = 2$.

4.6.2 Adjoint Actions

Finally, we would like to mention that the entropy of an expansive algebraic action is bounded from below by an action on the dual of the homoclinic group which is equipped with the discrete topology.

Let Γ be an amenable group which acts expansively on a compact abelian group X. Denote by α^* the dual action of $\alpha|_{\Delta_\alpha(X)}$ on $\widehat{\Delta_\alpha(X)}$; α^* is called the *adjoint action* and was introduced in [14, 29]. In [14, Proposition 4.3] it was shown that for expansive $\mathbb{Z}^d$-actions with completely positive entropy (with respect to the Haar measure of X) the adjoint action α^* is expansive, mixing and has completely positive entropy. Hence, α and α^* have very similar dynamical properties. Moreover, by the result of [10, Corollary 9.6], entropies of α and α^* satisfy

$$h(X, \alpha) \geq h(\widehat{\Delta_{\alpha_f}(X)}, \alpha^*).$$

4.7 Homoclinic Points and Applications

In this section we discuss various properties of algebraic dynamical systems which can be established using (summable) homoclinic points.

4.7.1 Symbolic Covers

4.7.1.1 Motivation

We present a (unique) construction of a map between two dynamical systems which respects the dynamics of these systems. Let us demonstrate this method by the following system which converts a signal.

Consider a sequence $s = (s_k)_{k \in \mathbb{Z}} \in \mathbb{R}^{\mathbb{Z}}$—the input signal produced by some stationary source. Moreover, denote by σ the time-shift, i.e., for $s \in \mathbb{R}^{\mathbb{Z}}$ and $l \in \mathbb{Z}$, $(\sigma^l s)_k = s_{k-l}$.

The signal only makes sense when $s \in \ell^\infty(\mathbb{Z}, \mathbb{R})$.

We send the signal through a channel, i.e., a linear map $\xi : \mathbb{R}^{\mathbb{Z}} \longrightarrow \mathbb{R}^{\mathbb{Z}}$ which sends the input signal s to the output signal $s' = \xi(s)$.

The map ξ has to fulfill two conditions:

(*) *shift-equivariance*: let s be the input signal which produces the output signal s', then the shifted signal $\sigma^l s$ has output signal $\sigma^l s'$, for every $l \in \mathbb{Z}$. Hence, for every $l \in \mathbb{Z}$ one has

$$\xi(\sigma^l s) = \sigma^l \xi(s) = \sigma^l s',$$

for every input signal $s \in \ell^\infty(\mathbb{Z}, \mathbb{R})$.

(**) *bounded energy*: a signal $s \in \ell^\infty(\mathbb{Z}, \mathbb{R})$ produces $s' \in \ell^\infty(\mathbb{Z}, \mathbb{R})$.

In other words, ξ is a map $\ell^\infty(\mathbb{Z}, \mathbb{R}) \longrightarrow \ell^\infty(\mathbb{Z}, \mathbb{R})$ such that the following diagram commutes

$$\begin{array}{ccc} \ell^\infty(\mathbb{Z}, \mathbb{R}) & \xrightarrow{\sigma^l} & \ell^\infty(\mathbb{Z}, \mathbb{R}) \\ \xi \downarrow & & \downarrow \xi \\ \ell^\infty(\mathbb{Z}, \mathbb{R}) & \xrightarrow{\sigma^l} & \ell^\infty(\mathbb{Z}, \mathbb{R}) \end{array}$$

for every $l \in \mathbb{Z}$.

Denote the standard basis of $\ell^\infty(\mathbb{Z}, \mathbb{R})$ by $\{\delta_k\}_{k\in\mathbb{Z}}$, i.e.,

$$\delta_k(n) = \begin{cases} 1, & n = k, \\ 0, & n \neq k. \end{cases}$$

Write $s^{(i)} \in \ell^\infty(\mathbb{Z}, \mathbb{R})$ as formal series:

$$s = \sum_{k\in\mathbb{Z}} s_k \delta_k = \sum_{k\in\mathbb{Z}} s_k \sigma^k \delta_0.$$

Then (*) implies that

$$\begin{aligned} (\xi(s))_l &= \left(\xi\left(\sum_{k\in\mathbb{Z}} s_k \sigma^k \delta_0\right)\right)_l \\ &= \left(\sum_{k\in\mathbb{Z}} s_k \sigma^k (\xi(\delta_0))\right)_l \\ &= \left(\sum_{k\in\mathbb{Z}} s_k (\xi(\delta_0))\right)_{l-k} \\ &= (s * \xi(\delta_0))_l\,, \end{aligned}$$

i.e., ξ is the convolution operator $C_{\xi(\delta_0)}$. In order to guarantee condition (**) we demand that $\xi(\delta_0) \in \ell^1(\mathbb{Z}, \mathbb{R})$, since for any $s \in \ell^\infty(\mathbb{Z}, \mathbb{R})$

$$\|s * \xi(\delta_0)\|_{\ell^\infty(\mathbb{Z},\mathbb{R})} \leq \|s\|_{\ell^\infty(\mathbb{Z},\mathbb{R})}\|\xi(\delta_0)\|_{\ell^1(\mathbb{Z},\mathbb{R})} < \infty.$$

Let us now consider two arbitrary dynamical systems (X, T) and (Y, S), where X, Y are metric spaces. Further, let $\xi : X \longrightarrow Y$ be a *factor map*, i.e., a continuous surjective map such that

$$\xi(Tx) = S\xi(x)$$

for all $x \in X$. If ξ is a homeomorphism, then the systems (X, T) and (Y, S) are said to be *conjugate*.

The study of such factor maps is one of the few internal problems in the field of Dynamics.

Since the conjugacy maps between two isomorphic topological dynamical systems are usually very complicated, Vershik raised the question whether there exists a *natural* construction of isomorphisms for algebraic systems.

4.7.1.2 Symbolic Covers of Principal Algebraic Actions

We present algebraic constructions of symbolic covers of principal Γ-actions.

Definition 4.7.1 Let $\Sigma_{\mathscr{A}}$ be the full-shift with finite alphabet $\mathscr{A}$ on the group Γ, i.e.,

$$\Sigma_{\mathscr{A}} = \{w = (w_\gamma)_{\gamma\in\Gamma} : w_\gamma \in \mathscr{A}\},$$

and Γ acts on $\Sigma_{\mathscr{A}}$ by translations σ, i.e., $(\sigma^\gamma w)_{\gamma'} = w_{\gamma^{-1}\gamma'}$. A closed σ-invariant subset Σ of Σ_N is called a subshift of Σ_N.

1. A subshift $\Sigma \subseteq \Sigma_N$ is a *symbolic cover* of an algebraic dynamical system (X, α) if there exists a continuous surjective map $\xi : \Sigma \longrightarrow X$ which is left-shift-equivariant, i.e.,
$$\xi \circ \sigma^\gamma = \alpha^\gamma \circ \xi, \tag{4.7.1}$$
for every $\gamma \in \Gamma$. Hence, the following diagram commutes
$$\begin{array}{ccc} \Sigma & \xrightarrow{\sigma^\gamma} & \Sigma \\ {\scriptstyle\xi}\downarrow & & \downarrow{\scriptstyle\xi} \\ X & \xrightarrow{\alpha^\gamma} & X \end{array}$$
for every $\gamma \in \Gamma$.
2. The symbolic cover Σ is an *equal-entropy symbolic cover* of X if the dynamical systems (Σ, σ) and (X, α) have the same topological entropy.

The first algebraic construction of symbolic covers is due to Vershik (1992). He constructed a Markov partition of a linear hyperbolic automorphism of $\mathbb{T}^2$, which is given by an equivariant map from an extension of certain β-shift. This construction makes use of homoclinic points (cf. [50–52]).

Vershik's approach was extended to algebraic expansive principal $\mathbb{Z}^d$-actions in [15]. There is a canonical way to define coding maps for principal actions as long as the group of summable homoclinic points is non-trivial.

Consider a homoclinic point $x^{(g)} \in \Delta^1_{\alpha_f}(X_f)$ which is obtained from a central multiplier g. The map $\xi_{x^{(g)}} : \ell^\infty(\Gamma, \mathbb{Z}) \longrightarrow X_f$, defined by

$$\xi_{x^{(g)}}(v) = \sum_\gamma v_\gamma \alpha_f^\gamma (x^{(g)}) \tag{4.7.2}$$

is a left shift-equivariant group homomorphism.

Let us discuss this in more detail. We will apply the ideas which we demonstrated in case of signal processing.

Note first that $w = g(f^*)^{-1} \in \ell^1(\Gamma, \mathbb{R})$ by definition. The convolution map $C_w : \ell^\infty(\Gamma, \mathbb{Z}) \longrightarrow \ell^\infty(\Gamma, \mathbb{R})$ given by $v \mapsto vw$ does commute with the left shift action on $\ell^\infty(\Gamma, \mathbb{R})$. Furthermore, $v(g(f^*)^{-1})f^* = vg \in \ell^\infty(\Gamma, \mathbb{Z})$. Hence, $(vg)_\gamma = 0 \mod 1$ for all $\gamma \in \Gamma$ and $\xi_{x^{(g)}} : \ell^\infty(\Gamma, \mathbb{Z}) \longrightarrow X_f$.

In most applications which are presented in this Section we make use of such coding maps.

Example 4.7.1 Let $f_N = N - x - y - x^{-1} - y^{-1} \in \mathbb{Z}[\mathbb{Z}^2]$ where $N \geq 4$. Then one can find $x_N \in \Delta^1(X_{f_N})$ such that $\xi_{x_N}(\{0, 1, \dots, N\}^{\mathbb{Z}^2})$ coincides with X_{f_N}.

Example 4.7.2 Consider $f = 2 - u_1 - u_2^{-1} \in \mathbb{Z}[\mathbb{Z}^2]$. The coding map $\xi_{x^{(g)}}$ with $g = (1 - u_1)^3$ serves as a coding from the full shift $(\Sigma_2, \sigma_{\Sigma_2})$ to X_f. Since both dynamical systems have entropy $\log 2$, the coding map is entropy-preserving.

The following example opens a completely different perspective on principal algebraic actions.

Example 4.7.3 The abelian sandpile model was introduced by Bak et al. (cf. [41, 47] for a mathematically rigorous treatment). It is a cellular automata which displays the phenomenon of self-organized criticality.

The model describes sandpiles which are placed on the lattice points of $\mathbb{Z}^d$ and on which randomly sand grains are dropped. If a sandpile reaches a critical height, the pile topples and distributes its grains to its nearest neighbours. Denote by $\mathscr{R}_d \subset \Sigma_{2d-1}$ the recurrent configurations of the d-dimensional abelian sandpile model on $\mathbb{Z}^d$.

The topological entropy of the d-dimensional abelian sandpile model $(\mathscr{R}^d, \sigma)$ is given by $\log M(l_{\mathbb{Z}^d})$, where $l_{\mathbb{Z}^d} = 2d - \sum_{i=1}^d (u_i + u_i^{-1}) \in \mathbb{Z}[\mathbb{Z}^d]$. Hence, $(\mathscr{R}_d, \sigma_{\mathscr{R}_d})$ has the same entropy as the so called harmonic model, i.e., the non-expansive principal algebraic system $(X_{l_{\mathbb{Z}^d}}, \alpha_{l_{\mathbb{Z}^d}})$ defined by $l_{\mathbb{Z}^d}$. The natural question was raised whether the abelian sandpile model is an equal-entropy cover of $(X_{l_{\mathbb{Z}^d}}, \alpha_{l_{\mathbb{Z}^d}})$. The answer is positive. In [47] it was shown that ξ_x, where x is summable homoclinic point, gives rise to a coding map.

Moreover, there is an interesting correspondence between other lattice models in Statistical Mechanics as well. Let $\mathscr{S}_d$ be the set of spanning trees of $\mathbb{Z}^d$, then

$$h(\sigma_{\mathscr{R}_d}) = h(\sigma_{\mathscr{S}_d})$$

Furthermore, let $(\mathscr{D}, \sigma_{\mathscr{D}})$ be the Domino model on $\mathbb{Z}^2$, which was introduced in Example 4.1.3. It is a well known fact that

$$4h(\sigma_{\mathscr{D}}) = h(\sigma_{\mathscr{R}_2}) = h(\sigma_{\mathscr{S}_2}).$$

Remark 4.7.1 More than twenty years ago Burton And Pemantle [8] raised the question about the link between the harmonic model $(X_{l_{\mathbb{Z}^d}}, \alpha_{l_{\mathbb{Z}^d}})$, on the one hand, and spanning trees on $\mathbb{Z}^d$ and the dimer model ($\mathbb{Z}^2$ case), on the other. Equality of their topological entropies suggests that symbolic models (spanning trees, dimers) form symbolic covers of $(X_{l_{\mathbb{Z}^d}}, \alpha_{l_{\mathbb{Z}^d}})$ [48]. Such strong correspondence could be used to answer some unresolved dynamical and probabilistic questions about lattice models, e.g., the uniqueness of the measure of maximal entropy of $\mathscr{R}_d$. In the ideal case such properties could be even traced back to algebraic properties of the Laurent polynomial.

Finally, we would like to mention that there are systems like irreducible ergodic non-expansive automorphisms of a compact connected abelian group X which have no non-trivial homoclinic point. Thus, the method for the algebraic construction of symbolic covers fails since it relies heavily on the existence of summable homoclinic points (cf. [33, 34]). Nevertheless, Lindenstrauss and Schmidt [34] were able to construct (with the help of one-sided homoclinic points) the so-called pseudo-covers of the corresponding principal actions.

4.7.2 Entropy and Periodic Points

In this section we discuss the growth rate of the number of periodic points of a principal action and its relation to entropy.

Let us start with toral automorphisms.

Example 4.7.4 Define A as in Example 4.1.5. A periodic point $\mathbf{a} \in \mathbb{T}^2$ of period n fulfills:

$$T_A(\mathbf{a}) = \mathbf{a} \quad \text{or equivalently} \quad A^n\mathbf{a} - \mathbf{a} = \mathbf{0} \mod 1.$$

Hence, in order to count the number P_n of periodic points with period n one has to find the preimage of (0, 0) under the map $L_n = A^n - I_2$, where I_2 denotes the 2×2 identity matrix. We treat L_n as a map from $\mathbb{R}^2$ to $\mathbb{R}^2$, then

$$A^n\mathbf{a} - \mathbf{a} = \mathbf{0} \mod 1 \quad \text{is equivalent to} \quad A^n\mathbf{a} - \mathbf{a} = \mathbf{k} \quad \text{for some } k \in \mathbb{Z}^2.$$

Pick's theorem tells us that the number of integer points in $L_n(\mathbb{T}^2)$ is given by the area of the parallelogram which is spanned by the columns of L_n which in turn agrees with

$$|\det L_n| = |(\lambda_+^n - 1)(\lambda_-^n - 1)|.$$

Hence, the growth rate of periodic points is given by

$$\lim_{n\to\infty} \frac{1}{n} \log |P_n| = \lim_{n\to\infty} \frac{1}{n} \log |\det L_n| = \log \lambda_+.$$

By Jensen's formula this value agrees with $\log M(f)$ and hence with the entropy of the dynamical system (X_f, α_f) where $f = u^2 - u - 1$.

Let us now consider the general case. Let Γ' be a subgroup of Γ and denote by

$$\mathrm{Fix}_{\Gamma'}(X_f) = \{x \in X_f : \alpha_f^{\gamma} x = x \text{ for every } \gamma \in \Gamma'\}$$

the Γ'-fixed points in X_f.

Assume for the moment that (X, α) is an algebraic $\mathbb{Z}^d$-action and let Γ_n denote a sequence of normal subgroups of $\mathbb{Z}^d$ with finite index such that

$$\langle \Gamma_n \rangle := \min\{\|\mathbf{m}\| : \mathbf{0} \neq \mathbf{m} \in \Gamma_n\}$$

goes to infinity as $n \to \infty$. For non-expansive actions there may be infinitely many Γ_n-fixed points, hence we have to make some modifications. Denote by P_Γ the index $\mathrm{Fix}_\Gamma(X)/\mathrm{Fix}_\Gamma(X)^\circ$, where $\mathrm{Fix}_\Gamma(X)^\circ$ is the connected component of the identity in $\mathrm{Fix}_\Gamma(X)$. In [46, Theorem 21.1] it was proven that

$$h((X, \alpha)) = \limsup_{n\to\infty} \frac{1}{|\mathbb{Z}^d/\Gamma_n|} \log \mathrm{P}_{\Gamma_n}(X).$$

For principal $\mathbb{Z}^d$-actions α_f we can say more. If the group $\Delta^1(X_f)$ is dense, then

$$\lim_{n\to\infty} \frac{1}{|\mathbb{Z}^d/\Gamma_n|} \log \mathrm{P}_{\Gamma_n}(X_f)$$

exists and is equal to the entropy $h(X_f)$.

The following result for expansive Γ-actions was established in [12]. In order to show that the entropy of α_f coincides with the growth rate of fixed points Deninger and Schmidt used symbolic covers and hence coding maps ξ_x, where $x \in \Delta^1_{\alpha_f}(X_f)$. They make use of the fact that periodic points of the symbolic shift are mapped to periodic points of X_f because of the equivariance (4.7.1) of the coding map. Let Γ be a residually finite amenable group and $f \in \mathbb{Z}[\Gamma]$ such that α_f is expansive. Let Γ' be a subgroup of Γ and denote by

$$\mathrm{Fix}_{\Gamma'}(X_f) = \{x \in X_f : \alpha_f^{\gamma} x = x \text{ for every } \gamma \in \Gamma'\}$$

the Γ'-fixed points in X_f. Moreover, let $\{\Gamma_n : n \geq 1\}$ be a sequence of finite-index normal subgroups of Γ such that for every finite subset F of Γ there exists $N \in \mathbb{N}$ such that for all $n \geq N$

$$\Gamma_n \cap (FF^{-1}) = e_\Gamma. \tag{4.7.3}$$

Then

$$h(\alpha_f) = \lim_{n\to\infty} \frac{1}{|\Gamma/\Gamma_n|} \log \mathrm{Fix}_{\Gamma_n}(X_f).$$

This result was used to prove that the logarithm of the Fuglede-Kadison determinant of the right-convolution operator coincides with the entropy of (X_f, α_f).

Remark 4.7.2 [29, cf., Example 7.5] Let us now consider the non-expansive principal algebraic $\mathbb{Z}^d$-action defined by the Laurent polynomial $f = 3 + u_1 + u_2 + u_1^{-1} + u_2^{-1}$. It is known that the homoclinic group of X_f is uncountable, but $\Delta^1_{\alpha_f}(X_f) = \{0\}$. Hence, the algebraic construction of a symbolic cover described above, does not apply to these system. However, X_f has many periodic points and we know that the upper growth rate of periodic points is equal to $h(\alpha_f)$. The hypothesis is that the (lower) growth rate of periodic points in X_f coincides with the entropy of (X_f, α_f) as well.

4.7.3 *Specification Property*

The concept of specification allows one to approximate given orbit segments or subconfigurations by periodic or homoclinic points as long as the time interval or distance between these subconfigurations is big enough for any fixed precision [6].

Definition 4.7.2 Let $f \in \mathbb{Z}[\mathbb{Z}^d]$ be an irreducible polynomial. Fix a translation-invariant metric d on X_f. Let $\varepsilon > 0$.

1. The system has *strong specification* if there exists a number $p(\varepsilon) > 0$ such that for every finite collection $\{Q_1, \dots, Q_r\}$ of finite subsets of $\mathbb{Z}^d$ with

$$(*) \quad \mathrm{dist}(Q_j, Q_k) := \min_{\mathbf{m}\in Q_j,\ \mathbf{n}\in Q_k} \|\mathbf{m} - \mathbf{n}\| \geq p(\varepsilon) \text{ for } 1 \leq j < k \leq r,$$

every collection $\{x^{(1)}, \dots, x^{(r)}\} \subset X_f$, and every finite-index subgroup G of $\mathbb{Z}^d$ with

$$\mathrm{dist}(Q_j + \mathbf{k}, Q_k) \geq p(\varepsilon) \text{ for } 1 \leq j < k \leq r \text{ and every } \mathbf{k} \in G \setminus \{\mathbf{0}\},$$

there is a $y \in \mathrm{Fix}_G(X_f)$ with

$$(**) \qquad d\left(\alpha_f^{\mathbf{n}}(y), \alpha_f^{\mathbf{n}}(x^{(j)})\right) < \varepsilon \text{ for } 1 \leq j \leq r \text{ and every } \mathbf{n} \in Q_j.$$

2. The principal system has *homoclinic specification* if there exists a number $p(\varepsilon) > 0$ with the following properties: every finite collection $\{Q_1, \ldots, Q_r\}$ of finite subsets of $\mathbb{Z}^d$ satisfying (*) and every collection $\{x^{(1)}, \ldots, x^{(r)}\} \subset X_f$ there is a point $y \in \Delta_{\alpha_f}(X_f)$ satisfying (**).

Let us explain these abstract definitions in more detail for the algebraic dynamical system (X, α). Assume that the system has one of the given specification properties. Then for a given $\varepsilon > 0$ one can find a constant $p(\varepsilon)$, which we interpret as time-span (d $= 1$) or distance. The definition says that for arbitrary points $x_1, x_2, \ldots, x_n \in X$ and subsets $Q_1, Q_2, \ldots, Q_n$ of $\mathbb{Z}^d$, which are separated by at least $p(\varepsilon)$ in the dist-norm, one can find a G-periodic respectively homoclinic point $y \in X$ such that the orbit segments $\{\alpha^{\mathbf{n}} x_i : \mathbf{n} \in Q_i\}$, $i \in \{1, \ldots, n\}$, are traced up to precision ε by this periodic or homoclinic point y for $i = 1, 2 \ldots, n$. It should be emphasized that in the definition of specification the size and form of the sets Q_i are completely arbitrary—which we interpret as uniformity.

Moreover, strong specification implies that the set of periodic points is dense and the system is topologically mixing. For expansive Γ-actions the strong specification is equivalent to homoclinic specification and both imply that the set of homoclinic points is dense [10, Theorem 6.3].

We know that expansive $\mathbb{Z}^d$ actions always have strong and homoclinic specification [29]. But even more is true: If for an expansive principal Γ-action the group $\Delta^1_{\alpha_f}(X_f)$ is dense in X_f, then the principal action has strong and homoclinic specification as well [30, Theorem 8.2].

Theorem 4.7.1 ([24]) *Let f be an irreducible element of $\mathbb{Z}[\Gamma]$ which has a formal inverse (fundamental solution) $w \in \mathbb{R}^\Gamma$. Suppose that there exists a central element $g \in \mathbb{Z}[\Gamma] \setminus \mathbb{Z}[\Gamma]f$ with $w \cdot g \in \ell^1(\Gamma, \mathbb{R})$ and a full shift Σ_N such that $\xi_{x^{(g)}} : \Sigma_N \to X_f$ is surjective. Then for every $\varepsilon > 0$, there exists a non-empty finite subset F_ε of Γ with the following property: if F_1, F_2 are subsets of Γ with*

$$F_\varepsilon F_1 \cap F_\varepsilon F_2 = \varnothing, \tag{4.7.4}$$

then there exists, for every pair of points x^1, x^2 in X_f, a point $y \in X_f$ with

$$\|x^j_\gamma - y_\gamma\| < \varepsilon \ \textit{ for every } \ \gamma \in F_j, \text{ and } j = 1, 2. \tag{4.7.5}$$

4.7.4 Local Entropy Theory

The local theory aims to characterize and study those dynamical systems which are *disjoint* from those with zero entropy—and which are therefore interesting from the view point of Ergodic Theory, since they have chaotic behaviour. In order to find a topological analogue of Kolmogorov automorphisms of ergodic theory, Blanchard [3] introduced non-equivalent notions of *completely positive entropy* (CPE) and *uni-*

formly positive entropy (UPE) (for topological dynamical systems with symmetry group $\mathbb{Z}$).

Let (X, T) be a topological dynamical system. An *entropy pair* is an element $(x, x') \in X \times X$ with $x \neq x'$ such that for every open cover $\{U_1, U_2\}$ of X with $x \in \operatorname{int} U_1^c$ and $x' \in \operatorname{int} U_2^c$, the topological entropy $h(\{U_1, U_2\})$ of this cover is positive. The system is UPE if and only if the set of entropy pairs coincides with $X \times X \setminus \operatorname{Diag}(X \times X)$ [4]. If the system (X, T) is UPE, then it is topologically weak mixing. A topological system has CPE if and only if all non-trivial topological factors have positive entropy.

The definition of entropy pairs was generalized to actions of discrete amenable groups in [27]. Moreover, a new combinatorial interpretation was given [4, Definition 3.1 and Theorem 3.16].

In [10] the following observations were made for algebraic dynamical systems (X, α). Let $\mathrm{IE}(X)$ be the set containing 0_X and all points $x \in X$ such that $(x, 0_X)$ is an entropy pair. The set $\mathrm{IE}(X)$ forms a closed normal subgroup of X. If $\hat{X}$ is finitely generated, then $\Delta_\alpha^1(X) \subseteq \mathrm{IE}(X)$, by combining this with Theorem 7.3 from [10] we obtain the following result:

Theorem 4.7.2 *If $\hat{X}$ is finitely generated and $\Delta_\alpha^1(X)$ is non-trivial, then $h(X) > 0$.*

The set $\mathrm{IE}(X)$ and X coincide if and only if (X, α) has CPE ([10, Corollary 7.5]). We list some other results. For general expansive algebraic actions the following result was established by Lind and Schmidt [29] for $\Gamma = \mathbb{Z}^d$, and has recently been generalized by Chung and Li [10] to all *polycyclic-by-finite* group, i.e., groups for which there exists a sequence of subgroups

$$\Gamma = \Gamma_1 \rhd \Gamma_2 \rhd \ldots \rhd \Gamma_n = \{e_\Gamma\}$$

such that Γ_i / Γ_{i+1} is finite or cyclic for every $i = 1, \ldots, n-1$.

Theorem 4.7.3 *Let Γ be a polycyclic-by-finite group, which acts expansively on a compact abelian group X. Then the following holds*

- *The group of (summable) homoclinic points is non-trivial if and only if the action has positive entropy.*
- *The group of (summable) homoclinic points is dense in X if and only if the action has completely positive entropy. Moreover, if the entropy of the Γ-action is finite, then the Haar measure is the unique measure of maximal entropy if and only if the (summable) homoclinic points are dense in X (cf. [10, Theorem 8.6]).*

4.7.5 Central Limit Theorems

Ergodic theory studies various probabilistic properties of measure-preserving dynamical systems such as ergodicity, mixing, decay of correlations, large deviations, and limit theorems. One of the most important and studied theorems in probability theory is the Central Limit Theorem (CLT): a stationary sequence of random variables $\{X_n\}$ satisfies the CLT if

$$\frac{\sum_{i=0}^{n-1} X_i - n\mathbb{E}X_1}{\sqrt{n}} \to \mathcal{N}(0, \sigma^2)$$

in distribution, where $\mathcal{N}(0, \sigma^2)$ is the Gaussian distribution with mean 0 and variance $\sigma^2 > 0$. Typically, one requires that $\{X_n\}$ is a sequence of *weakly dependent* random variables. Central Limit Theorems have also been established in the context of dynamical systems, i.e., for sequences $X_n = \phi(T^n(\omega))$, where $T : \Omega \longrightarrow \Omega$ is a measure-preserving transformation of some measure space $(\Omega, \mathcal{B}, \mu)$, and $\phi : \Omega \longrightarrow \mathbb{R}$ is some observable. Typically, in order to obtain a CLT for dynamical systems, one has to consider dynamical systems with sufficient hyperbolic behaviour and relatively smooth observables. Under these conditions, the corresponding sequence $\{X_n = \phi \circ T^n\}$ has *summable decay of correlations*

$$\sum_{n=0}^{\infty} \rho(X_0, X_n) < \infty,$$

where $\rho(\cdot, \cdot)$ is the correlation coefficient. For a hyperbolic toral automorphism $T : \mathbb{T}^d \longrightarrow \mathbb{T}^d$ (c.f., Example 4.1.5) and any Hölder continuous observable $\phi : \mathbb{T}^d \longrightarrow \mathbb{R}$, decay of correlations is exponential

$$\rho(\phi, \phi \circ T^n) \leq C\lambda^n,$$

for some $\lambda \in (0, 1)$ and $C > 0$, and the CLT follows easily. Rather surprisingly one can obtain CLT's using homoclinic points.

M. Gordin pioneered the homoclinic approach to Central Limit Theorems [18, 19]. In fact, he proposed a more general notion of a *homoclinic transformation*. Suppose $(X, \mathcal{F}, \mu)$ is a probability space, and $\alpha : X \longrightarrow X$ is a measure-preserving automorphism, i.e., we consider an measure-preserving action of $\mathbb{Z}$ on X. A non-singular invertible transformation S is called *homoclinic* if $\alpha^n S \alpha^{-n} \to \mathsf{Id}$ as $n \to \pm\infty$. Observe that S itself is a homoclinic point in the classical sense under the map $S \mapsto \alpha S \alpha^{-1}$ acting on the group of invertible measurable transformations. This definition extends in an obvious fashion to actions of general discrete groups Γ. In the context of Algebraic Dynamics, if $x \in \Delta_\alpha(X)$ is a homoclinic point, then $S : X \longrightarrow X$, $S(z) = z + x$, is a homoclinic transformation. Note also that S preserves the Haar measure in this case.

Gordin [19] noted a link with coupling methods of Probability Theory: homoclinic transformations provide a form of a *deterministic coupling* which is sufficient to establish an asymptotic independence property for a stationary processes $\{X_n = \phi(\alpha^n x) | n \in \mathbb{Z}\}$, when the observables $\phi : X \longrightarrow \mathbb{R}$ is of a special form, which in turn is sufficient to obtain a Central Limit Theorem for $\{X_n\}$.

In particular, as a corollary, one obtains the following version of CLT for algebraic $\mathbb{Z}$-actions.

Theorem 4.7.4 ([19]) *Assume that $\phi \in L^\infty(X, \lambda_X)$ is real-valued and can be represented in the form*

$$\phi(z) = \Phi(z + x) - \Phi(z) \tag{4.7.6}$$

for some $\Phi \in L^1(X, \lambda_X)$ and $x \in \Delta_\alpha(X)$. If, in addition,

$$\sum_{n=-\infty}^{\infty} \|\Phi(z + \alpha^n x) - \Phi(z)\|_\infty < \infty, \tag{4.7.7}$$

then for $n \to \infty$, the sums

$$\frac{1}{\sqrt{n}} \sum_{k=0}^{n-1} \phi(\alpha^k z)$$

have an asymptotic normal distribution with mean 0 and variance $\sigma^2 \geq 0$,

$$\sigma^2 = \lim_{n\to\infty} \frac{1}{n} \left|\left| \sum_{k=0}^{n-1} \phi(\alpha^k z) \right|\right|_2^2 = \sum_{n=-\infty}^{+\infty} \int \phi(z)\phi(\alpha^n z)\, \lambda_X(dz). \tag{4.7.8}$$

Expression (4.7.8) is precisely the one expected for systems with rapid mixing (decay of correlations). It is relatively easy to find functions Φ satisfying (4.7.7): e.g., if x is a summable homoclinic point ($x \in \Delta_\alpha^1(X)$), then any Lipschitz function Φ satisfies (4.7.7). A non-trivial and interesting question is to identify the class of *smooth homoclinic coboundaries* ϕ, i.e., functions satisfying (4.7.6) with Φ sufficiently "smooth". When $X = \mathbb{T}^d$ this question has a long history (see [42] for a good overview of known results): the answer depends on the Diophantine properties of x as a vector in $\mathbb{R}^d$. For more general groups, the method of Moore and Schmidt [39] might offer a practical approach to the problem.

Gordin and Weber [21] proposed another homoclinic approach to Central Limit Theorems. When compared with the previous Theorem, the main result of [21] is valid for a more restrictive class of algebraic dynamical systems, namely, expansive $\mathbb{Z}^d$-actions, but at the same time, the class of suitable observable functions ϕ is more transparent.

Theorem 4.7.5 (Theorem 2.2, [21]) *Suppose α is an expansive $\mathbb{Z}^d$-action on an a compact second countable abelian group X. Suppose also that α has completely positive entropy. Let $\phi \in L^2(X, \lambda_X)$ be a real-valued function with mean 0, which satisfies for every $x \in \Delta_\alpha(X)$,*

$$\sum_{\mathbf{n}\in\mathbb{Z}^d} ||\phi(z) - \phi(z + \alpha^{\mathbf{n}} x)||_2 < \infty.$$

Then the CLT holds for ϕ. Moreover, ϕ satisfies the Almost Sure Central Limit Theorem as well: for all $t \in \mathbb{R}$

$$\frac{1}{\log m}\sum_{k=1}^{m}\frac{1}{k}\mathbb{I}\Big(\frac{1}{|D_m|^{1/2}}\sum_{\mathbf{n}\in D_m}\phi(\alpha^{\mathbf{n}}z)\le t\Big)\to\frac{1}{\sqrt{2\pi\sigma_\phi^2}}\int_{-\infty}^{t}\exp\Big(-\frac{\tau^2}{2\sigma_\phi^2}\Big)d\tau,$$

where $\{D_m\}$ is a sequence of finite sets in $\mathbb{Z}^d$ with $D_m \to \mathbb{Z}^d$ and

$$\frac{|\partial D_m|}{|D_m|}\to 0.$$

Note that by a result of Lind and Schmidt [29], the requirement of complete positivity of entropy is equivalent to the density of $\Delta_\alpha(X) = \Delta_\alpha^1(X)$ in X. For other interesting probabilistic applications of homoclinic points see [20, 22, 23].

References

1. Andersson, K.G.: Poincare's discovery of homoclinic points. Arch. Hist. Exact Sci. **48**(2), 133–147 (1994)
2. Berger, R.: The undecidability of the domino problem. Mem. Amer. Math. Soc. No. **66**, 72 (1966)
3. Blanchard, F.: Fully positive topological entropy and topological mixing. Symbolic dynamics and its applications (New Haven, CT, 1991), Contemp. Math. Amer. Math. Soc. Providence, RI **135**, 95–105 (1991)
4. Blanchard, F.: A disjointness theorem involving topological entropy. Bull. Soc. Math. France **121**(4), 465–478 (1993)
5. Bowen, L., Li, H.: Harmonic models and spanning forests of residually finite groups. J. Funct. Anal. **263**(7), 1769–1808 (2012)
6. Bowen, R.: Periodic points and measures for axiom-A diffeomorpisms. Trans. Americ. Math. Soc. **154**, 377–397 (1971)
7. Bunimovich, L.A., Sinai, Y.G.: Spacetime chaos in coupled map lattices. Nonlinearity **1**(4), 491–516 (1988)
8. Burton, R., Pemantle, R.: Local characteristics, entropy and limit theorems for spanning trees and domino tilings via transfer-impedances. Ann. Probab. **21**(3), 1329–1371 (1993)
9. Capocaccia, D.: A definition of Gibbs state for a compact set with Zn action. Commun. Math. Phys. **48**(1), 85–88 (1976)
10. Chung, N.-P., Li, H.: Homoclinic groups. IE groups, and expansive algebraic actions. ArXiv e-prints (2011), available at 1103.1567
11. Clark, A., Fokkink, R.: On a homoclinic group that is not isomorphic to the character group. Qual. Theory Dyn. Syst. **5**(2), 361–365 (2004)
12. Deninger, C., Schmidt, K.: Expansive algebraic actions of discrete residually finite amenable groups and their entropy. Ergodic Theory Dynam. Systems **27**(3), 769–786 (2007)
13. Einsiedler, M., Rindler, H.: Algebraic actions of the discrete Heisenberg group and other non-abelian groups. Aequationes Math. **62**(1–2), 117–135 (2001)
14. Einsiedler, M., Schmidt, K.: The adjoint action of an expansive algebraic Zd-action. Monatsh. Math. **135**(3), 203–220 (2002)
15. Einsiedler, M., Schmidt, K.: Markov partitions and homoclinic points of algebraic Zd-actions. Proc. Steklov Inst. Math. **216**(3), 259–279 (1997)

16. Fisher, M.E.: Statistical mechanics of dimers on a plane lattice. Phys. Rev. (2) **124**, 1664–1672 (1961)
17. Fowler, A.C.: Homoclinic bifurcations for partial differential equations in unbounded domains. Stud. Appl. Math. **83**(4), 329–353 (1990)
18. Gordin, M.I.: A homoclinic version of the central limit theorem. Zap. Nauchn. Sem. Leningrad. Otdel. Mat. Inst. Steklov. (LOMI) **184** (1990), no. Issled. po Mat. Statist. **9**, 80–91, 321 (Russian); English transl. J. Math. Sci. **68**(4), 451–458 (1994)
19. Gordin, M.: Homoclinic approach to the central limit theorem for dynamical systems. Doeblin and modern probability (Blaubeuren, 1991). Contemp. Math. Amer. Math. Soc. Providence, RI **149**, 149–162 (1993)
20. Gordin, M.I.: Homoclinic processes and invariant measures for hyperbolic automorphisms of tori, Zap. Nauchn. Sem. S.-Peterburg. Otdel. Mat. Inst. Steklov. (POMI) **368** (2009), no. Veroyatnost i Statistika. **15**, 122–129, 284 (Russian, with English and Russian summaries); English transl., J. Math. Sci. (NY) **167**(4), 501–505 (2010)
21. Gordin, M., Weber, M.: On the almost sure central limit theorem for a class of Zd-actions. J. Theoret. Probab. **15**(2), 477–501 (2002)
22. Gordin, M.I.: A criterion, based on homoclinic sums, for the vanishing of a spectral density. Zap. Nauchn. Sem. S.-Peterburg. Otdel. Mat. Inst. Steklov. (POMI) **228** (1996), no. Veroyatn. i Stat. **1**, 94–110, 358 (Russian, with English and Russian summaries); English transl., J. Math. Sci. (New York) **93**(3), 311–320 (1999)
23. Gordin, M.I.: Some remarks on homoclinic groups of hyperbolic automorphisms of tori. Zap. Nauchn. Sem. S.-Peterburg. Otdel. Mat. Inst. Steklov. (POMI) **223** (1995), no. Teor. Predstav. Din. Sistemy, Kombin. i Algoritm. Metody. I, 140–147, 339 (Russian, with English and Russian summaries); English transl., J. Math. Sci. (New York) **87**(6), 4067–4071 (1997)
24. Göll, M., Schmidt, K., Verbitskiy, E.: Algebraic actions of the discrete Heisenberg group: expansiveness and homoclinic points. Indag. Math. **25**(4), 713–744 (2014)
25. Göll, M., Schmidt, K., Verbitskiy, E.: Wiener's Lemma for the discrete Heisenberg group: invertibility criteria and applications to algebraic dynamics. In preparation (2014)
26. Golse, F.: On the dynamics of large particle systems in the mean field limit. This contribution
27. Kerr, D., Li, H.: Independence in topological and C*-dynamics. Math. Ann. **338**(4), 869–926 (2007)
28. Li, H., Thom, A.: Entropy, determinants, and L2-torsion. J. Amer. Math. Soc. **27**(1), 239–292 (2014)
29. Lind, D., Schmidt, K.: Homoclinic points of algebraic Zd-actions. J. Amer. Math. Soc. **12**(4), 953–980 (1999)
30. Lind, D., Schmidt, K., Verbitskiy, E.: Entropy and growth rate of periodic points of algebraic Zd-actions, dynamical numbers-interplay between dynamical systems and number theory, vol. 532. Contemporary Mathematics American Mathematical Society Providence, RI, (2010)
31. Lind, D., Schmidt, K., Verbitskiy, E.: Homoclinic points, atoral polynomials, and periodic points of algebraic Zd-actions. Ergodic Theory Dynam. Syst. **33**(4), 1060–1081 (2013)
32. Lind, D., Schmidt, K., Ward, T.: Mahler measure and entropy for commuting automorphisms of compact groups. Invent. Math. **101**(3), 593–629 (1990)
33. Lindenstrauss, E., Schmidt, K.: Invariant sets and measures of nonexpansive group automorphisms. Israel J. Math. **144**, 29–60 (2004)
34. Lindenstrauss, E., Schmidt, K.: Symbolic representations of nonexpansive group automorphisms, Israel J. Math. **149**, 227–266 (2005). Probability in Mathematics
35. Mackey, G.W.: On induced representations of groups. Amer. J. Math. **73**, 576–592 (1951)
36. Miles, R.: Expansive algebraic actions of countable abelian groups. Monatsh. Math. **147**(2), 155–164 (2006)
37. Mielke, A.: Lecture Notes on Evolutionary G-convergence. This contribution
38. Mielke, A., Zelik, S.V.: Infinite-dimensional hyperbolic sets and spatio-temporal chaos in reaction-diffusion systems in $\mathbb{R}^n$. J. Dynam. Different. Eqns. **19**(2), 333–389 (2007)
39. Moore, C.C., Schmidt, K.: Coboundaries and homomorphisms for non-singular actions and a problem.H. Helson, Proc. London Math. Soc. (3) **40**(3), 443–475 (1980)

40. Petersen, K., Schmidt, K.: Symmetric Gibbs measures. Trans. Amer. Math. Soc. **349**(7), 2775–2811 (1997)
41. Redig, F.: Mathematical aspects of the abelian sandpile model. Mathematical Statistical Physics, pp. 657–729. Elsevier B. V, Amsterdam (2006)
42. Rozhdestvenskii, A.V.: On nontrivial additive cocycles on a torus, Mat. Sb. **199**(2), 71–92 (2008). (Russian, with Russian summary); English transl., Sb. Math. **199**(1–2), 229–251 (2008)
43. Rudolph, D.J., Schmidt, K.: Almost block independence and Bernoullicity of Zd-actions by automorphisms of compact abelian groups. Invent. Math. **120**(3), 455–488 (1995)
44. Ruelle, D.: Thermodynamic formalism, 2nd edn. Cambridge Mathematical Library, Cambridge University Press, Cambridge (2004). The mathematical structures of equilibrium statistical mechanics
45. Shilnikov, L.P.: Homoclinic trajectories: from poincare to the present. Mathematical Events of the Twentieth Century, pp. 347–370. Springer, Berlin (2006)
46. Schmidt, K.: Dynamical systems of algebraic origin. Progress in Mathematics, vol. 128. Birkhauser Verlag, Basel (1995)
47. Schmidt, K., Verbitskiy, E.: Abelian sandpiles and the harmonic model. Commun. Math. Phys. **292**(3), 721–759 (2009)
48. Schmidt, K., Verbitskiy, E.: New directions in algebraic dynamical systems. Regul. Chaotic Dyn. **16**(1–2), 79–89 (2011)
49. Timberlake, T.: A computational approach to teaching conservative chaos. Am. J. Phys. **72**(8), 1002–1007 (2004)
50. Vershik, A.M.: The fibadic expansion of real numbers and adic transformations. Preprint, Mittag-Leffler Institute **26**(3), 22–27 (1991/1992) (Russian); English Transl. Funct. Anal. Appl. **26**(3), 170–173 (1992)
51. Vershik, A.M.: Arithmetic isomorphism of hyperbolic automorphisms of a torus and of sofic shifts. Funktsional. Anal. i Prilozhen. **26**(3), 22–27 (1992) (Russian); English Transl., Funct. Anal. Appl. **26**(3), 170–173 (1992)
52. Vershik, A.M.: Locally transversal symbolic dynamics. Algebra i Analiz **6**(3), 94–106 (1994) (Russian); English Transl., St. Petersburg Math. J. **6**(3), 529–540 (1995)
53. Walters, P.: An introduction to ergodic theory. Graduate Texts in Mathematics, vol. 79. Springer-Verlag, New York (1982)
54. Zelik, S.V.: Spatial and dynamical chaos generated by reaction-diffusion systems in unbounded domains. J. Dynam. Different. Eqns. **19**(1), 1–74 (2007)

Index

A. Muntean et al. (eds.), *Macroscopic and Large Scale Phenomena: Coarse Graining, Mean Field Limits and Ergodicity*, Lecture Notes in Applied Mathematics and Mechanics 3, DOI 10.1007/978-3-319-26883-5

FSC
www.fsc.org
MIX
Papier | Fördert gute Waldnutzung
FSC® C083411

Kolibri 360 GmbH
Ferdinand-Jühlke-Straße 7
99095 Erfurt, Deutschland
produktsicherheit@kolibri360.de